Landscaping
Principles and Practices

5TH EDITION

Landscaping

Principles and Practices

5TH EDITION

JACK E. INGELS

State University of New York
College of Agriculture and Technology
Cobleskill, New York

DELMAR

™

THOMSON LEARNING

Africa • Australia • Canada • Denmark • Japan • Mexico • New Zealand • Philippines
Puerto Rico • Singapore • Spain • United Kingdom • United States

NOTICE TO THE READER

Cover art courtesy of Jack E. Ingels — Cover design by McKinley Griffen Design

Delmar Staff

Publisher: Tim O'Leary
Acquisitions Editor: Cathy L. Esperti
Senior Project Editor: Andrea Edwards Myers

Production Manager: Wendy A. Troeger
Marketing Manager: Maura Theriault

COPYRIGHT © 1997 Delmar, a division of Thomson Learning, Inc. The Thomson Learning™ is a trademark used herein under license.

Printed in the United States of America
8 9 10 XXX 02 01 00

For more information, contact Delmar, 3 Columbia Circle, PO Box 15015, Albany, NY 12212-0515; or find us on the World Wide Web at http://www.delmar.com

International Division List

Asia
Thomson Learning
60 Albert Street, #15-01
Albert Complex
Singapore 189969
Tel: 65 336 6411
Fax: 65 336 7411

Japan:
Thomson Learning
Palaceside Building 5F
1-1-1 Hitotsubashi, Chiyoda-ku
Tokyo 100 0003 Japan
Tel: 813 5218 6544
Fax: 813 5218 6551

Australia/New Zealand:
Nelson/Thomson Learning
102 Dodds Street
South Melbourne, Victoria 3205
Australia
Tel: 61 39 685 4111
Fax: 61 39 685 4199

UK/Europe/Middle East
Thomson Learning
Berkshire House
168-173 High Holborn
London
WC1V 7AA United Kingdom
Tel: 44 171 497 1422
Fax: 44 171 497 1426

Latin America:
Thomson Learning
Seneca, 53
Colonia Polanco
11560 Mexico D.F. Mexico
Tel: 525-281-2906
Fax: 525-281-2656

Canada:
Nelson/Thomson Learning
1120 Birchmount Road
Scarborough, Ontario
Canada M1K 5G4
Tel: 416-752-9100
Fax: 416-752-8102

Library of Congress Cataloging-in-Publication Data

Ingels, Jack E.
    Landscaping principles and practices / Jack E. Ingels. — 5th ed.
      p. cm.
    Rev. ed. of: Landscaping, 4th ed. c1992.
    Includes index.
    ISBN 0-8273-6735-X
    1. Landscape architecture. 2. Landscape gardening.
3. Landscape industry. I. Ingels, Jack E. Landscaping.
II. Title.
SB472.153   1997   96-22507
712 — dc20   CIP

Contents

SECTION
2

Landscape Contracting

Landscape Maintenance

SECTION 3

Preface

America's landscapes are rich, diverse, and dynamic. They include unspoiled natural regions, farmlands, urban and suburban population centers, recreational and commercial parks, historic settings, mountains, prairies, and seascapes. Our national landscape is a mix that is never quite the same from one decade to the next. Each segment of the American landscape serves a population that is as diverse as the nation itself. The individual who pursues a career in landscaping is a service professional who seeks to understand the needs of those who use and enjoy the land. Meeting those diverse needs and satisfying people without harm to others or to the environment is the challenge of the modern landscape professional.

Landscaping: Principles and Practices, 5th edition, is intended as an overview of a complex and rapidly growing industry. The text acquaints the reader with the three major branches of the industry, design, installation, and management, by organizing the chapters into three main sections. The first section, "Landscape Designing," will acquaint the reader with the intellectual and physical processes that result in the creation of a new outdoor environment. It will also offer training in the graphic skills needed to present ideas on paper so that others may understand them. The second section, "Landscape Contracting," addresses the office and field skills needed to bring the landscape project to fruition. Contractual relationships between landscapers and the others with whom they work in the construction trades are discussed. Calculative techniques needed to turn landscape drawings into quantitative measurements are

described, and opportunities for the reader to practice while reading are presented. In addition, basic methods of planting and lawn installation are discussed, as well as consideration of landscape irrigation and interior plantscapes. Section three covers "Landscape Maintenance," the key to modern landscape management. Plant maintenance, lawn care, winterization, and pricing are among the topics covered in the chapters of this section.

As an introductory text, Landscaping: Principles and Practices, 5th edition, is directed to traditional and nontraditional learners seeking to begin their training as landscape professionals. While much of the material covered is unquestionably horticultural, the text presents the profession first and foremost as a business. By following the text, readers progress from the basic principles of landscape design and graphics, through methods of installation and maintenance, to specific business methods and techniques vital to successful estimating, bidding, and budgeting.

The text aids the reader's learning through the use of specific learning objectives, end-of-chapter reviews, extensive illustrations, and supplemental activities. Technical terms are defined as they are introduced within the chapters and again in the glossary at the back of the text.

Features of the Fifth Edition

❧ Chapters have been organized in accordance with the three branches of the industry.

❦ All plant references include both botanical and common names.

❦ Graphic training has been expanded and includes additional lettering styles, a more comprehensive description of both tools and techniques, and added examples to aid the reader.

❦ New coverage is given to the selection and evaluation of plants for the landscape.

❦ The Landscape Process is described and discussed in detail, including the development of functional diagrams and preliminary designs prior to actual master planning.

❦ Xeriscaping is described and discussed, complete with details for designing, installing, and sustaining a water-wise landscape.

❦ New photos have been added to the color section to illustrate modern landscape planning.

❦ New examples have been added to the portfolio of landscape drawings.

❦ An extensive section on landscape calculations enables the reader to accomplish a complete landscape take-off once the calculative skills are mastered. The chapter is written in a manner that permits the reader to begin at whatever level is suited to his/her experience with mathematics.

❦ An overview of the construction industry, including the similarities and differences between regular contractors and landscape contractors, is given.

❦ Landscape specifications and their importance in the bidding and construction process are described in a comprehensive new chapter.

❦ The documents of the contractive process are described in detail.

About the Author

Jack E. Ingels holds a Board of Trustees' Distinguished Teaching Professorship at the State University of New York's College of Agriculture and Technology at Cobleskill. He is responsible for the College's Landscape Program, which prepares associate and baccalaureate degree students for career-track positions with landscape firms throughout the nation.

Professor Ingels completed his undergraduate degree at Purdue University and his graduate studies at Rutgers University. Postgraduate training was at Ball State University. His fields of specialization include ornamental horticulture, landscape design, landscape garden history, and plant pathology. He is an experienced university educator with a diverse practical background in industry as well. He is also the author of *Ornamental Horticulture: Science, Operations, & Management*, another successful text in Delmar's agricultural series.

Acknowledgements

The author wishes to express appreciation to the following for their help in the preparation of this text:

Stewart Allen, Director,
The Allen Organization

Robert C. Bigler Associates, Architects

Eric Blamphin, Designer

Michael Boice, Designer

Richard Centolella

Cooperative Extension Service
Cornell University, Ithaca, New York

Vickie Davis

Edward Dennehy, Designer

Albert Glowacki Landscaping

Vicki Harris

Russell Ireland, Landscape Contractor

Thomas Kenly, Designer

John Krieg, L. A.

James Lancaster, Designer

Peter Lee

Lied's, Inc.

Mike Lin

Mark Magnone, L. A.

Joye Noth

Lawrence Perillo, Designer

Edward J. Plaster

Robert Rodler, Landscape Contractor

Smallwood Design Group

Ricky Sowell

State University of New York
College of Agriculture and Technology
Cobleskill, NY

A.J. Tomasi Nurseries, Inc.

Landscaping: Principles and Practices was classroom tested at the State University of New York College of Agriculture and Technology at Cobleskill, New York.

The following individuals devoted their time and considerable professional experience to reviewing the manuscript for this edition of the text:

Arlene Corbin
Seminole, Florida

Mark Schusler
Frost, Texas

Jan Traynor
Randolph, New Jersey

David Twente
Wellington, Missouri

Hilton Webb
Eden, North Carolina

Edward L. Wright
Graves County High School
Mayfield, Kentucky

Carl E. Mitchell
Lee-Davis High School
Mechanicsville, Virginia

Don C. Leibelt
Green Bay East High School
Green Bay, Wisconsin

Bryan Starr
Jefferson High School
Jefferson, Oregon

David Briggs
Jefferson-Scranton & East Greene
Jefferson, Iowa

Introduction

An Overview of the Landscape Industry

When William Shakespeare posed the question, "What's in a name?", he could have been referring to the term *landscaper*. Few job titles conjure up such a variety of definitions and descriptions from well-meaning people, each certain that his or her definition is correct. Each would also probably believe that the term is not complex, because the industry itself is not complex. The same thing happens when people are asked to define the duties of a conservationist, a doctor, or a teacher. What seems so simple and straightforward at the outset grows in complexity upon realizing that there are endless things to conserve, dozens of ways to practice medicine, and students requiring educational services from nursery school through post-graduate programs.

Landscaping is a service industry. It serves people by fabricating environments where they can live, work, play, or just pass time. These environments are primarily in the outdoors or are in interior settings that seek to suggest the outdoors. Landscapers use the products of others to create habitat areas that can range from utilitarian to fantastic. While humans are the predominant clientele, others with whom we share the planet may also benefit from the landscapers' efforts.

Landscapers serve residential clients, civic clients, corporate clients, and others. They work as one-person, part-time operations and as international corporate giants, and in companies that fit somewhere in between those extremes. Landscape firms are listed in the yellow pages of small villages, cities, and world capitals. In short, there is no one way to be a landscaper or to practice landscaping. There are many different levels of practice and differing degrees of educational preparation required, depending upon how a person opts to work within the industry.

The popular image of the landscaper as a small, one- or two-person operation, with one truck, a lawnmower, and some hand tools with which to serve a nearby residential clientele, is a valid one. There will probably always be that type of landscaper because there will always be a demand for simple plant installation and maintenance services made by a clientele who would rather hire someone else to do the work than do it themselves. It isn't that those clients couldn't do the work, but rather that they do not want to do it. As long as there are people who can hire others to do work that they do not enjoy doing, there will be a market for those services. The small, one-truck landscape operation fills a need. The array and quality of their services vary, and the requirements for entry into the industry at that level are minimal. There is little or no requirement for investment in land or buildings. There is no industry accreditation necessitating formal schooling at this level of practice. Many of the needed skills are easily learned, often by apprenticing with another firm for a season or two. Companies at this level of practice have the greatest turnover rate of any in the industry. While statistics are sketchy, there is no reason to believe that their record of success is better than small businesses at large, where mini-companies with five or fewer employees fail more frequently than they succeed. Still, because it is an easy business to get into, with little financial risk even if it should fail, there is no reason to forecast the demise of this segment of the industry.

The larger segment of the industry is a multi-billion dollar industry that serves a population that is increasingly discerning and demanding in its requirements for technical competency and service quality. Demographic trends in the America of the 1990s indicate that more people are living in non-traditional ways, such as singles, single parents, married without children, or married with fewer children. That, coupled with accelerating land and building costs, is resulting in homes being built on smaller lots, often clustered close together while sharing a common green acreage nearby. Landscapes are predicted to become increasingly smaller but more luxurious as the baby boom generation, which has seen the sights of Europe and elsewhere, begins to gray and settle down to a quieter lifestyle at home. Many have good incomes, and good pension programs, and it is expected that they will create a new luxury residential landscape market in the United States. The opportunities for employment within the industry will be good, but will require a higher level of professionalism than was typical of the industry in the past.

Concurrent with rising standards for residential landscaping are comparable expectations by commercial and civic clients. They use landscaping as mirrors to the public, hoping to reflect the quality of their products or services via the positive image conveyed through their corporate or institutional setting. While they may value weed-free lawns and properly pruned trees for entirely different reasons than a homeowner, their appreciation of companies and individuals capable of providing these features is as genuine.

As the demand for landscape services has grown in volume, client expectations, and financial commitment, traditional small companies have been unable to meet the needs of the expanding client pool. Larger companies have come into existence, most since the 1970s, knowing at the outset which client type they would serve, what specialty services they would offer, and how to provide those services in a qualitative and cost-effective manner. These companies are as attuned to business management techniques as they are to ornamental horticulture. They seek employees who know how to work with plants and people. The expectations for employee education are growing rapidly, and the competition between companies for trained potential employees is often fierce. At all levels within the organizational structure of larger companies, the Caucasian male, long the dominant component of the labor and management team, is being joined by men and women of other ethnic origins and from other countries. These nontraditional workers will soon represent the majority of workers in the industry. To facilitate and accelerate the training of the workers, many companies offer in-house educational sessions on topics that their workers need to understand. Using videos and/or instructor-led training sessions, frequently conducted in Spanish or other non-English languages, the companies encourage their employees to grow professionally and personally. Many companies offer their employees full access to corporate planning objectives, financial statements, marketing strategies, and records. Some companies will subsidize the costs of their employees taking courses at nearby colleges. All of this and more is done by companies to help their employees become better educated workers who understand their roles in the company. By feeling part of the organization, rather than estranged from it, employees are better able to understand the significance of their work in satisfying the needs of the clients and conveying a positive image of the company.

The landscape industry of the 1990s continues to offer the services that it has in the past, structured both traditionally and nontraditionally. The three branches of the industry are still the same as they have been for many years: landscape architecture, landscape contracting, and landscape management. **Landscape architects** are the professionals who conceptualize the outdoor spaces.

They seek a perfect match between the desires of a client, the capabilities of a site, and the best interests of the environment. If perfection is unattainable, then they strive to come as close to it as practical. New landscapes have their origins in the mind's eye of the landscape architect or an allied professional, the **landscape designer**. In most states landscape architects are registered professionals. Landscape designers may be, too, depending upon individual state laws. Registration usually requires the accomplishment of formal education, a specified period of apprenticeship, and successful completion of a certification examination. Certification as a landscape architect is a more difficult achievement than designation as a landscape designer, although there is considerable variation among the states. Usually, the greatest apparent distinction between a landscape architect and a landscape designer is in the scope of their projects. Landscape designers usually work with residential clients, while landscape architects are more likely to be involved with institutional or corporate clients or very affluent residential clients seeking complex assistance that is beyond the capabilities of most designers.

Using the traditional tools of the draftsman and/or the new technology of computers, the landscape architect or designer transfers ideas and concepts from his or her mind onto paper where it can be seen and understood by clients and others. By using different views of the design, printed text, colored renderings, computer imaging, and old-fashioned persuasion to get the client excited about the proposal, the landscape architect or designer seeks to gain client acceptance of the ideas and approval to begin. Once approval is gained, the design professional must then prepare and present the design proposal in a way that will explain its intent to those who will eventually build it. Construction drawings, cost estimates, contract signing, and supervision of the construction on behalf of the client all follow as the designer's ideas come off the drawing board and go into the ground. Landscape architects are the client's representative from beginning to end if the client so wishes. When the client is not one person, but instead is a corporate board or governmental agency, the landscape architect is especially valued as the one who provides a face and a voice to the others involved with completing the project.

Most states now require that a registered landscape architect receive formal training from a university program accredited by The American Society of Landscape Architects. There are about forty accredited programs across the nation, although the number changes from time to time.

The actual construction of landscapes requires skill and tools that are different from those of the landscape architect. It is the **landscape contractor** who accepts responsibility for transforming the landscape architect's or landscape designer's plans and ideas into reality. Today's landscape contractor wears several hats: horticulturist, construction craftsman, business manager, service specialist. Landscape contracting companies are the most complex branch of the landscape industry, and the branch which has moved farthest and fastest from its traditional roots in recent years. Formerly a small landscape contracting firm would have attempted to provide all of the skills and services needed to accomplish the installation of a landscape project. Some companies still do. However, the economic realities of doing business for cost-conscious and service-savvy clients today has convinced some companies to only do what they can do better and more cost effectively than anyone else. While they may accept responsibility for complete construction of a landscape project, they will only personally do a part of it, such as plantings and pavements, and **subcontract** the remainder of the work to other specialty firms that can do the work better, faster, and cheaper. Pools or other water features, electrical work, stonework, or other requirements of the design may be subcontracted. It is often disappointing to young people who enter the industry with an enthusiasm for everything to discover that their companies' involvement with big projects has a more narrow focus than they were expecting.

Due to the diverse nature of landscape contracting, employees must possess a wide range of skills, with no one person likely to embody them all. The person most skilled in sales solicitation is unlikely also to be an expert at paver installation or fence construction. The educational requirements needed for employment as a landscape contractor are not as definitive as for a landscape architect. Few states have certification standards, although there is some movement in that direction currently being promoted by The Associated Landscape Contractors of America. They have established curriculum standards for colleges and universities that choose to offer an ALCA-approved major in landscape contracting. Numerous two-year and four-year college programs in the nation offer training in landscaping, but only four or five have sought and received the ALCA accreditation. ALCA is also promoting the certification of landscape contractors at a state or regional level, which would meet national standards for competency established by ALCA. Certification would require periodic renewal to assure that landscape professionals stay up to date.

For employees working at the crew level in a landscape contracting company, the required horticultural skills can usually be learned on the job. Many people enter the industry by apprenticing. Should an opportunity or personal ambition cause the crew-level worker to move into a supervisory position, there may be a need for greater skills than can be learned on the job. As an employee moves through the ranks of a landscape contracting organization, the daily use of horticultural knowledge is often supplanted by the need for human resource and business skills. Therefore, someone seeking to maximize career opportunities in landscape contracting should be trained not only in ornamental horticulture, but in business and personnel skills as well. While it is still possible to have a career in landscape contracting without acquiring a college degree somewhere along the way, the opportunities are greater for men and women who bolster their personal credentials with solid academic training before or during the development of their careers.

Landscape management involves the extended care of existing landscapes. It begins with the expiration of the guarantees given by the landscape contractor, at which time the landscape architect has gone on to seek other clients and the landscape contractor has moved trucks, tools, and workers to break ground at another site. The landscape that they leave behind is far from complete. It is not yet grown, not yet weathered, not yet used ... and abused. The new landscape must be sustained and managed, just as a building, farm, or office is managed. A budget must be established and a schedule put into place that will predict the number of mowings, snow plowings, prunings, litter sweeps, flower rotations, etc., that are needed to guide the design toward mature fruition at the standard of maintenance desired by the client. While landscape management may sound like the grown-up version of the neighbor's child who cut lawns for pocket money, it is the fastest growing area of landscaping today. It requires horticultural skills different from those needed by the landscape designer and contractor, and often necessitates equally different tools and equipment. The subcontracting of landscape management services is much less common than the subcontracting of landscape contracting services, but management companies may divide their workforce for specialization with either residential or commercial clients.

While there are numerous successful companies that offer only landscape contracting or landscape management services, even more companies offer both. They see a successful construction performance for a client as the way to acquire the long-term management contract. Over a period of several years, the latter may prove more financially lucrative than the former. Construction income may be a one-time profit, but the management contract represents ongoing income. As the economy fluctuates, companies that offer both construction and management can often alter their balance of work to meet the needs of the marketplace and remain profitable even as the more specialized firms begin to flounder. Aside from the differences in the task performed by the field crews, the organization and operation of landscape companies offering construction and management services are so similar that the inclusion of both under one corporate umbrella makes sense. The educational preparation required for career development within such companies is nearly identical as well.

Even landscape architects who have long regarded themselves as professionally distinctive from the rest of the industry are recognizing the benefits of closer collaboration with landscape contractors. Some landscape contractors now employ their own registered landscape architects, permitting them to offer a full range of services to a client and thereby relieving the client of the need and bother to deal with multiple companies in order to get a landscape project completed. One-stop shopping is desired by many private clients, and the large landscape firms are able to offer that convenience. Such companies also provide employment opportunities for landscape architects, many who value the complete control over a project that is provided when one company handles everything from initial contact with the client, through the design stage, and on to construction and long-term maintenance. Frequently referred to as **design-build firms**, such companies are becoming increasingly common nationwide.

In summary, the landscape industry in America is changing. Once composed entirely of small companies that served a local market and tried to provide all of the services required by their residential clientele, the industry is evolving into one that serves larger markets, often crossing state lines, and serving a more diverse clientele. It employs people from across the ethnic, gender, and educational spectrum. It actively strives to improve its professional image and price tag by hiring, training, and retaining talented people who are, in turn, seeking positions with companies that offer them good starting positions and the opportunity for advancement. It is an industry growing so quickly that there are few wise old men and women to learn from. As a result, company executives are teaching themselves and passing on what seems to work to their subordinates. Active national organizations such as the American Society of Landscape Architects, the Associated Landscape Contractors of America, the Professional Grounds Management Society, the Professional Lawn Care Association of America, and the National Landscape Association are contributing to the education of their members through newsletters, conferences, video productions, and book sales that bring new ideas and techniques to the forefront. As the industry upgrades its self image, it should follow that it will upgrade its public image. That is essential if the industry is to attract and retain the talent it needs from the college campuses of the nation. An upgraded public image is also essential to permit companies to charge fairly for their services. Customers who wouldn't flinch at bills for medical services or lawyers fees must be persuaded that the talent and skills applied to the development of their landscapes warrant a professional charge as well. That acceptance will only come from customers when it is warranted; thus the continuing need for professionalism at all levels within the landscape industry. It is an interesting point in time to observe the landscape industry. It still has a foot in the past, but it is both looking and moving toward a future that is promising, yet undefined.

Landscape
Designing

Using Drawing Instruments

CHAPTER 1

Objectives:

Upon completion of this chapter, you should be able to

🌿 identify and use properly the traditional tools of the landscape designer

🌿 measure and duplicate angles

🌿 measure and interpret dimensions to scale

The drawing tools of the landscape architect and the landscape designer have traditionally been the same as those of the engineering draftsperson. With very little alteration over many years, traditional tools are still highly important to modern designers, and a knowledge of their proper use is essential to understanding how landscape design plans are created. However, in the 1980s landscape designers began supplementing traditional drawing methods with computer-aided methods. Compu-terized drawing technology has grown rapidly, necessitating a knowledge of both traditional and high-tech methods if the full spectrum of drawing options is to be available to the designer. This chapter will focus on the traditional tools. The use of computers in the landscape industry will be discussed in a later chapter.

The Purpose of the Drawing Tools

When an idea begins to form in the mind of a designer, it is seldom clearly defined. It is usually necessary for the designer to make some preliminary sketches for his or her own benefit in order to prepare something for eventual presentation to the client. As the design evolves from fuzzy idea to final presentation, it becomes important to represent the idea in a form that will enable the client to understand the proposal, envision how it will look, and later present it to the landscape contractor in sufficient detail to permit realistic cost estimation and actual construction. To be most beneficial to all users, a landscape design must be drawn to scale, representing actual dimensions and placements. Through its graphic styling it must create an image of how the new landscape will look and what it will be like to live within it. The plan must also convey an impression of professional competency on behalf of the landscape firm if the client is to be persuaded to entrust that designer and that company to steer the project to completion.

The Tools and How They Work

The **T-square** is a long straightedge that derives its name from its shape, Figure 1-1. When used with a smooth-surfaced **drawing board** or **drawing table** that has four 90-degree corners, the T-square can be used to draw a series of parallel horizontal or vertical lines, Figure 1-2. It is important that the T-square be kept flush with the edge of the drawing board at all times when parallel or perpendicular lines are being drawn. Common uses of the T-square include the representation of property lines, roads and drives, fences, and utility lines, as well as adding borders around drawings. The T-square is also used as a base of support for other tools such as triangles and lettering guides.

The **drawing pencil** is the most frequently used tool of the landscape designer, and probably the most important. It is the link between the designer's mind and the paper. It may be inexpensive and

short-lived, as with wooden pencils, or more costly but longer-lasting, as with lead holders, Figure 1-3. Wood pencils have a graphite core (the lead), while lead holders have a chamber that holds replaceable leads. Both wood pencils and lead holders require some type of sharpening device to create the desired point. **Sandpaper pads** are often used, as are **specialized pencil sharpeners**, Figures 1-4, 1-5. Millimeter pencils hold thin leads of prescribed diameters, which create lines of comparable widths and never require sharpening. The consistent width can be a convenience as long as varied line widths are not required.

Drawing leads are given H or B ratings, which are measures of their comparative hardness or softness. The harder leads have H ratings, and softer leads have B ratings. The higher the H rating is, the harder the lead; while the higher a B rating is, the softer the lead. Leads may be rated as hard as 9H or as soft as 6B. While no rule exists, most landscape designers work with 2H or 3H pencils or leads for basic drawing. However, other leads are frequently employed for special needs, especially for interpretation techniques. The harder a lead is, the lighter will be the line and the longer the lead will retain its point. The softer a lead is, the darker and more easily smudged will be the line, and the more often will the lead require sharpening.

Figure 1-1 T-square

Figure 1-2 The T-square may be used to create parallel horizontal or vertical lines.

Figure 1-4 A wood pencil being sharpened using a sandpaper pad

Figure 1-3 Types of drawing pencils. From bottom to top: Wood pencil, lead holder, and millimeter pencil

Figure 1-5 The lead in the lead holder is sharpened using a mechanical lead sharpener.

Technical pens are used when ink is needed or preferred over pencil for the drawing. While there is sharper line definition and greater permanency with ink, it is more difficult to correct errors or make changes to inked drawings. Pens are available in both disposable and refillable forms. Refillable pens have a reservoir that contains the ink. Like millimeter pencils, technical pens create lines of a prescribed width. That necessitates a set of pens, rather than one, if the drawing requires lines of varying widths, Figure 1-6. The tips of the pens are constructed of different metals. Which type of tip is appropriate depends on the type of drawing surface being used. Some papers and plas-

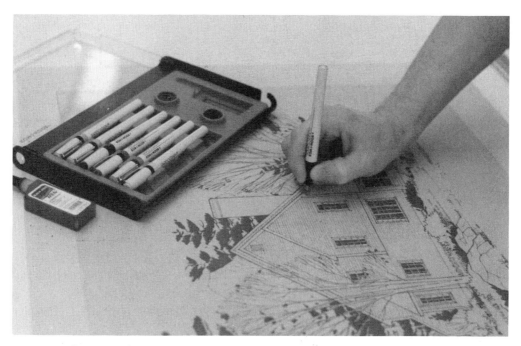

Figure 1-6 Technical pens. Each one creates a different line width.

tics are more abrasive to certain metals than to others, causing some pen tips to wear out sooner than others.

Lettering guides or templates are used to create a stenciled lettering style that may be appropriate for certain kinds of drawings. Whether using pencils or technical pens, the lettering guide must be matched to the width of the lead or pen tip. The result is lettering that appears somewhat stiff and mechanical, Figure 1-7. Lettering guides are used in combination with the T-square to keep letters evenly aligned. When ink is used, the guides must have raised edges to allow the template to slide back and forth along the T-square without smearing the letters.

The **Ames lettering guide®** is an inexpensive plastic device used in association with the T-square to produce guidelines for hand lettering. It can be set to create lines in varying numbers and of varying widths, which can then be replicated endlessly without requiring additional measurement, Figure 1-8. The Ames guide also has straightedge sides, which allow it to be used in actual creation of lettering as well.

Figure 1-7 Lettering template

Triangles serve many functions in landscape design and illustration. Commonly constructed of clear or colored plastic, the triangles will have either 30-60-90 or 45-45-90 angle combinations,

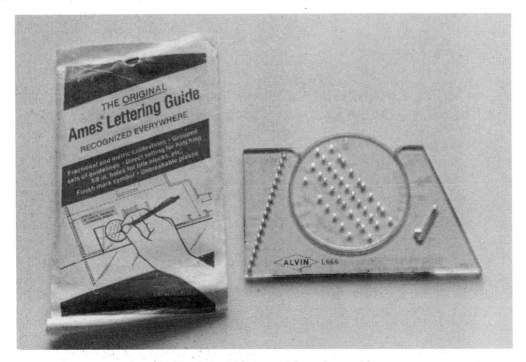

Figure 1-8   The Ames® lettering guide

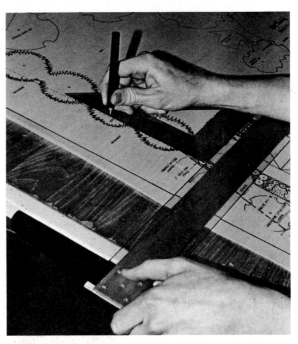

Figure 1-9   The 45 degree – 90 degree triangle (left) and the 30 degree – 60 degree triangle (right).

Figure 1-10   The triangle can be used with the T-square to create angles of 30, 45, 60, or 90 degrees.

Figure 1-9. The triangles may be used as straight-edges by themselves or in combination with the T-square to create lines at consistent angles of 30, 45, 60, or 90 degrees, Figure 1-10.

A **compass** is used to create circles, Figure 1-11. Circles are the basis for the symbolization of many landscape features, including plants, patios, pools, and lawn areas. The metal pointed leg of the compass is placed at the center of the circle, while the leaded leg creates the arc. When drawing circles, remember that the distance between the two legs of the compass should be half the desired diameter of the circle. For example, if a circle with a 4-inch diameter is desired, the compass legs should have a spacing of 2 inches, Figure 1-12.

The **protractor**, Figure 1-13, measures the relationship between two joined lines. The relationship

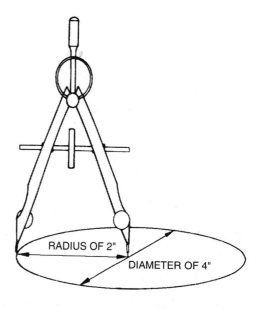

Figure 1-12 To create a circle with a diameter of 4 inches, the legs of the compass are spaced 2 inches apart. The radius of this circle measures 2 inches.

Figure 1-11 Compass

Figure 1-13 A 360-degree protractor. It is also available in a 180-degree model.

is known as an **angle**; the unit of measurement is a **degree**. To measure an angle, the center of the protractor's base is placed at the point where the two lines of the angle join. The 0-degree mark of the protractor's baseline is aligned along the lower line of the angle. The angle is determined by reading **up from 0 degrees** to the point where the second line intersects the protractor. In Figure 1-14,

Figure 1-14 The protractor is placed along one line of the angle and aligned with zero (0-degree). The angle is read at the point of intersection of the second line.

the protractor is measuring a 25-degree angle. It is important to remember that the reading is always taken between two existing lines, starting at 0 degrees. In this way, confusion over the protractor's double scale is avoided.

Curves are hard plastic or flexible plastic supports for curvilinear lines, Figure 1-15. The hard plastic curves, known as French curves, are available in assorted shapes. They are moved freely to whatever placement best supports the pencil as it creates freeflowing lines for the design. The flexible curves allow nearly limitless support for the designer's pencil, since they can be twisted into practically any shape. Both French curves and flexible curves are available in different sizes and lengths.

The **scale** is a tool used to represent actual dimensions in a reduced size. While the scale resembles an ordinary ruler, it has many more uses than merely measuring inches, Figure 1-16. Two types of scales are available: the engineer's scale and the architect's scale. While both are commonly used, beginning designers may find the engineer's scale easier to use initially, since it divides the inches into units that are multiples of ten; whereas the architect's scale uses multiples of four and is

Figure 1-15 Tools for guiding curvilinear lines. Left, the French curve; Right, the flexible curve

not as easy to read until the user is more familiar with it.

Depending upon the desired size of the drawing or the size of the property that must be represented on paper, a scale will be selected for the drawing, to correspond with one of the six sides on the scale instrument. For example, if a property line is 100 feet long, it can be represented on the designer's drawing as a line 10 inches long if the scale of the drawing is 1" = 10'. The same property line length would only be 5 inches long on the drawing if the scale of the drawing was changed to 1" = 20'. In Figure 1-17, the same measurement of 60 feet is located on each of

Figure 1-16 The scale. The triangular shape of this instrument creates six measuring sides.

Figure 1-17 Reading the scale. The six views show where one measurement (60 feet) appears on each side of the scale instrument. Individual foot units are not shown in the scales on the bottom line. (Drawing not to scale)

the six sides of the scale instrument. Note that each unit represents 1 linear foot regardless of the side of the instrument that is being used, as long as the scale of the drawing matches the side of the scale tool.

Templates are available in many styles and sizes to assist the designer with the creation of a neat, attractive plan. Circle templates can be helpful for uses such as plant symbolization; irrigation templates for representing the components of irrigation systems; tree and shrub templates for use when original styling is unwarranted; geometric shape templates to represent assorted non-plant features in a design; burnishing guides to permit rapid representation of textured pavement surfaces; and assorted other templates for cars, people, arrows, etc. The variety is almost endless.

Erasers need no explanation of purpose; but it is worth noting that there are different kinds available, depending upon the designer's need. **Plastic erasers** have all but completely replaced the traditional rubber eraser as the workhorse in the drawing room. The plastic eraser wears away more slowly and is washable, permitting it to be cleaned of graphite and used again. It does not become hard and unusable over time. Should a sharp edge be required on the eraser, the plastic can also be carved or whittled with a sharp knife or razor blade. Plastic erasers are produced by a number of companies and in a variety of forms, including plugs that fit into electric erasing machines. **Kneaded erasers** are soft and pliable and can be used to lift graphite from a drawing by placing the eraser on the line, pressing down, and then removing it. As a technique for lightening a line, it is very helpful. **Special purpose erasers** for the removal of inked lines or colored marker are available, but many do not perform as well as they claim. **Liquid erasers** are also available for removing old, dried ink lines or those created by copy machines.

An **erasure shield** can be helpful when erasing graphite lines, especially those made with soft B leads, which are prone to smearing, Figure 1-18. The thin metal shield is filled with punched out slits, which can be positioned over the line or word to be removed, thereby protecting the rest of the drawing from being smeared by the eraser.

Drafting powder also prevents smudging and

Figure 1-18   An erasure shield reduces graphite smears

keeps the drawing clean. The powder is available in bottles, shakers, and bags. A light dusting is sprinkled over the drawing as the designer works, absorbing skin oil and preventing the drafting tools from making direct contact with the graphite. As hands, arms, and tools move back and forth across the paper's surface, smudging is minimized.

Drafting tape is a low-adhesive paper tape produced in continuous rolls or as individual paper dots. The low adhesion is intentional and desirable, since its purpose is to hold the drawing on the board while the designer works, then release the drawing upon completion without tearing off the corners. Other paper tapes, such as masking tape, should not be used. Although they often look like drafting tape, they have much greater adhesion and can tear paper when removal is attempted. **Transparent mending tape** is also useful when drawings do get torn, and most designers keep a roll nearby.

The **rolling ruler** is a multi-use tool, and as such is favored over the more traditional ones by some designers. It is literally a straightedge on a roller. That permits the designer to make a series of parallel lines at any angle of choice without using another straightedge for base support, Figure 1-19. The tool can also be used to make circles, Figure 1-20, although there is not the infinite choice of

Figure 1-19   The rolling ruler being used to create parallel lines

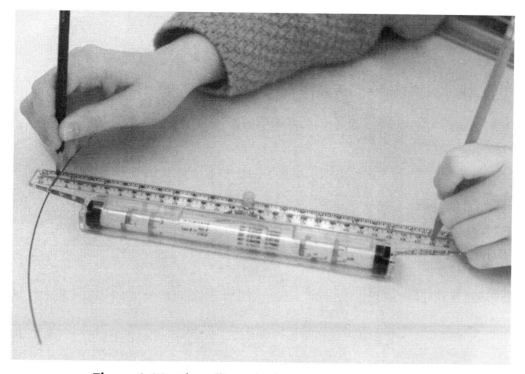

Figure 1-20   The rolling ruler being used to create circles

diameters offered by the compass. Skillful use of the rolling ruler requires some practice, because it is prone to slipping on the paper.

Rendering tools, such as colored pencils and color markers, are usually not required or used during the design stage of a landscape project. They are more commonly used when preparing the design for presentation to the client. Therefore the description and discussion of rendering tools and uses will be reserved for a later chapter.

Getting Started

Tools are manufactured in a wide range of sizes and qualities. Less expensive and lower quality ones are commonly available in neighborhood malls. High-quality, professional tools can be purchased in a drafting supply or art supply store or from a catalogue that carries professional instruments. As with most things, the value of the initial investment usually determines the long-term value of the purchase.

Practice Exercises

A. Tape a piece of drawing paper onto a drawing board. Practice drawing parallel horizontal lines and vertical lines using a T-square. Use a 2H pencil first and then a 3H or 4H pencil. Which pencil marks smear most easily? Which are easiest to erase?

B. With your protractor, measure the angles in the figure below. Duplicate the angles on a separate sheet of paper.

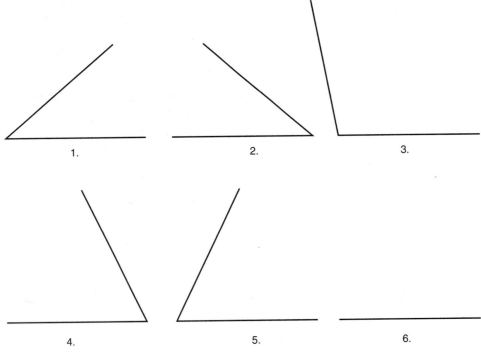

C. Using a scale instrument to measure and a triangle as a straightedge, draw a line 30 feet long to the scale of 1" = 10'. Draw another 30-foot line to the scale of 1" = 50'. Draw an 87-foot line to the scale of 1" = 40' and another to the scale of 1" = 30'.

D. Use a compass and a scale instrument to create a circle with a 20-foot diameter drawn to the scale of 1" = 10'. What is the radius setting for the compass? Draw another circle with a 20-foot diameter to the scale of 1" = 20'.

Achievement Review

A. Explain the function of each drawing instrument listed.

   1. T-square
   2. triangle
   3. compass
   4. protractor
   5. scale
   6. curve
   7. drafting powder
   8. rolling ruler
   9. technical pen
   10. Ames lettering guide

B. From the choices offered, select the best answer to each question.

   1. A compass set for a 2-inch radius forms a circle of what diameter?
      a. 2 inches
      b. 4 inches
      c. 1 inch
      d. 6 inches

   2. If the scale of the drawing is 1" = 5', the completed circle described in question 1 would be how many feet wide?
      a. 5 feet
      b. 10 feet
      c. 15 feet
      d. 20 feet

   3. Which of the following indicates the hardest pencil lead?
      a. 2H
      b. 3H
      c. 4H
      d. 5H

   4. Which two instruments combine most easily to make a 90-degree angle?
      a. compass and scale
      b. T-square and triangle
      c. protractor and scale
      d. compass and T-square

   5. Why is masking tape unsuitable for securing a drawing to the drawing table?
      a. Paper tape discolors the drawing paper.
      b. The adhesion is too strong.
      c. The adhesion is too weak.
      d. Cloth tape is better.

   6. Why are plastic erasers preferable to rubber erasers?
      a. They are less expensive.
      b. They are disposable.
      c. They last longer.
      d. They eliminate the need for erasure shields.

   7. If a driveway is 50 feet long in reality, how long will its symbol be on a drawing with a scale of 1" = 20'?
      a. 2-1/2 inches
      b. 5 inches
      c. 7-1/4 inches
      d. 10 inches

   8. If the scale of the drawing is changed to 1' = 5', how long will be the symbol of the drive?
      a. 2-1/2 inches
      b. 5 inches
      c. 7-1/4 inches
      d. 10 inches

   9. If the scale of the drawing is changed to 1" = 10', what will be the actual length of the drive?
      a. 2-1/2 inches
      b. 5 inches
      c. 10 inches
      d. 50 feet

   10. Which is the softest lead?
      a. 5B
      b. 4B
      c. 3B
      d. 2B

CHAPTER 2

Lettering

Objectives:

Upon completion of this chapter, you should be able to

- ❦ explain the importance of designer lettering
- ❦ describe different methods of lettering designs
- ❦ create a variety of lettering styles

The Importance of Lettering

Today's sophisticated graphic technology has made everyone consciously or subconsciously aware of the importance of lettering as a support element in the overall visual presentation of a product or service. Many corporations obtain trademark protection for not only the names of their products, but for the lettering styles as well. Seemingly endless variations of style can transform the twenty-six characters of the alphabet into words that lend interest and excitement to graphic displays. Land-scape plans and drawings also require lettering that matches the sophistication and importance of the work they are intended to explain. While never overshadowing the plan or drawing, neither should the size and style of letter-

ing used by a designer diminish the worth of the project being proposed. As illogical and unfair as it seems, many clients will associate the graphic quality of a presentation with the quality of the design. Therefore anyone who would be a good landscape designer should also strive to be a good graphic artist. The landscape designer must master enough graphic art skills to permit all aspects of a landscape proposal to be conveyed fully to a client. Foremost among those graphic skills is the development of a good hand lettering style, as well as knowledge of other ways that letters can be applied to drawings.

Methods of Lettering

Several factors affect the selection of a lettering method by a designer, not the least of which is the choices available. Not every firm possesses a full state-of-the-art range of selections. That aside, the choice of method is most likely to be affected by the importance of the immediate drawing and the time schedule for its production. Preliminary drawings that will stay in-house and will not be seen by the client require lettering that is fast and functional. Presentation drawings for the client and the public require lettering that is carefully styled and sized to fit the presentation graphics. In nearly every case, however, a landscape designer will seek and select a lettering method that is easy to create. There is little justification for spending a long time on lettering. Time is money, even for artists and designers.

The methods of applying letters to landscape drawings include

❦ lettering templates.

❦ waxed press-on letters.

❦ lettering machines.

❦ transfer film.

❦ hand lettering.

Each method has certain characteristics that make it suitable for certain circumstances or types of drawings.

Lettering Templates

Lettering templates are guides that reproduce the same letter size and style over and over. They may be used with a drawing pencil, lead holder, millimeter pencil, or technical pen. The drawing instrument must be selected to match the width of its tip with the width of the lines punched into the template; otherwise the lines of the letters will be wobbly. If technical pens and ink are used, lettering templates must have raised edges so they can be used without smearing the ink, Figure 2-1.

Lettering templates in various sizes and styles are available on the commercial market. A single landscape drawing may require letters of various sizes. Thus, a selection of lettering templates may be needed to complete a drawing. Using mixed styles of letters within the same drawing is usually avoided because it creates visual complexity.

The use of lettering templates by landscape architects and designers is limited. Templates

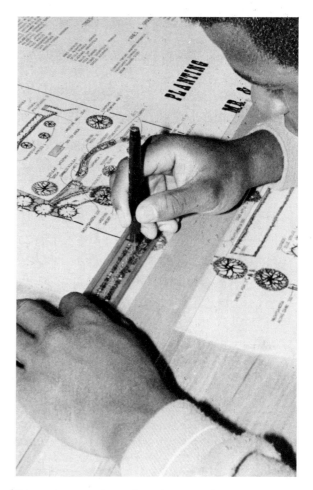

Figure 2-1   Using a lettering template

are time-consuming to use, and they create stiff, mechanical-looking letters, lacking the visual interest that hand lettering creates or the style that more sophisticated technology can now produce.

Waxed Press-on Letters

When a hand-drawn landscape drawing requires stylized lettering to accompany it, one of the methods available is the use of **waxed press-on letters**. Quick and easy to use, the letters are commercially manufactured in an assortment of styles and sizes, Figure 2-2. The letters are mounted and

Folio Extra Bold	**Gill Extra Bold**	News Gothic Bold	*Berling Italic*
Folio Bold Condensed	Grotesque 7	News Gothic Condensed	**Berling Bold**
Franklin Gothic	GROTESQUE 9	**Pump Medium**	Beton Medium
Franklin Gothic Italic	*Grotesque 9 Italic*	**Pump**	**Beton Bold**
Franklin Gothic Cond.	Grotesque 215	**Simplex Bold**	**Beton Extra Bold**
FRANKLIN GOTHIC EX COND	Grotesque 216	Standard Medium	Bookman Bold
FRANKLIN GOTHIC COND	Helvetica Ex Light	**Standard Extra Bold Cond.**	*Bookman Bold It.*
Futura Light	Helvetica Light	Univers 45	**Carousel**
Futura Medium	Helvetica Medium	Univers 53	Caslon 540
Futura Medium Italic	**Helvetica Bold**	Univers 55	**Caslon Black**
Futura Bold	*Helvetica Light Italic*	Univers 57	**Century Schoolbook Bold**
Futura Bold Italic	*Helvetica Med. Italic*	Univers 59	Cheltenham Old Style
Futura Demi Bold	*Helvetica Bold It.*	Univers 65	**Cheltenham Med**

Figure 2-2   Styles and sizes of press-on letters

sold on sheets of plastic or waxed paper, Figure 2-3. They are transferred to the drawing surface by the heat of friction created when a pencil is rubbed over the surface of the sheet, Figure 2-4. To improve their adhesion, the letters are then rubbed again with a pencil through a sheet of waxed paper that usually accompanies each sheet of letters, Figure 2-5.

Advantages of Waxed Press-on Letters

- easy to apply
- create a professional graphic image
- easily removed with cellophane tape, permitting correction of errors

Disadvantages of Waxed Press-on Letters

- expensive
- crack with age
- wasteful; vowels commonly depleted before the consonants

Figure 2-3   A sheet of one style of press-on letters

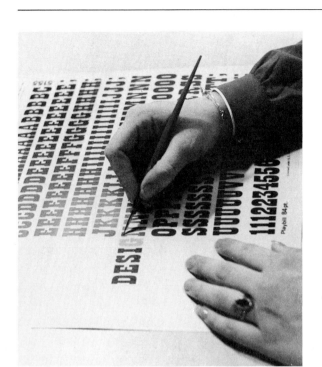

Figure 2-4 Press-on letters are transferred onto the design by the heat of friction.

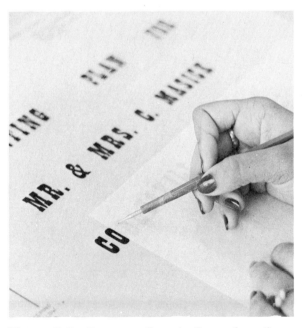

Figure 2-5 To secure them to the surface, the letters are rubbed again through a sheet of waxed paper.

Lettering Machines

Designers who prepare large numbers of presentation drawings are apt to favor a **lettering machine** over waxed press-on letters. While the initial expense is greater, over time the machine is more cost effective. The lettering machine is simple to operate, since it functions much like a typewriter. A lettering style and size are selected and placed on the machine. Some models use plastic wheels; others use plastic chips. Regardless, the operator selects the letter desired, and the machine imprints it onto a clear tape that is contained in a cartridge. When the desired words are completed, the tape is clipped off and applied to the drawing. Because the tape is transparent and thin, it is nearly invisible after application, Figures 2-6 and 2-7. Unlike waxed press-on letters, the lettering machine involves no waste. Machines and lettering fonts are purchased separately, as are tape cartridges. Prices vary greatly between manufacturers and models.

Transfer Film

Transfer film is paper-thin, transparent, plastic sheeting with a clear adhesive backing and a sheet of protective waxed paper over it. It enjoys a number of uses in graphic art since anything that is drawn on it can then be placed and affixed anywhere desired by the designer. Commonly referred to as "sticky back" by landscape architects and designers, transfer film is used for plant lists, construction notations, client names and addresses, corporate logos, and dozens of other design components. The initial lettering can be prepared on a typewriter or word processor and printed on a standard sheet of paper. It is then replicated onto the transfer film by passing it through a photocopy machine, using the film in place of conventional copy paper. Lather the protective backing is pulled away and the film positioned where desired on the design, Figure 2-8. The use of transfer film by landscape designers is common. It allows lettering that is used repeatedly, such as the company name and logo, scale indicators, slogans, page markings, and standard notations, to be added quickly whenever and wherever they are needed by simply pulling the master sheet from a file and passing it through a copy machine. Transfer film is also an easy

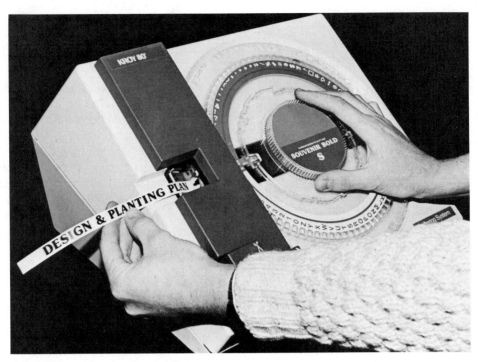

Figure 2-6 The lettering machine permits many styles and sizes of letters to be produced without muscle.

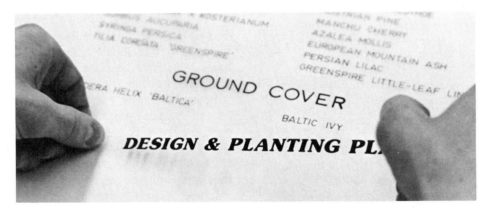

Figure 2-7 The lettered tape is peeled from its backing and applied to the tracing surface.

way to merge the modern technology of computer-generated lettering with traditional forms of designing and drawing. Selecting from the many font styles of modern word processors and adjusting the size of the letters to match the drawing being labeled, designers can use transfer film in the same way they used waxed press-on letters in the past.

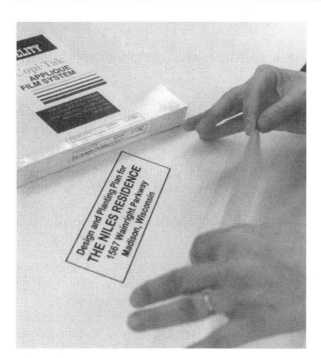

Figure 2-8 Transfer film being applied

Hand Lettering

One of the hallmarks of professionally trained designers and graphic artists is a distinctive style of **hand lettering**. While it does not make the design better nor does it cause the proposed project to cost a penny more or less, good hand lettering offers visual evidence of designer competency to clients. Perhaps it is because most people do not have good lettering skills that they view favorably those who do. Whatever the reason, the development of a lettering style that is interesting, lively, and rapidly executed should be a goal of anyone who aspires to practice professionally as a landscape architect or landscape designer.

Hand-drawn letters are created with single strokes of the pen, pencil, or marker. To attempt more complex styles would be counterproductive because the lettering would take too long and the letters would probably by inconsistently styled. Instead, designers use styles that are simple variations of the basic block style that they learned as children in school, Figure 2-9. In that style, letters are constructed with single stroke lines, and their

heights and centers are controlled by the location of three **guidelines**. The guidelines may be pre-printed on the paper or may need to be drawn on lightly before lettering begins. If that early school training allowed any variation of the basic block form, it was to slant letters to the right at approximately 60 degrees to suggest a more natural, humanistic interpretation of the style, Figure 2-10.

To create letters that spell out words while simultaneously contributing to the overall appearance of the design presentation requires that the basic block style be altered. Alterations take one of two forms:

❦ changing the location of the center guideline.

❦ changing the number of guidelines from three to five.

Figure 2-9 The alphabet in basic block style

Figure 2-10 Slanted block alphabet

Changing the Center Guideline

Maintaining three guidelines, but raising or lowering the center one, causes the weight of the letter to shift down or up accordingly. Figures 2-11 and 2-12 illustrate the standard block style of lettering

Figure 2-11 Distorted block alphabet with midline raised

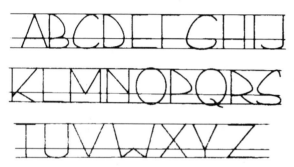

Figure 2-12 Distorted block alphabet with midline lowered

Figure 2-13 Slanted, distorted block alphabet with guidelines. Half of the alphabet has the midline above center, and half has the midline below center.

with the center lines changed. Figure 2-13 illustrates the changed appearance of slanted block letters when the center guideline is either raised or lowered. These styles require some practice, but are easily learned.

Changing the Number of Guidelines

Since the purpose of guidelines is to aid the designer in proper shaping of the letters, it follows that changing the number of guidelines from three to five will significantly alter the appearance of the letters. In Figure 2-14, two guidelines are added to the three used in the standard block style. Note that line 2 is very close to line 1 and that line 4 is very close to line 5. The main purpose of the two additional lines is to introduce a slight upward slant to the horizontal strokes of the letters and numerals. In the three-guideline style, all horizontal strokes are aligned directly on the guidelines. The five-guideline style creates an intentional horizontal angle while keeping the degree of angling consistent throughout, Figure 2-14. The five-guideline style has a looser appearance due to the slight horizontal angle of the lines and the inclusion of several deliberately distinctive letters, such as the *K* and *R*. It is only distinctive when done as upper case (capital) letters. It is not adaptive to lower case (non-capital) letters.

Hand Lettering Techniques and Tips

Good hand lettering is a graphic skill that too many beginning designers fail to master, largely because of their impatience and their desire to get on to bigger and better challenges. With practice and adherence to a simple but effective regime, everyone can develop a rapid and attractive lettering style that will enhance the total graphic presentation.

    a. **Always use guidelines.** Whether drawn with a straightedge or with a tool such as the Ames lettering guide (see Figure 1-8), all lettering should be constructed using guidelines. They will keep the lettering consistently sized and styled throughout the plan. They will also keep the lettering straight as long as the

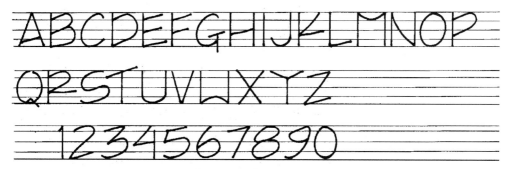

Figure 2-14 Lettering and numerals styled using five guidelines

drawing is secured to the drawing board and properly aligned.

b. **Keep the guidelines light.** It is usually unnecessary to erase lettering guidelines if they are drawn lightly. They need only to be dark enough to be visible to the person doing the lettering. A pencil with very hard lead is suitable for drawing guidelines. If the work is to be copied, guidelines can be made with a light blue pencil or pen, since photocopy and ozalid process machines do not reproduce blue well.

c. **Use straightedge support for all vertical lines.** All letters look best when their vertical alignment is parallel. A predictable mistake made by beginners and impatient designers is the belief that they do not need to use a straightedge for lettering once the guidelines are drawn. With the exception of very small letters, all letters need the sharpness and uniformity that straightedges provide. Figure 2-15 illustrates the use of a T-square and triangle to assure consistent vertical alignment during lettering.

d. **Continually monitor line quality.** Different types of drawing tools create different types of lines, and the designer must know what constitutes quality for each type. Maintaining an even line tone can be difficult when pencil is the drawing instrument. Lines drawn confidently with the support of a straightedge may be darker than curving lines drawn more cautiously. Curving lines often tempt the beginning designer to sketch the line rather than drawing it confidently with a single stroke.

Figure 2-15 The triangle, resting on the T-square, supports the vertical lines of the letters.

Pencil leads also widen if not sharpened frequently. That causes varying line widths within words. Unless deliberate line-width variation is being introduced, single-stroke lettering should also be single line-width lettering. Also, keep line intersections free of overlaps to avoid creating an inconsistent, shaggy appearance.

❦ Maintain even line tone.

❦ Do not sketch lines.

❦ Sharpen pencils to maintain consistent line width.

❦ Do not overlap line intersections.

e. **Space letters and lines to permit comfortable reading.** Letters that are crowded together or spaced too far apart create words that are difficult to read, Figure 2-16. The spacing between letters should be consistent throughout the word. The spacing between words should be greater than that between letters within a word. There must also be space between lines of words. The spacing common to printed lettering, such as this text, applies to hand lettering. If it is to communicate, the lettering must be easy to read.

f. **Maintain case consistency.** When upper and lower case lettering is used within a single presentation, the cases should not be mixed within words and sentences unless it is the first word of a sentence or is a word that warrants capitalization, Figure 2-17.

Introducing Line-Width Variation

An additional variation on the hand-lettering styles described above can be incorporated with the deliberate addition of **line-width variation**. The word *deliberate* is significant because the designer is intentionally creating a style, not just making a mistake in violation of the earlier concern for line quality.

Figure 2-16   Incorrectly spaced letters, words, and lines

Figure 2-17   Incorrectly mixed upper and lower case letters

To introduce deliberate line-width variation, the designer must either select or create an instrument with a chiseled tip. If pencil is the lettering instrument, the point needs to be widened and flattened with a sandpaper pad, Figure 2-18. If markers are used, those with broad nibs are the appropriate choice. Technical pens and millimeter pencils cannot be used for lettering when line-width variation is the style of choice, because their tips cannot be modified.

Using the chiseled tip, lines that are either thin or wide can be made by the same instrument, depending upon how it is held and the direction of the stroke, Figure 2-19. The designer must make a decision about whether to make the vertical strokes thin and the horizontal strokes wide, or vice-versa. Then, as long as there is no alteration of that decision, a distinctive and consistent lettering style will result. Figure 2-20 illustrates an alphabet and set of numerals created with thin vertical lines and wide horizontal strokes, using straightedge support for all vertical lines.

REGULAR POINT   CHISEL POINT

CHISEL POINT—SIDE VIEW

Figure 2-18 Wedge-point drawing pencil

Figure 2-19 With a chisel-tipped lead, the pencil can make both broad and thin lines.

Figure 2-20 Line-width variation in lettering can be accomplished using a chisel-tipped pencil.

Practice Exercises

A. On a sheet of drawing paper, lay out three horizontal guidelines, each 1/4-inch apart. Following the examples shown in this chapter, create a full upper case alphabet, 1/2-inch height. Use a hard lead for the guidelines and a softer lead for the lettering. (Drafting powder can be used to prevent smearing.) Use a T-square to support your triangle for straightedge support of the vertical lines. Letter the alphabet several times and repeat as often as necessary those letters that you find difficult.

B. Repeat the exercise above using a 60-degree triangle to support the vertical lines, thus creating a slanted block alphabet. Repeat as necessary to improve your skill.

C. Repeat exercises A and B using hard lead pencil for the three guidelines and a felt-tipped pen to create the letters. Repeat as necessary to improve your skill.

D. Lay out five guidelines for lettering a complete alphabet and set of numerals. The letters and numerals should be 1/2-inch in height. Make lines 1 and 2 and lines 4 and 5 close together, with line 3 at the midpoint. Refer to the example in the chapter for guidance. Using the T-square to support the triangle for straightedge guidance of the vertical lines, create the letters and numerals as shown in Figure 2-14.

E. Repeat exercise D using a chisel-tipped soft lead pencil. Remember to chisel the tip frequently to retain the consistency of the style.

F. Prepare a sheet of drawing paper for the lettering of a paragraph with words that will be 1/4-inch in height. Select one of the styles of lettering that you have practiced in the above exercises. Copy the first paragraph under the heading *Hand Lettering* in this chapter. Remember to space properly within words and between words. Also remember to leave space between lines of words.

Achievement Review

A. List five methods of lettering commonly used by landscape architects and designers.

B. Select the best answer to the following questions.

1. Assuming that all lettering choices were available to a landscape designer, what would most likely determine the method selected?
   a. the cost of the lettering method
   b. the level of graphic sophistication needed
   c. the preferences of the client
   d. the type of project being designed

2. What characteristic must all methods of lettering possess if used in landscape designing?
   a. colorful
   b. fancy
   c. rapid
   d. large

3. Which lettering method is most wasteful of material?
   a. lettering machine
   b. waxed press-on letters
   c. templates
   d. transfer film

4. What is the purpose of guidelines in hand lettering?
   a. to control the size and style of the letters
   b. to monitor line quality
   c. to match the lettering with the presentation style
   d. to assure correct placement of all labels and wording

5. What happens when the middle guideline of a three-line style is raised or lowered from the center?
    a. Nothing happens.
    b. The letters become wider.
    c. More space is needed between the letters.
    d. The weight of the letter shifts.

6. What does the deliberate introduction of line-width variation do to the letters created?
    a. It makes the letters wider.
    b. It destroys line quality.
    c. It creates a desirable difference in the lettering style.
    d. It eliminates the need for straightedge support.

Graphic Materials and Techniques

Objectives:

Upon completion of this chapter, you should be able to

- distinguish between plan views, elevations, perspective views, and axonometric views of landscape proposals
- create plan view symbols for major landscape features and label them correctly
- lay out a complete landscape design plan
- compare different levels of graphic presentations and the proper uses of each
- describe the characteristics and uses of graphite pencils, colored pencils, felt pens, markers, pastels, pressure graphics, foam board, and the papers and films used for both original and copied work
- arrange the components of a graphic presentation

Same Design, Different Views

The interpretation of a designer's proposal for a project advances through numerous levels of refinement as it moves from abstract idea to acceptance by the client and eventual implementation by the contractor. Each person or group that has input into the evolution of the project, from vague concept to finished product, possesses a different level of design expertise and/or understanding. Each person also is involved with the project in a different way. For some the involvement is emotional; that is usually the client. For others the involvement is technical; that is most likely the contractor. The client wants to love the project. The contractor wants to know how to build it. Along the way, the designer also is continually questioning his or her own thinking. Will this idea work? Will this feature fit the space? What will be the result if I put these two features together? The landscape designer must be able to prepare drawings that will speak to each constituent.

Plan Views

The most common drawing prepared by landscape designers is the **plan view**. It is drawn to scale and is therefore measurable. In a typical plan view, objects are seen as though the viewer is above the drawing, viewing it at a perpendicular angle to the ground. It is not unlike that experienced when viewing the ground from an airplane.

The plan view drawing is an excellent way for a designer to show the proposed placement of plants and other elements of the design in relation to each other and to the existing features such as the buildings, roadways, utility poles, and similar items. Horizontal measurements are readily made. The exact width and length of objects or the distance between objects can be determined by anyone who understands how to use a scale instrument, Figure 3-1. Landscape architects, landscape designers, and landscape contractors all understand how to read plan views. For them to correctly use the plan, it need not be attractive, only accurate. Clients, however, frequently do not understand how to read plan views, since they have only horizontal dimensions. The lack of vertical measurements means that the third dimension is missing. People untrained in reading plans or who lack knowledge of plant heights are unable to envision how the design proposal will look when completed. For such clients, the designer may need to embellish the plan view with graphic tricks such as symbols that suggest how a plant or other feature will actually look. Other techniques such as the use of colors and shadows may aid client understanding. In many instances it may be necessary to prepare a different type of drawing that is better suited to the client's understanding.

Elevation Views

An **elevation view** is also a scaled drawing, thereby making its dimensions measurable. It has two dimensions, horizontal and vertical, Figure 3-2. Although lacking one of the horizontal dimensions possessed by plan views, elevations are generally more helpful to clients for visualizing a design proposal. They are well suited for displaying size relationships between plants, buildings, and people. They are also helpful in showing vertical relationships between various features of the landscape, such as raised decks and recessed patios, or hills and valleys. For contractors, the combination of plan views and elevations of the same project provides them with most of the physical measurements and explanations they need to get the design built. As with the plan view, the elevation view may be done with simple, minimal graphic effort or with elaborate presentation techniques. The designer makes the decision of how much time to spend on the graphics, depending on the importance of the project, its current stage of development, and the viewer for whom it is intended.

Perspective Views

Other than an actual model of the proposed landscape project, nothing allows a better visualization of the designer's intentions than a **perspective** drawing. A perspective allows multiple sides of a proposed project to be seen in one drawing, Figure 3-3. While it can be scaled, the dimensions are not measurable because most of the lines of the drawing are seen to recede away from the surface of the paper. Receding lines in a perspective are termed *foreshortened*, meaning that they are not drawn to their correct length. To do so would cause the dimensions to appear unrealistic in the drawing. Also in a perspective drawing, lines that are actually parallel nevertheless appear to converge if extended far enough. Controlling the angle of line convergence requires the designer to have sufficient knowledge of drawing technique to create a realistic interpretation. As a drawing to aid construction, the perspective has only limited value. However, as a means of helping both the designer and the client visualize the proposal's finished appearance, the perspective has great value. Ranging in graphic sophistication from a quick sketch to a highly detailed interpretation, the perspective drawing plays an important role in landscape design graphics. Frequently, however, it is the one least used because the designer fails to recognize that the client is not seeing the proposal as clearly as he or she sees it. Other times, the project's value does not justify the time required for construction of a perspective view.

DESIGNED BY RUSSELL IRELAND
EAST NORWICH, NY

Figure 3-1 Example of a landscape plan

Figure 3-2 A front elevation of a residence and plantings (Courtesy of Lawrence Perillo, Smallwood Design Group, Naples, Florida)

PERSPECTIVE DRAWING by MARTIN BOZAK

Figure 3-3 A perspective view permits easier visualization of how the proposed design will appear.

In some cases the lack of a perspective view may be due to the designer's inability to construct the drawing. With the continual advances in computer graphic technology, however, lack of hand drawing skills may not serve as an excuse much longer.

Axonometric Views

While not as realistic to the observer as perspective drawings, **axonometric views** offer a look at multiple sides of objects while also providing measurable dimensions to scale. Lines of the design that recede are not foreshortened to create the realism of a perspective, nor do parallel lines appear to converge. Object lines that are parallel in reality are left parallel in the drawing. With no perspective tricks, axonometric drawings are less helpful to the client who wants to see how the proposal will appear when built; but the drawings are of much greater assistance to contractors, who can take measurements directly from them, Figure 3-4. Since all lines are drawn as actual scaled dimensions, and since the technique for constructing the drawing is much easier than with perspectives, many landscape architects and designers favor axonometrics as a rapid alternative to perspectives. Although axonometrics often look distorted, they still offer a semblance of realism and in many ways are much more usable by a greater number of people than are perspectives.

Figure 3-4 Axonometric drawing. All dimensions are measurable to scale.

Landscape Symbols

When preparing landscape designs in plan view, designers use symbols that offer a suggestion of how the proposed or existing features would appear if seen from above. To be functional, the symbols must be drawn to their true size. The size is actual in the case of existing objects such as buildings and streets, and eventual with plants, which are usually drawn to their mature size rather than their size at the time of installation. For preliminary drawings that will not be seen by clients, but help the designers work their way through their own thought processes, or for construction plans that provide installation information for landscape contractors, the symbols need only be of correct size and placement. However, for presentation plans that will be used to help the client understand the design proposal and aid in selling the proposal, more elaborate and realistic styling of the symbols is needed.

To make the client's understanding of the design as easy as possible, the symbols need to be suggestive of the features they represent, yet not so graphically complex that they distract the client from full appreciation of the design. Whether drawn by hand or selected from a computer-generated menu, the symbols become the principal way that the designer explains a proposal to the client.

Needled Evergreens

When recalling the appearance of a pine or spruce tree, it is easy to understand the symbols commonly used to represent needled evergreens, Figure 3-5. The symbols suggest the spiny leaves and rigid growth habit of these plants, which are green throughout the year. These symbols can be used to represent both trees and shrubs as long as they are needled and evergreen.

Broadleaved Evergreens

Another group of plants are green throughout the year, but have wider and usually thicker, fleshy leaves. Plants such as the hollies, rhododendrons, and camellias exemplify the category of broad-leaved evergreens. The symbols used to represent them suggest the larger leaf size and semi-rigid growth habit of the plants, Figure 3-6. As with the needled evergreens, these symbols can be used for both broadleaved evergreen trees and shrubs.

Deciduous Shrubs

Deciduous shrubs drop their leaves in the autumn. As a category of plants, they are much more numerous and diverse than either of the evergreen categories; therefore the symbols used to represent them tend to be more generic or general. The edge of the symbol is loose and irregular, suggestive of the less rigid growth habit of most deciduous shrubs, Figure 3-7. With this symbol, as with all plant symbols, a prominent dot in the center marks the spot on the plan where the plant is to be set into the ground.

Figure 3-5   Symbol styles for needled evergreens

Figure 3-6   Symbol styles for broadleaved shrubs

   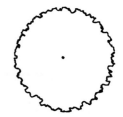

Figure 3-7   Symbol styles for deciduous shrubs

Deciduous Trees

Like deciduous shrubs, deciduous trees also lose their leaves during the winter season. Unlike most shrubs, trees usually have only a central trunk rather than multiple stems. Also, trees are generally taller than shrubs. All of these differences are suggested in the symbols selected by designers to represent deciduous trees, Figure 3-8. The symbols for trees are usually wider since trees are commonly larger than shrubs. Their line width may be greater, too, reflecting their greater height and prominence within the design. Depending upon how much detail must be shown beneath them, the tree symbols may be simple and plain, or intricately detailed to show branches and/or leaves.

Vines

Trees and shrubs tend to grow radially out from their centers. That is why the compass or circle template is so useful in forming their symbols. However, vines grow in a linear manner, and they do not hold to a predictable shape. Their symbols are shaped to suggest that rambling linearity, Figure 3-9.

Trailing Groundcovers

Groundcovers are those plants, usually eighteen inches or less in height, that fill the planting bed beneath the trees and shrubs. Like vines, many groundcovers are shapeless, linear plants that would have little impact on the design if used alone; but when grouped in masses and given time to fill in, they become an important textural component of the design. The symbols used to represent trailing groundcovers are therefore more textural than structural, Figure 3-10. The texture is applied to all areas in the design where groundcover plants will be used.

Hardscape Materials

Design materials that are not living plant materials are often referred to as **hardscape**. They include such things as pavings, fencing and wall materials, furnishings, lighting, and water features. Like plant

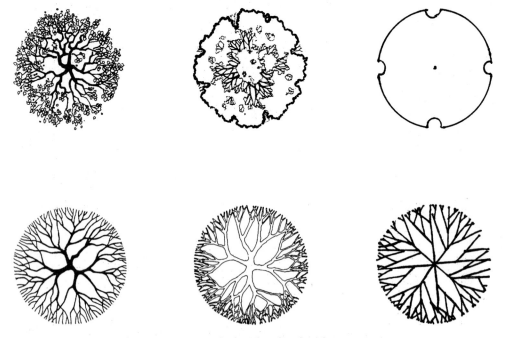

Figure 3-8 Symbol styles for deciduous trees

VINES AGAINST A BUILDING WALL

VINES ON A FENCE

VINES ON A FENCE

Figure 3-9 Symbol styles for vines

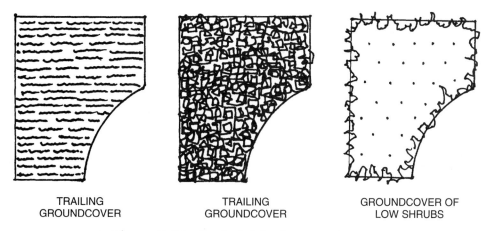

TRAILING
GROUNDCOVER

TRAILING
GROUNDCOVER

GROUNDCOVER OF
LOW SHRUBS

Figure 3-10 Symbol styles for groundcovers

symbols, hardscape symbols attempt to suggest how the materials or objects will actually appear in the landscape. They need to be scaled appropriately to allow the graphic to look realistic. When the scale of the plan is so small that it becomes impos-sible to draw each brick, stone, or similar feature to its exact size, then a textural interpretation may be used to suggest the hardscape rather than interpret it literally. Figure 3-11 illustrates typical hardscape features as represented in landscape plan views.

STONEWORK

LOOSE AGGREGATES AND MULCHES
(CRUSHED STONE, MARBLE CHIPS,
WOOD CHIPS)

DECKING

STEPS AND LANDING

PICNIC TABLE

BRICK

OUTDOOR GRILL

FENCING WITH GATE

Figure 3-11 Some common construction symbols

Explaining an Idea With Symbols

Once designers have learned to style symbols that others can interpret, they can begin using that skill to explain design concepts to others. While there are few hard-and-fast rules in graphics, over time certain techniques have been proven effective in assuring that the viewer is not misled or confused when interpreting the plan. Some of these techniques are more easily accomplished by hand drawing than by computer drawing, Figure 3-12.

❦ When one landscape object passes beneath another, the upper symbol is drawn with a wider, darker, and/or solid line, while the symbol for the object hidden beneath it is drawn with a thinner, lighter, and/or broken line.

❦ Plants that are exactly alike and are massed will overlap slightly, and the area of overlap is not drawn. That permits a single label to identify all of the plants in that massed grouping.

❦ Plants that are not exactly alike, but are massed, are not overlapped. Each is drawn completely and separately, since separate labeling will be required.

Labeling

To make the landscape plan as useful and understandable as possible, it is usually necessary to add labeling and other notations to the plan. The goal is to use only as much labeling as necessary to convey the required information, while assuring that

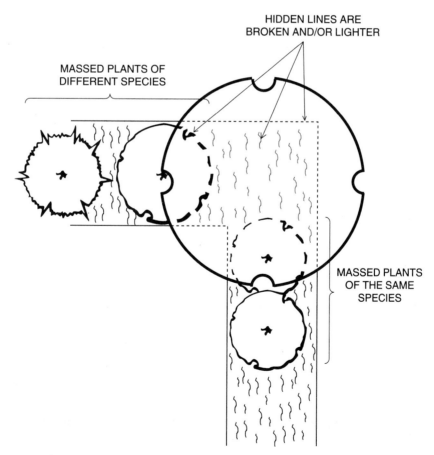

Figure 3-12 Techniques for explaining the designer's intent

the plan reads easily and quickly. Several types of labels should appear on the finished plan. Included are the symbol labels, the plant list, a directional arrow, the scale, the client's name and address, and the designer's name and firm.

Symbol Labeling

All of the graphic symbols used on the landscape plan must be labeled, and the labels must be on or as close to the symbols as possible. The closer the labels are to the symbols, the easier the plan will be to read. When the scale of the drawing results in very small symbols, it may be necessary to code the labels. That means that a number is used on or near the symbol that corresponds to a lettered label placed elsewhere on the plan, Figure 3-13. When more than one plant is being labeled, as in a grouping, the plant label or code number is followed by the number of plants in the grouping, Figure 3-14. By using one label for several plants or groupings of the same species, the number of labels can be minimized and clutter avoided. An excess of labeling lines (leader lines) can be confusing, often mistaken for object lines. They should be eliminated whenever possible.

Hardscape materials are labeled similarly to plant materials, but are seldom coded. There is no need to indicate specific numbers of bricks, stones, or boards. Instead, the feature to be constructed should be briefly described by such factors as height, color, or special treatment. For example, a label might read "Vertical louvered fence–5' tall/ grey stain." Another might be "Concrete with exposed aggregate surface."

Plant Lists

From the plan's symbols and labels, a tally can be made of how many of each species of plant are required in the total plan. The tally permits the development of a **plant list**, Figure 3-15. An important component of the layout sheet, the plant list always includes the first two items (in three columns) and frequently includes the last two:

❦ Two listings of the plant species; one column listing them in alphabetical order by botanical names and an accompanying column listing the common names

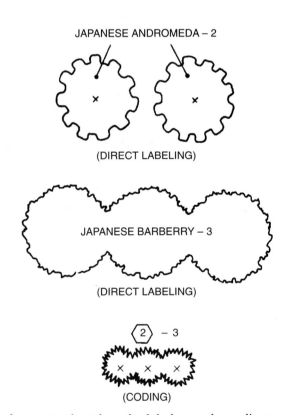

Figure 3-14 When the label or code applies to more than one plant of that species, the number should be indicated.

PLANT SPECIES CODE

#1	COMMON LILAC	SYRINGA VULGARIS
#2	CREEPING JUNIPER	JUNIPERUS HORIZONTALS

Figure 3-13 Coding the labeling of plant symbols is less desirable than direct labeling, but is sometimes necessary. Never use the techniques of direct labeling and coding on the same landscape plan.

❦ A column noting the total numbers of each plant species used

❦ A column noting the sizes of the plants at the time of installation

❦ A column containing special notations about certain species that the designer does not want overlooked

Scale Indicator and Directional Arrow

To permit the landscape plan to be measured and eventually installed, the scale to which it was drawn must be known and noted on the layout sheet. Figure 3-16 illustrates typical scale indicators.

PLANT LIST

BOTANICAL NAME	COMMON NAME	QTY	SIZE	COMMENT
ACER GINNALA	AMUR MAPLE	8	6'	B&B
BETULA LENTA	SWEET BIRCH	2	8'	B&B/SPRING INST.
CARPINUS BETULUS	EUR. HORNBEAM	14	6'	B&B/MATCHED
DEUTZIA GRACILIS	SLENDER DEUTZIA	25	18"	CONTAINERS
EUONYMUS ALATUS	WINGED EUONYMUS	25	3'	B&B
ILEX GLABRA	INKBERRY	19	3'	BR
MAGNOLIA STELLATA	STAR MAGNOLIA	3	4'6"	B&B/TAGGED
PINUS NIGRA	AUSTRIAN PINE	8	8'	B&B

Figure 3-15   A typical plant list. One of the components of the final layout

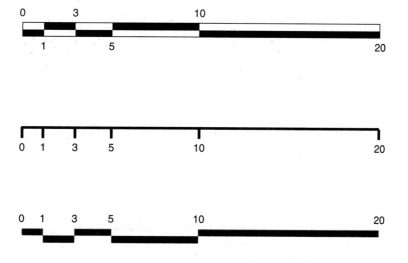

EXAMPLES ARE NOT TO SCALE

Figure 3-16   Examples of scale indicators

Also important to full comprehension of the design is an awareness of its directional orientation. That is provided by the directional arrow, Figure 3-17. While it only needs to be accurate in order to be functional, it is traditional for designers to make their directional arrows somewhat ornate. Some firms copyright their arrows as one way of lending distinction to their graphic presentations. The arrows should be neither too ornate nor too large, but they can be fun to design.

Client and Designer Names

The inclusion of the names of the clients and the designer is important. Not just words, these labels should be regarded as important components of the plan layout. The client's name usually includes the address of the project and may include additional descriptive text. It should be done using the largest and/or most ornate lettering on the plan. It is important to the customization of the plan and should be prominently featured.

The designer's name may include additional information, such as the firm name and address. It should not be formed with letters as large as those used for the client's name, and it should be located away from the client's name on the layout sheet.

Design Sheet Layout

There is no single correct way to arrange the design and various lettering components on the sheet. Certainly the design is the most important component and needs to be featured. The other components will vary greatly in size and prominence from layout to layout.

Traditionally, landscape designers do their actual designing on opaque paper. The paper may be gridded or plain. Then a sheet of thin, transparent tracing paper called **vellum** is placed over it and the entire work traced, Figure 3-18. At the same time, the plant list and other labeling components are added, either by tracing or by the use of waxed press-on letters or transfer film. At this stage in the development of the layout the designer must decide what to place where on the sheet. If a computer creates the layout, the same decisions must be made. While use of the computer negates the need to trace the plan onto vellum, the need still remains for the designer to decide where to

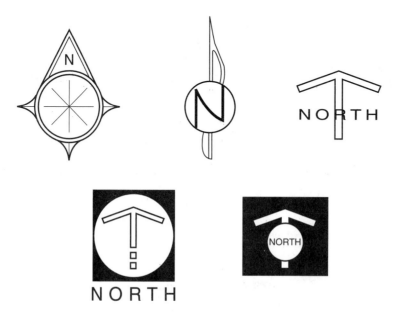

Figure 3-17 Examples of directional arrows

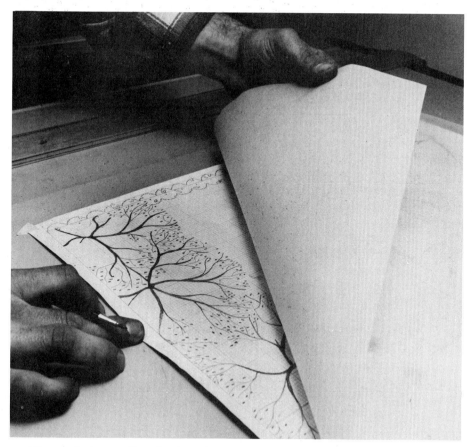

Figure 3-18 Vellum is placed over the original design for tracing.

place the various components and what prominence to assign to each. Figures 3-19, 3-20, and 3-21 illustrate different sheet layouts. While different, there are some similarities:

- The design is placed on the left side or at the top of the sheet.
- The plant list is sized to help counterbalance the design on the sheet.
- The client's name and address block, with the largest lettering, is placed near the bottom to give stability to the layout.
- The scale indicator and directional arrow are placed near the design and located to balance the white space on the sheet.
- The designer's name and/or firm name are placed where they can help balance the white space.

NOTE: Some firms will group all of the labels together in a box, which is repeated on each sheet of a multiple-sheet design proposal.

- An imaginary line drawn down the center of the sheet would find the graphic weight equal on each side of center.
- A border line is often drawn around the sheet to tie together all of the components on the sheet.

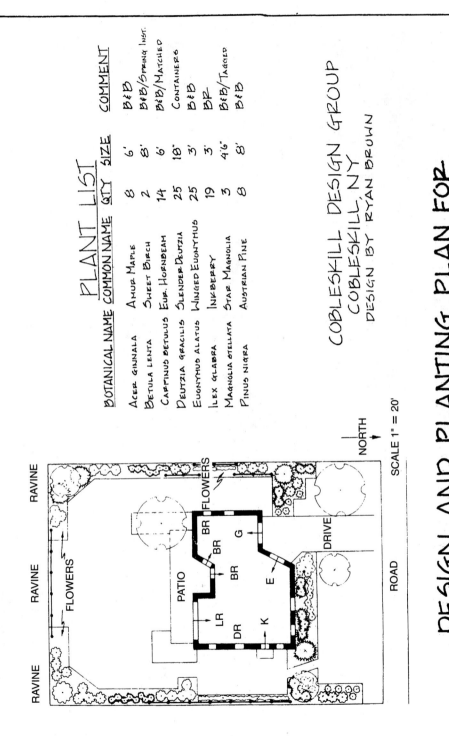

PLANT LIST

BOTANICAL NAME	COMMON NAME	QTY	SIZE	COMMENT
Acer Ginnala	Amur Maple	8	6'	B&B
Betula Lenta	Sweet Birch	2	8'	B&B/Spring Inst.
Carpinus Betulus	Eur. Hornbeam	14	6'	B&B/Matched
Deutzia Gracilis	Slender Deutzia	25	18"	Containers
Euonymus Alatus	Winged Euonymus	25	3'	B&B
Ilex Glabra	Inkberry	19	3'	BR
Magnolia Stellata	Star Magnolia	3	4'6"	B&B/Tagged
Pinus Nigra	Austrian Pine	8	8'	B&B

COBLESKILL DESIGN GROUP
COBLESKILL, NY
DESIGN BY RYAN BROWN

NORTH

SCALE 1" = 20'

RAVINE RAVINE RAVINE RAVINE

FLOWERS

FLOWERS

PATIO

BR BR

LR BR

DR G

K E

DRIVE

ROAD

DESIGN AND PLANTING PLAN FOR
THE WHITESTONE RESIDENCE
DUANESBURG, NEW YORK

Figure 3-19 A typical plan view sheet layout

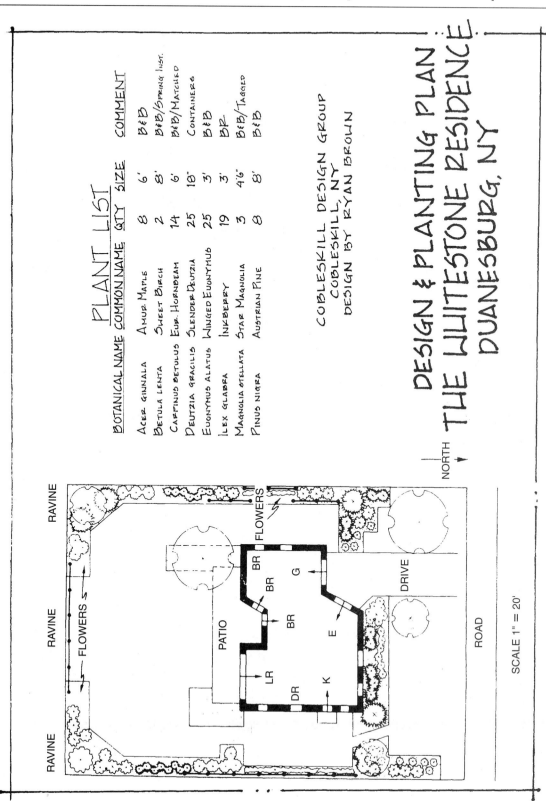

Figure 3-20 A variation on the horizontal sheet layout

RAVINE   RAVINE   RAVINE

FLOWERS

PATIO

FLOWERS

LR   BR
        BR   BR
DR   G
    K   E
              DRIVE

ROAD

SCALE 1" = 20'

NORTH

PLANT LIST

BOTANICAL NAME	COMMON NAME	QTY	SIZE	COMMENT
Acer ginnala	Amur Maple	8	6'	B&B
Betula lenta	Sweet Birch	2	8'	B&B/Strong lean
Carpinus betulus	Eur. Hornbeam	14	6'	B&B/Matched
Deutzia gracilis	Slender Deutzia	25	18"	Containers
Euonymus alatus	Winged Euonymus	25	3'	B&B
Ilex glabra	Inkberry	19	3'	BR
Magnolia stellata	Star Magnolia	3	4-6'	B&B/Tagged
Pinus nigra	Austrian Pine	8	8'	B&B

DESIGN AND PLANTING PLAN
THE WHITESTONE RESIDENCE
DUANESBURG, NY

COBLESKILL DESIGN GROUP ··· RYAN BROWN

Figure 3-21 A vertical layout of the same design. As with the horizontal layouts, this too is balanced on the sheet.

Different Drawings for Different Requirements

Graphic presentations range from quick sketches to elaborate, detailed drawings. Some emphasize design concepts, while others explain construction methods. Some are seen by no one but the designer, while others appear in full-color brochures distributed to the public. It is foolish to spend more time on a graphic presentation than is warranted by the importance or value of the project. Figures 3-22, 3-23, and 3-24 illustrate a complete sheet layout, a perspective, and a construction detail drawing, respectively. All have required precision by the designer, and all are intended for viewing and/or use by someone other than the designer. They justify the time required for

their preparation. Figure 3-25 illustrates two common elements of landscape graphics: a plan view plant symbol and a human figure. Each is shown as it might be interpreted in a quick sketch, a simple presentation plan, and a refined presentation for a large project. The graphic style used should be consistent throughout the piece of work. If the building and plants in a drawing are being done with elaborate detail, it would be inappropriate to insert a human figure done as a loose sketch. The use of color is also an arbitrary decision that the designer must make. The first level of the decision deals with whether to use color at all. The second level of the decision concerns what medium to use, since some media are faster or easier to apply than others. Familiarity with the media of the graphic artist is important for landscape architects and many landscape designers.

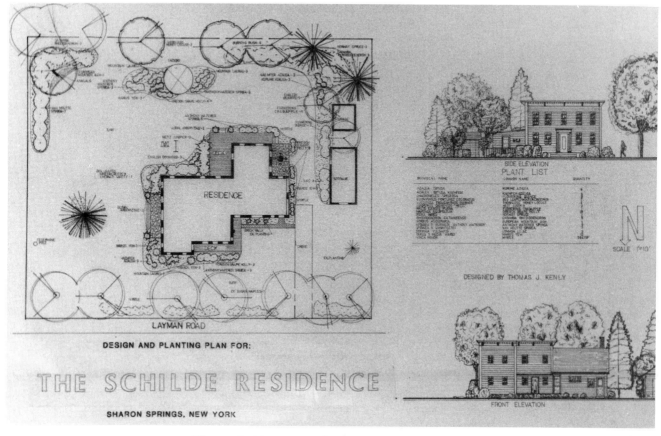

Figure 3-22 A complete presentation drawing

PERSPECTIVE DRAWING
SCOTT RIVES   MARCH 1977

Figure 3-23 A perspective view can be time-consuming to prepare if done in this detail, but provides a client with the best understanding of what the designer is proposing.

- ALUMINUM RE-
  FLECTOR HOOD
- CAST ALUMINUM
  LOUVERS PAINTED
  MATTE BLACK
- INCANDESCENT
- TO BE USED ON
  WALKWAYS AND
  PEDESTRIAN COURT
  AREAS

FINISH GRADE

BASE ANCHOR

2" FIBRE CONDUIT

3/4" STEEL CONDUIT

CAP TYPE

- 14" DIAMETER
  PLEXIGLAS
  INCANDESCENT
  GLOBE MATTE
  WHITE
- MOUNTED ON 2½"
  DIAMETER STEEL
  POST MATTE
  BLACK FINISH
- MOUNTING HEIGHT
  6'-6" ABOVE GRADE
- POST EMBEDDED
  8" SITTING OVER
  ENTRANCE CON-
  DUIT FOR CABLE
- TO BE USED ON
  WALKWAYS AND
  PEDESTRIAN
  COURT AREAS

FLUSH HANDHOLE
WITH COVER

FINISH GRADE

CONCRETE
BASE

3/4" STEEL
CONDUITS

2" FIBRE
CONDUITS

GLOBE
TYPE

2'-0"

WALK LIGHTING
DETAIL   NO SCALE

Figure 3-24 A detail drawing for walk lights on a college campus (Courtesy James Glavin, Landscape Architect, Syracuse, New York)

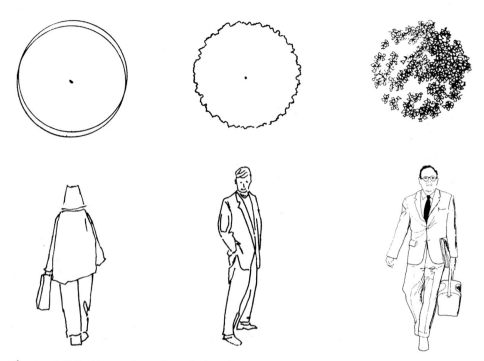

Figure 3-25 Examples of symbol styling and figure drawings for (left) quick sketches, (center) simple presentations, and (right) refined presentations

Graphic Media

The media of graphic art fall into two broad categories: those used to draw and delineate and those used to render and interpret. There is no sharp line of distinction between these two categories, since some media can be used either way.

Graphite Pencil

Graphite pencils were described as drawing instruments previously. Their center cores of graphite (lustrous carbon) are manufactured with varying degrees of hardness. Firmer leads have high H numbers and are more suitable for drawing than rendering. Softer leads, with high B numbers, are better suited for sketching and creating tonal variations needed to render drawings realistically.

Softer leads (such as 4B to 6B) can create a thin, fine line if kept sharpened, but sharpening is needed frequently. If chiseled to a wedge, soft leads will create wider, softer-appearing lines. Flat sketch pencils cannot be sharpened to a point, but are excellent for making wide strokes when tones of grey are needed, Figure 3-26. The apparent softness of graphite tones is also influenced by the

Figure 3-26 Soft leads with chiseled tips and flat sketch pencils can make wide graphite lines.

drawing paper. Hard-surfaced papers will make the graphite tone appear even and uniform. Softer, textured papers will introduce tonal variation. Thin papers that permit light to pass through (translucent) will cause graphite tones to appear weaker than they would if used on heavier paper that light cannot penetrate (opaque).

Pencil has long found favor with landscape designers because it is the most versatile medium, being usable for both drawing and rendering. It is also the least expensive, most commonly available, and least intimidating of the art media. Used alone or in combination with other media, the graphite pencil is capable of tonal subtleties that are difficult to represent with other media.

In switching from writing use to sketching or rendering use, the pencil needs to be held differently by the designer. As held for writing, the pencil is controlled mainly by the fingers. For graphic use, the pencil needs to lie lower in the hand, thereby permitting greater wrist action with less dependency on the fingers, Figure 3-27. As the angle of the pencil changes, the width and the crispness of the lines change, becoming wider and more fuzzy as more of the lead's edge makes contact with the paper. Also, by allowing a soft graphite lead to rotate slightly and/or by varying the pressure applied to the pencil, the designer is able to introduce additional interest into a pencil line. The resulting width and tonal variations within a single line suggest that the line is turning, advancing, or receding from the surface. That is in contrast to the static appearance of a straight-edge-supported line. Both types of lines have their uses, and the graphite lead is capable of producing both. A pencil sketch commonly uses both types of lines plus a range of tones, all resulting from skillful use of the inexpensive graphite drawing pencil, Figure 3-28.

Colored Pencil

Several brands of colored pencils are available from art supply stores. Prismacolors (reg. trademark) are perhaps the best known and enjoy the widest use. Colored pencils are wax based and

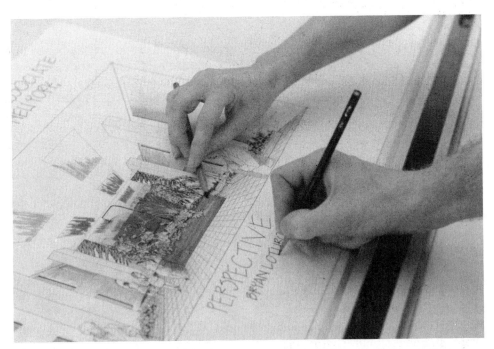

Figure 3-27 The drawing pencil in two common positions: Left, for rendering, and Right, for lettering

WIDER, FUZZIER, SOFT LINES ARE USED FOR ORGANIC OBJECTS

THE PENCIL IS TWISTED TO CREATE THE TEXTURE OF THE TREE TRUNK

VERTICAL LINES SUGGEST THE SKY

HORIZONTAL LINES SUGGEST THE GROUND PLANE

THIN, HARD LINES ARE USED FOR ARCHITECTURAL OBJECTS

LINES AND TONE ON THIS SIDE EXPLAIN THE LOCATION OF THE LIGHT

TONE IS VARIED TO SHOW THE TOP AND UNDERSIDE OF THE TREES

Figure 3-28 Pencil sketches use the full range of graphite capabilities. (Courtesy of Mike Lin, The Mike Lin Graphic Workshop, Manhattan, Kansas)

sold in sets that can include as many as 120 different colors. They are available in traditional pencil form, with round color cores enclosed in wood, also with thinner color cores enclosed in wood, and as long, square solid color sticks with no wood. The pencils are best sharpened with a pocket knife or single-edged razor blade and pointed with a sandpaper pad or knife, since their soft wax core is quickly chewed up by conventional graphite pencil sharpeners.

With little exception, colored pencils are used for the rendering of design plans, elevations, and perspective illustrations, rather than actual creation of designs. While designers may find them helpful for quick design studies, their principal value is to add color excitement to works that are to be used to sell the design to clients. Both beginners and experienced designers appreciate colored pencils because they allow the designer maximum control in the application of the color. Unlike markers that create brilliant color the instant they touch the paper, colored pencils can be applied slowly and lightly, allowing the designer to build color intensity to the desired level.

Colored pencils can be blended by overlaying different colors. As long as the designer is careful to apply each layer lightly so that light can penetrate it, the colors will mix, resulting in different tones and frequently in new colors. Wise designers carefully note and record which pencils were used to create desirable mixes so that they can duplicate the effect at a later time, perhaps on a different piece of work.

To gain the smoothest application of colored pencils, they should be applied uniformly in a single direction, much like a paint roller is used when painting a wall. Additional colors or layers should be applied in the same way, but at 90 degrees to the previous direction of application.

When smoothness is not the objective, colored pencils can create textures in numerous ways. With the point sharpened, textures can be created using short, repetitive strokes. While the lack of color in this text makes it impossible to illustrate the color mixing features of colored pencils, Figure 3-29 does illustrate various textures created with colored pencils. These textures can be used to render plant materials, hardscape surfaces, clothing, and many

Figure 3-29 Colored pencil textures. Top left: The line creates endlessly repeating box shapes. Top right: The texture is formed from tiny circles. Lower left: The line is like a flock of birds in flight. Lower right: The line creates repeating *W*'s.

other features in landscape illustrations. Colored pencil works effectively to add color and textures when used alone or when used in combination with other media. It is wise to apply a spray fixative over the completed rendering to prevent it from smearing.

Felt Marker Pens

These instruments are fine-point markers that deliver lines similar to those created by technical pens, yet marker pens are disposable and much less expensive. Line widths vary from medium to extra fine. The inks used may be water soluble, water resistant, or waterproof, depending upon the manufacturer. All dry instantly upon contact with the surface, and most are applicable to a variety of drawing surfaces, Figure 3-30. Felt pens are gradually replacing technical pens as the instrument of choice for creating plans and other drawings that require ink. They are good for lettering, bordering, outlining, and symbol styling. They can also be

Figure 3-30 Felt pens can create varied lines widths.

Figure 3-31 Different types of marker nibs. Each creates lines of different widths.

used for sketching and texturing, although the required skill level of the artist is greater since erasing is nearly impossible.

Color Markers

Technically, the tool being described here is just another felt marker. However, color markers are used differently from the point-tipped instruments described as marker pens.

Color markers contain a reservoir of pigment that is liquified in assorted solvents that vary between manufacturers. They are available in a wide range of colors, tips, and qualities. It is important for the beginning landscape designer to realize that the inexpensive color markers used by children are unsatisfactory for use in professional rendering. Here is one case where the higher price does translate into better quality. Design, Berol, Faber Castell, and Pantone are among the most popular markers used for landscape rendering today, and there are many other equally good

brands. The markers are available with a variety of tips (nibs) that include broad chisel tips, pointed and ultra-fine pointed tips, and flexible foam tips suggestive of a paint brush, Figure 3-31. Some markers are available with two different nibs, one at each end of the same marker. The color selections are extensive, but there are great inconsistencies between manufacturers. Though called by the same name, a marker color from two different manufacturers can be two different shades. That sometimes makes it difficult to mix marker brands. Also, differences in the pigment solvents can make it difficult to use different brands of markers for a single rendering.

The first-time use of color markers can be a bit intimidating. Unlike colored pencils, which can be applied lightly and, if necessary, erased, markers soak in, dry instantly, and defy erasure. Color markers are a wet medium, with similarities to water colors, leaving novice users with the same feeling of reduced control over the outcome of the rendering. They require some practice and adhering to some basic rules of application in order to apply them effectively, but it is worth the effort. Used incorrectly, marker renderings are streaked, splotchy, gaudy, and amateurish. Used correctly, markers bring brilliance and realism to renderings that no other media can create as quickly.

Color markers can be used on all drawing surfaces common to landscape illustration. However the resulting color and the level of brilliance are directly related to the color of the paper, its degree of translucency or opaqueness, and the type of light being cast upon the illustration, Figure 3-32. In general, this could be said of all color media, not just markers.

Most landscape renderings rely upon the broad-nib color marker as the workhorse, with the finer pointed markers used as supplements. It is important to understand how to use broad-nib markers in order to maximize their use and longevity. The markers should be used with only an edge touching the paper, not the full stub end. The chisel tip offers the option of making thin or broad lines, depending upon which way it is drawn across the paper, Figure 3-33. Like colored pencils, markers should be applied evenly in a single direction across the surface to reduce streaking. A straightedge may be used for support, but it is better to use something disposable, such as heavy cardboard or chipboard, and change it frequently. Plastic triangles are not satisfactory because they eventually begin to leave streaks of past colors on the new drawings. Also, the solvents in some markers will dissolve the plastic edges and make the triangles unusable later for line work with pencils and pens.

Color markers should be used with the lightest colors applied first, then gradually build to darker colors and darker tones. Like colored pencils, color markers can be layered to create darker tones of the same color or new colors from mixing different colors. A colorless blender, which is a marker containing only solvent, can dilute a marker color applied over it while both are still wet. The result will be a lighter tone of the color than would otherwise be produced. Also, colored pencils can be applied over markers to change colors and to lighten or darken the marker colors. Many mistakes made with markers can be hidden by applying colored pencils over the marker. Pencils also work well to add textures to marker surfaces. In landscape graphics, the combined use of colored pencils and color markers is more common than the exclusive use of either.

Pastels

Pastels are chalky crayons made of powdered pigment. They are available in a variety of colors, in round or square sticks. Some have an oil additive that reduces some of the chalky quality. For landscape renderings, the best pastels are square sticks without an oil base. The messiness of chalky pastels is exactly what makes them useful for land-

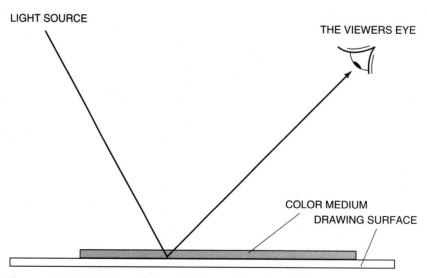

Figure 3-32 The three factors in determining how the eye perceives color are the quality of light, the choice of medium, and the paper selected.

Figure 3-33 The broad nib marker can make wide or thin lines depending upon the edge used and the direction of movement.

Figure 3-34 Pastel squares can create both wide and thin lines.

scape graphics. They allow the designer to add color quickly to large areas such as lawns, skies, and bodies of water, where the shape of the areas is indeterminate and a subtle, blended color effect is desired.

Application techniques for pastel vary. Using a square stick rather than a round one provides the designer with some control when a line is desired rather than a tone. By breaking the pastel stick into two uneven pieces and applying the color with the edge of the stick, not the end, varied line widths can be drawn, Figure 3-34. Another application technique takes advantage of the tendency of pastels to smear by scraping off particles of pigment from the stick, then using fingers or paper blending stumps to create a smooth texture in the area desired, Figure 3-35. Once applied, pastels require the application of a spray fixative to prevent unwanted smearing.

Pastels are comparatively inexpensive and require some practice to use effectively. Minimal application is always better than heavy-handed application. Pastels can be applied over dried marker colors and over colored pencil, but pencils and especially markers do not work well when applied over pastels.

Figure 3-35 Pastels can be applied as powdered pigment by scraping with a coin and then blending with a paper stump or finger.

Pressure Graphics

A vast assortment of material is available to the landscape graphic artist who wishes to speed up the development of a presentation while maintaining a professional appearance. Some of the materials have been described in earlier chapters, such as waxed press-on letters, lettering machines, and transfer film. To that list can be added printed border tapes, dry transfer sheets of plant symbols, cars, and people, as well as sheets of patterns and tones that can be used to create textures or shades on the surfaces of landscape features. Rubbed off from sheets with a pencil or stylus much like the press-on letters, which are also a dry transfer product, pressure graphics are a fast, though somewhat expensive, means of instantly enhancing a landscape presentation. Extensive use of pressure graphics is usually reserved for presentation designs.

Foam Board

For the presentation of designs to clients and in any other situation where it becomes necessary to mount drawings temporarily or permanently, foam board is frequently the mounting board of choice. It is a lightweight but sturdy board composed of two sheets of thin cardboard permanently bonded to a central sheet of styrofoam. Drawings can be mounted using glue, dry wax, pins, or other devices. Unless the mounting is permanent, the foam board sheets can be stored and used again for other presentations. Foam board is also used in landscape model construction. A sharp knife, metal straightedge, and protected cutting surface are needed to cut foam board properly.

Drawing Papers and Films

Many different drawing and copy surfaces are available for use by landscape designers. The choice of a surface depends on one or more of the following factors:

a. Whether the drawing is temporary or permanent
b. What drawing instrument is being used
c. Whether the drawing will be copied and, if so, with what process
d. Whether the drawing will be rendered and, if so, with what rendering media

Some of the papers used most commonly for landscape illustration are described below. Many other specialized art papers may be selected, but these are the ones most frequently used.

Quadrille Paper

- Soft-surface ruled paper divided into 4, 5, 8, or 10 squares per inch.
- Inexpensive. It is available in pads or large sheets.
- Used for plan view designs, usually in first-draft form. Simplifies drawing to scale when selected to match with the drawing's scale.
- Most often used with drawing pencils.
- Opaque.

Newsprint Paper

- Rough surface. Absorbent.
- Inexpensive. It is available in pads or large sheets.
- Useful for sketching and preliminary design work.
- Accepts soft pencil and pastel. Markers bleed.
- Opaque.

Sketch Paper

- Smooth, hard surface. Less absorbent.
- Top brands such as Strathmore can be expensive.
- Available in pads or sheets.
- Useful for original drawings.
- Accepts most drawing and rendering media with very little bleeding. Markers will soak through.
- Opaque.

Marker Paper

- Smooth, hard surface. Little to no absorbency.
- Comparatively expensive. Available in pads.
- Used for marker renderings. Does not bleed or soak through.
- Also accepts other drawing and rendering media.
- Semi-translucent. May be used with laser printers and copiers.

Vellum

- Smooth, hard surface.
- Available in weights from light to heavy. Lightweight vellum is commonly used for sketching. Heavier weights are used for tracing and for making copies. Price varies by weight and size of roll.
- Accepts drawing pencils and pens.
- Transparent to translucent, depending upon weight.

Mylar Film

- Polyester drafting film in .003 and .005 gauge.
- Comparatively expensive. Available in sheets and rolls.
- Used for pencil and pen drawings where durability of the drawing is important or copies are required.
- Matte finished on one or both sides.
- Translucent.

Print Papers

Frequently the paper that the design is initially drawn on is not the paper used for the finished work. While it may be best to draw on a paper or film that accepts pencil or pen, it may be preferable to transfer the drawing to different paper for rendering. For some projects, it may be necessary to reproduce the original for use by other parties. Sometimes dozens of copies are required as a complex project moves through the design and approval process.

Two duplication processes are commonly used to reproduce landscape drawings in large numbers: the **diazo process** and the **photocopy process**. Both processes make direct positive copies, unlike the old blueprint process that made negative copies. The diazo process requires a specially treated paper, ultraviolet light, and an ammonia vapor developer to make copies, Figure 3-36. The photocopier operates much like any office copier. It works best with high-quality bond paper and uses black toner to reproduce the image, Figure 3-37. Each process has some advantages and some limitations. The diazo process is less expensive and less prone to problems common to photocopiers, such as streaking, jamming, or depleted toner. However, the diazo process necessitates periodic changing of the ammonia to maintain its strength, and it is unusable if the special paper is out of stock or has been stored too long or exposed to light (it is light sensitive). In general the photocopy process requires a greater initial financial investment, ongoing maintenance contracts, and periodic service calls. Diazo copiers usually give longer periods of uninterrupted service. Photocopies are somewhat sharper copies than diazo copies. Some marker solvents will dissolve the toner lines of the photocopies, so caution is necessary if doing marker renderings on photocopies.

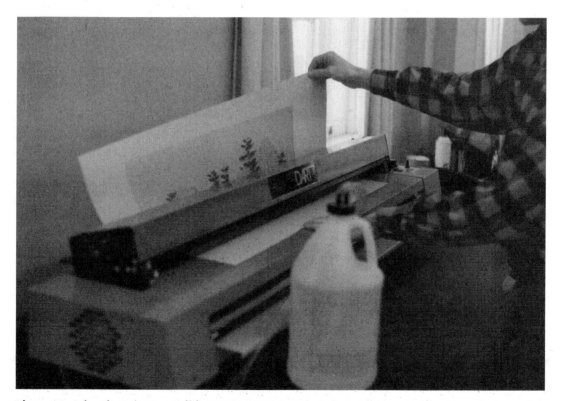

Figure 3-36 The Diazo Ozalid copy process requires specially treated paper and ammonia (contained in the bottle) to make a copy. The original drawing must be on translucent paper.

Figure 3-37 The photocopier works much like an ordinary office copier. It requires toner to reproduce black line copies. The original drawing need not be on translucent paper.

The Layout of Graphic Presentations

Much of what was described earlier in this chapter about the layout of a design presentation applies to the layout of any graphic presentation. The various drawings, lists, enlarged views, name blocks, and informational items are the components of the layout. Some are more important than others and should be featured prominently. All have shapes, usually some type of rectangular shape, although components such as directional arrows or even property lines may not be rectilinear. In deciding where components should be placed within the total presentation, an attempt should be made to arrange the components so that they compliment the rectilinearity of the other components and collectively create a larger rectangle. Figure 3-38 illustrates a presentation comprised of several different components. Notice how the long side of each component parallels the long side of the presentation layout, and the short sides of the components and the layout sheet are also parallel. The outside edges of the components are kept in alignment so

that the viewer's eye connects them, and the rectangle is again repeated.

Parallel lines and shapes are an important part of the presentation layout, and so is the ratio between the length of the layout sheet and its width. Layout sheets that are long and thin or perfectly square do not work well. The preferred shape for a layout presentation is a rectangle that is two-thirds as wide as it is long. For example, a sheet that is 18 inches wide should be 27 inches long in order to create the desired ratio. Another example would be a sheet that is 24 inches × 36 inches. Some designers favor a slightly different ratio that gives a little more length to the layout. Their preferred width to length ratio is 3 to 5, such as 18 inches × 30 inches or 36 inches × 60 inches.

The third element of a successful layout presentation is the balance of white space between the components. Equally as important as the components to the success of the layout, white space must be evenly distributed through the presentation. It need not occupy as much area as the components, but it should be evenly distributed throughout the layout. Figure 3-38 illustrates an even distribution of white space.

Figure 3-38 Careful balance of the white space around the components is important to proper sheet layout. (Courtesy of A. J. Tomasi Nurseries, Inc., Pembroke, Mass.)

Practice Exercises

A. On a sheet of drawing paper, use a compass or circle template, a scale, a straightedge, and a graphite pencil to create the symbols described below. Copy the examples illustrated in this chapter.

  1. A needled evergreen tree that is 20 feet wide, drawn to the scale of 1" = 10'.
  2. A needled evergreen shrub that is 5 feet wide, drawn to the scale of 1" = 5'.
  3. A broadleaved evergreen shrub that is 8 feet wide, drawn to the scale of 1"= 4'.
  4. A deciduous shrub that is 10 feet wide, drawn to the scale of 1"= 20'.
  5. A deciduous tree that is 30 feet wide, drawn to the scale of 1"= 10'.
  6. Three needled evergreens of the same species, each one 10 feet wide, massed and drawn to the scale of 1"= 10'.
  7. Three deciduous shrubs, each one 10 feet wide and each one a different species, massed and drawn to the scale of 1"= 10'.
  8. A span of fencing 15 feet long with posts every 5 feet, drawn to the scale of 1"= 5'. Growing along it is a deciduous vine that spreads 10 feet wide. Draw the vine to the same scale as the fence.
  9. A deciduous tree that is 25 feet wide, drawn to the scale of 1"= 10'. Beneath the tree are three 10-foot-wide broadleaved evergreen shrubs, drawn to the same scale. Position them so that part of each shrub is hidden by the tree.
  10. Go back and label each plant or group of plants in the previous nine exercises. Make the lettering 1/8-inch tall. Use guidelines, straightedges, and the lettering style of your choice. Remember that when there is more than one plant to which the label applies, the number of plants becomes part of the label. The plant names to be used are are follows:

    In #1: Colorado Spruce
    In #2: Savin's Juniper
    In #3: Carolina's Rhododendron
    In #4: Showy Border Forsythia
    In #5: Silver Maple
    In #6: Globe Arborvitae
    In #7: Vanhoutte Spirea, Beautybush, and Ninebark
    In #8: Shadow fencing, 5', and Boston Ivy
    In #9: Flowering Dogwood and Japanese Andromeda

B. Draw scale indicators for the following: 1"= 5', 1"= 10', 1"= 20'.

C. Draw four different directional arrows to indicate where north is located on a plan view. Make each one 2 inches long. Use pencil for two of them and felt pen for two of them.

D. Prepare a client name and address component. Use a five-guideline lettering style and straightedges for line support. Make light guidelines with a 3H or 4H pencil. Do the letters in 2H or H pencil. Use the following wording and letter sizes.

DESIGN AND PLANTING PLAN FOR	[1/2" letters]
MR. AND MRS. JOHN DOE	[1" letters]
1234 MAIN STREET	[1/2" letters]
HEARTLAND, INDIANA	[1/2" letters]

E. Practice applying different rendering media to develop skill in both smooth and textural application. Apply large swatches of graphite and color tones to sketch paper or white drawing paper and practice until the following techniques are mastered:

1. Smooth, consistent, streakless tones
2. Smooth application with gradual value change from very light to very dark
3. Lighten a dark Prismacolor tone with an overlay of white pencil
4. Lighten a dark color marker tone with an overlay of white pencil
5. Create a grainy texture with colored pencil applied over a wash of smooth colored pencil
6. Create an architectural texture with colored pencil applied over a wash of color marker

Achievement Review

A. Indicate if the following characteristics describe

  a. Plan views
  b. Elevations
  c. Perspectives
  d. Axonometrics

1. Measurable drawing that allows comparison of building heights with heights of nearby plants and other features.

2. Measurable drawing showing the spacing between all features used in the landscape design.

3. Measurable drawing displaying multiple sides.

4. The vantage point of the viewer is directly above the drawing.

5. Receding parallel lines are not made to converge.

6. Receding parallel lines apppear to be converging.

7. The most realistic view.

8. The dimensions are not measurable.

9. Two measurable dimensions: horizontal and vertical.

10. All measurable dimensions are horizontal.

B. Select the best answer to the following questions.

1. Which of the following is frequently used as both a drawing tool and a rendering tool?
   a. graphite pencil
   b. colored pencil
   c. color marker
   d. pastels

2. Which of the following is applied wet?
   a. colored pencil
   b. pastels
   c. color marker
   d. waxed press-on letters

3. Which of the following is the softest and messiest to apply?
   a. graphite pencil
   b. colored pencil
   c. color marker
   d. pastels

4. Which of the following is most costly per unit?
   a. graphite pencil
   b. colored pencil
   c. color marker
   d. pastels

5. Which of the following is used for finished presentation drawings?
   a. mylar film
   b. bond print paper
   c. newsprint
   d. quadrille paper

6. Which of the following is used for tracings?
   a. vellum
   b. diazo paper
   c. quadrille paper
   d. sketch paper

7. Which of the following is used to mount and display presentation drawings?
   a. diazo paper
   b. marker paper
   c. bond print paper
   d. foam board

8. Which of the following is non-absorbent?
   a. newsprint
   b. marker paper
   c. sketch paper
   d. quadrille paper

9. Which is the most durable drawing surface?

   a. marker paper
   b. vellum
   c. mylar film
   d. diazo paper

10. Which of the following has a chemically treated surface that must be developed in order to see the drawing?
    a. marker paper
    b. foam core board
    c. mylar film
    d. diazo paper

CHAPTER 4

Analyzing the Landscape Site

Objectives:

Upon completion of this chapter, you should be able to

- define the word site, and explain its significance in the development of a landscape
- list the features by which a site can be evaluated
- describe the limitations that the terrain imposes upon human activities
- understand the basic concepts of land grading

The word **site** refers to a piece of land that has the potential for development. The evaluation of a site involves any or all of the following factors:

- A comparison of the characteristics of two or more possible land areas
- An analysis of the features of a single piece of land
- A comparison of the potential uses for a site
- Selection of a site, and a final detailed analysis of it

Site comparisons and analyses are part of the responsibilities of professional site planners.

The following examples illustrate how site analysis functions in the landscape process.

Comparing Sites

A religious organization plans to build a facility that will serve as a summer camp for its members and their families. The facility is intended to provide space for thirty residential cabins, a center for worship services, assorted woodland recreational areas, adequate parking, and all necessary public utilities. Privacy is also deemed to be a necessity.

Two sites are available, both of which are within the financial means of the organization. Following are listed the principal characteristics of the two sites.

Site A	Site B
The site consists of 500 acres; 400 are wooded and rolling, 100 are cleared and level.	The site consists of 550 acres, all of which are wooded and rolling.
The site adjoins a major state highway.	The site is near a state highway, but actually adjoins a secondary road.
Land adjacent to the site is presently wooded, but is zoned commercial.	Land adjacent to the site is state forest land, which is not likely to be rezoned.
A small stream runs through the property.	The property has no surface water, but it borders on a large lake.
Utility lines are already installed near the property.	Utility lines are farther from the site than they are in Site A.
The soil is heavy clay, and one section of the wooded acreage has major rock outcroppings.	The soil is loamy to sandy and without rock formations.

Once the characteristics are noted, the site planner can begin a comparative analysis to determine which site will best satisfy the client's needs. This process is not like fitting a pair of shoes or solving a puzzle; there is never a perfect solution. The 100 percent perfect site does not exist among the available choices. Site analysis often results in reshaping the client's needs. Though failing to meet certain needs, the analysis may introduce new possibilities for the land that had not occurred to the client originally.

In this example, neither site is totally right or wrong for the organization. Site A would permit the construction of cabins and other buildings, plus parking lots, on the 100 acres of cleared flat land,

without the need for extensive and costly clearing and grading of the land. Site B would need to be cleared and leveled before building could begin. Site A is reached more easily than Site B. However, the better site could depend upon the client's attitude toward remoteness and privacy.

Both sides offer sizeable woodlands, but the commercial zoning of the land around Site A could result in a great change in the character of the area over a period of years. Site B, surrounded by state-owned forest land, is unlikely to change as much as Site A, if at all, through the years. As a financial investment, Site A, because of the nearby commercial zoning, would increase in value faster than Site B. The preferable site could depend upon the client's concern or lack of concern about the land as an investment.

Finally, the lake in Site B would probably be considered as a more desirable feature for water recreation than the small stream in Site A. Whether the client would favor Site B over Site A would depend upon the importance of the lake as a positive site feature, as compared to the negative features of Site B and the positive features of Site A.

Analyzing One Site

A farmer wants to utilize 400 acres of marginally productive farmland in a way that will increase its profitability. The farmer seeks suggestions for potential use of the site as a recreational facility.

The principal features of the site are as follows:

❦ The site consists of 400 acres; 200 acres are cleared and level, and 200 acres are wooded but open beneath the canopy. The wooded acres were used as a grazing woodlot for cattle. The two 200-acre portions are separated by a stream where the trout fishing is good.

❦ The stream flows into a nearby public reservoir where recreational boating and swimming are allowed.

❦ Small wildlife abounds on the site.

❦ The soil is somewhat rocky and drains well.

❦ One corner of the woodlot is swampy and contains a growth of vegetation rare to the area.

❦ The site receives substantial snow accumulation during the winter.

❦ The land around the site is agricultural in nature.

❦ The area is zoned for agriculture and small business.

❦ The site and adjacent areas were the scenes of many local skirmishes during the Civil War.

❦ The site is a rich source of American Indian artifacts, as well as prehistoric fossils.

In this example, the site analysis becomes an itemized account of the property's characteristics. Most of them cannot be rated as positive or negative, since the client has not yet decided how to use the property. The site's characteristics themselves may suggest the best use. The terrain is suitable for all types of camping. Canoeing and boating are possibilities. Nature studies or hunting could take place in the swampy areas and habitats (living and growing areas). The historical and archaeological features could be developed as tourist attractions.

In short, the site analysis can help to suggest uses for the land that the client had not thought of before.

The Characteristics of a Site

No two sites are identical. Although some may appear to be quite similar, each has a combination of qualities that sets it apart from other properties. A landscape planner should have an organized method to assess the hundreds of characteristics pertaining to each site.

Separating the factors into categories is a logical place to begin. Some of the site's characteristics are natural features, while others are man-made. Other features are cultural (associated with human society). Still others are basically physical features. Some other characteristics are most important as visual features. Some features are unmistakably positive factors, and others are definitely negative in impact. Many have a neutral quality until they are judged in the context of the proposed design.

A checklist is one way to compile the factors of a site analysis, Figure 4-1. In this example, the information, when filled in, will be both complete and concise. The checklist can then be taken to the drawing table to provide essential input for the design process.

The analysis checklist is a flexible tool for the designer. An item can be simply checked (✔) to indicate its presence, or additional notations can be made to indicate its size, direction, quality, state of repair, importance, prominence, possible uses, and so on. The more notations there are, the more helpful is the data in producing a landscape plan that will fit the site.

Reading the Terrain

The rise and fall of the land describes its **terrain**. The record of an area's terrain is its **topography**. Topography is charted and recorded nationally by the United States Geological Survey, in the form of topographic maps, Figure 4-2.

The maps are drawn to a scale of 1 inch = 2,000 feet. The broken lines on the map are **contour lines**, representing a vertical rise or fall of 10 feet over the horizontal distance measured from the map's scale. Each contour line connects all of the points of equal elevation on that map, and is labeled to indicate its elevation. The vertical distance between contour lines, the **contour interval**, is always stated on the map (10 feet in the U.S. Geological Survey topographic maps). Steep slopes are identified by closely spaced contour lines. Gradual slopes are denoted by more widely spaced contour lines. See Figure 4-3.

For large sites, the U.S. Geological Survey maps may provide satisfactory data for a designer to use in planning the landscape. These maps are available for most areas of the United States, and may be purchased at a nominal cost from regional offices of the Survey. For smaller sites, the scale of the map may need to be 1 inch = 50 feet, or smaller, and the contour interval as precise as 1 foot between the lines. To obtain such precise data, the landscape planner may have to hire a private surveyor.

Site Analysis for the Property of

Client's Name *Mr. & Mrs. John Doe*

Client's Address *1234 Main Street*

Tucson, Arizona

Taken By *JCS* Date Taken *April 12, 1982*

Site Characteristics	Physical Importance	Visual Importance	Pos. +	Neg. –	Neutral ?
NATURAL FACTORS					
Existing vegetation	2 SHADE TREES IN R.A. 1 FL. TREE OFF S.W REAR CORNER	ALL IN GOOD HEALTH AND ATTRACTIVE	✓		
Stones, boulders, rock outcroppings	NONE	NONE			
Wind, breezes	WESTERLY / GUSTY			✓	
Surface water features	NONE	NONE			
Groundwater	TOO DEEP TO BE USABLE				✓
Soil conditions	CALICHE LAYER			✓	
Birds and small game	✓	✓	✓		
Large game	NONE				
Existing shade	FAIRLY GOOD		✓		
Turf plantings	NONE OF QUALITY				✓
Terrain features	LEVEL				✓
Direct sunlight	LOTS OF IT / ALL DAY				✓
Off-site views		MOUNTAINS IN DISTANCE	✓		
Others					
Hardiness zone 9					
Soil pH 8.4					
Soil texture SANDY DOWN TO THE CALICHE					
MAN-MADE FACTORS					
Architectural style of building(s)		SPANISH			✓
Presence of outbuildings	NONE				✓
Existing patios	10 × 15 CONCRETE	NOTHING SPECIAL			✓
Existing walks, paths, steps, ramps	DRIVEWAY IN / WALKS IN /CONCRETE		✓		
Swimming pool	YES	OVAL / 30' LONG	✓		
Fountains, reflecting pools	NONE	NONE			
Statuary	NONE	NONE			
Fences, walls	NONE	NONE			
Existing lighting	NONE	NONE			
Off-site features		HOMES ON ALL SIDES			✓
Others FIRE HYDRANT	FRONT CORNER OF LOT			✓	
CULTURAL FACTORS					
Power lines (aboveground)	YES	UNATTRACTIVE		✓	
Power lines (belowground)	NO	NO			
Telephone lines	YES			✓	
Water lines	INSTALLED				✓
Historical features	NONE	NONE			
Archaeological features		ATTRACTIVE SPANISH STYLING	✓		
Nearby roadways	IN FRONT OF HOUSE				✓
Neighbors	GOOD RELATIONSHIP /	ELDERLY ON N SIDE FAMILIES + KIDS ON E + W	✓		
Off-site benefits		GOOD DISTANT VIEWS	✓		
Off-site nuisances	WIND IS TOO STRONG			✓	
Zoning regulations	YES / PROTECTIVE		✓		
Nearby public transportation	NO				✓

Figure 4-1 Site analysis checklist

Figure 4-2 U.S. Geological Survey topographic map (Courtesy of U.S. Geological Survey)

SECTION VIEW

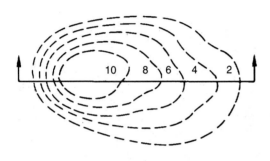

PLAN VIEW

Figure 4-3 The contour interval is 2 feet. Closely spaced contour lines represent a steep slope. Widely spaced lines show a gradual slope.

Figure 4-4 illustrates some of the land forms recognizable from a topographic map. In order to interpret the map fully, students should know the following points regarding contours and contour lines:

🌿 Existing contours are always shown as broken lines.

🌿 Proposed contours are always shown as solid lines.

🌿 Contours are labeled either on the high side of the contour or in the middle of the line.

🌿 Spot elevations are used to mark important points.

🌿 Contours lines neither split nor overlap (except in overhangs).

🌿 Contour lines always close on themselves. The site map may not be large enough to show the closing, but it does occur on the land.

🌿 Run-off water always flows downhill along a line that is perpendicular to the contour lines.

Once the contours of a site are known and plotted, slopes can be measured and analyzed. **Slopes** are measurements that compare the horizontal length (measured from the map's scale) to vertical rise or fall (as determined by the contour lines and contour interval). Slopes may be stated as ratios or percents. As a **ratio**, the horizontal space required for each foot of vertical change in elevation is commonly expressed as 3:1, 4:1, and so forth, Figure 4-5. As a **percent**, the vertical distance is divided by the horizontal distance, and the answer is expressed as 33 percent, 25 percent, and so forth. Another way to visualize percent of slope is to picture the slope extending along a horizontal distance of 100 feet. The vertical distance then becomes comparable to the percent of slope, Figure 4-6.

The Need for Terrain Information

The ease or difficulty of development depends upon whether the land is level or rolling, rocky or sandy, forested or open. A study of the terrain also supplies answers to such basic questions as: Where does the surface water flow? Will water collect in puddles anywhere? What types of human activities can take place? Will grass grow on that slope? Can a car be parked safely on that slope?

Most human activities require that the land be flat or nearly flat. Land that is 5 percent or less in slope is perceived by users as flat. Flat land is the easiest terrain to develop, but it may be difficult to move off surface water. A slope of at least 1 percent is usually necessary to drain off surface water on turf and other planted landscape areas.

Figure 4-4 A topographic map labeled to show different land forms

RATIO: COMPARING THE HORIZONTAL AND VERTICAL
MEASUREMENTS OF A SLOPE

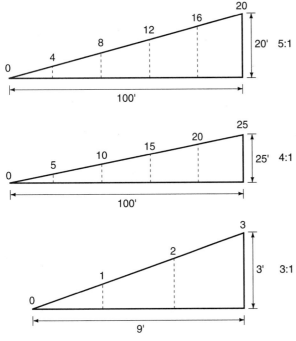

Figure 4-5 Ratio of slope

$$\text{PERCENT OF SLOPE} = \frac{\text{VD (VERTICAL DISTANCE)}}{\text{HD (HORIZONTAL DISTANCE)}}$$

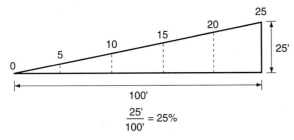

$$\frac{25'}{100'} = 25\%$$

Figure 4-6 Percent of slope

Human activities can usually take place on a
slope of 5 percent to 10 percent, but users will
sense the nonlevel footing. Land that slopes more
than 10 percent may require alteration (grading) to
make it more usable. Figure 4-7 lists the acceptable
slopes for various landscape components.

Grading the Land

When the terrain is not suitable for the activities
planned for the site, it may be necessary to reshape
it. The form of the land is changed by a process
called **grading**. Grading can be as simple as one
worker leveling and smoothing a small area of
earth with a spade and a rake, and hauling away
the leftover soil in a cart. It also can be so extensive
that massive bulldozers and dump trucks are
required to chew up and haul away entire mountains. Regardless of the extent of the project, grading is usually done for one of four reasons:

🌿 To create level spaces for the construction of
buildings

🌿 To create the level spaces required for activities and facilities such as parking lots, swimming pools, and playing fields

🌿 To introduce special effects into the landscape,
such as better drainage, earth berms, tree
wells, and ponds

🌿 To improve the rate and pattern of circulation
by means of better roads, ramps, tracks, or
paths

When earth is removed from a slope, the grading
practice is called **cutting**. When earth is added to a
slope, the practice is called **filling**. On a contour
map, a cut is shown as (1) a solid line diverting
from and then returning to an existing contour line,
and (2) moving in the direction of a higher contour.
A fill is shown as (1) a solid line diverting from and
then returning to an existing contour line, but (2)
moving in the direction of a lower contour line. See
Figure 4-8. A typical graded slope is illustrated in
Figure 4-9.

Since the grading process can involve the movement of tons of soil and rocks, designers should
approach such specifications cautiously. In cut and
fill operations, the soil that is removed from the cut
should be used to create the fill whenever possible.
This practice minimizes the need for hauling. If
possible, the topsoil layer should be stripped away
and stockpiled before grading begins. The topsoil
can then be spread over the finished grade before
the site is replanted.

ILLUSTRATED EXAMPLE						
2%	3%	4%	10%	30%	33%	50%

PERCENT OF SLOPE							
IDEAL	1/2% TO 2%	2% TO 3%	1% TO 4%	1% TO 11%	16% TO 33%	20% TO 33%	33% TO 50%
ALLOWABLE	1/2% TO 3%	1% TO 5%	1/2% TO 8%	UP TO 11%	UP TO 33%	UP TO 50%	UP TO 65%

LANDSCAPE COMPONENT						
SITTING AREA, PATIOS, TERRACES AND DECKS	LAWNS	WALKS	DRIVEWAYS AND RAMPS	BANKS PLANTED WITH GRASS	BANKS PLANTED WITH GROUND-COVERS AND SHRUBS	STEPS

Figure 4-7   Recommended slopes for common landscape components

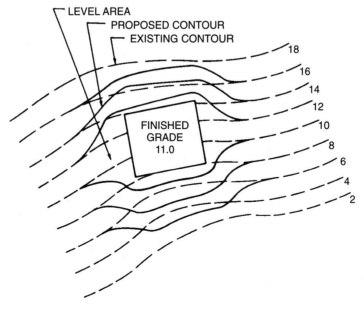

Figure 4-8   Cut and fill as shown on a topographic map

Figure 4-9 Cross-sectional view of a typical graded slope

When land is graded, not only is the topsoil disturbed, but the surface water drainage and vegetation are disturbed as well. Water must drain away from the buildings, not toward them. Freshly graded slopes must be stabilized to guard against erosion. The surface roots of valuable trees must be protected from the destruction of cutting and the suffocation of filling. Figure 4-10 shows a typical slope before and after grading. Figures 4-11 and 4-12 show common techniques for dealing with trees that exist prior to the grading of a site.

Figure 4-10 A typical slope before and after grading

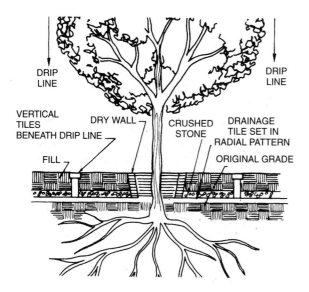

Figure 4-11 Raising the grade around an existing tree

Figure 4-12 Although the level of the lawn has been lowered, the tree's roots remain at the original level because of the retaining wall.

Achievement Review

A. Indicate if the site characteristics listed are natural (N), man-made (M), or cultural (C).

1. condition of the turf
2. nearness of public transportation
3. rock outcroppings
4. swimming pool
5. terrain features
6. off-site views
7. buildings on the site
8. style of the architecture
9. existing shade
10. prevailing breezes
11. soil conditions
12. historical features
13. provisions for parking
14. nearness of neighbors
15. presence of wildlife
16. traffic sounds
17. zoning regulations
18. presence of large, old trees
19. existing lighting
20. surface water patterns

B. Define the following terms.

1. terrain
2. topography
3. contour line
4. contour interval
5. slope

C. Complete the following sentences that describe the characteristics of contours and contour lines.

1. Existing contours are always shown as _____ lines on topographic maps.
2. Proposed contours are always shown as _____ lines on topographic maps.
3. Contours are always labeled on the _____ side of the contour or in the _____ of the line.
4. Important points on topographic maps are marked by _____.
5. A 3:1 or 4:1 comparison is the _____ of a slope.

D. Label the parts of a typical cut and fill in the following diagram.

E. Identify the land forms A through E on the contour map on the following page.

Suggested Activities

1. Fill several greenhouse flats with loamy soil and plant with grass seed. Tip the flats to create various slopes up to 50 percent. Water all flats with the same amount of water, applied from a sprinkling can. Record the erosion noted over the weeks at each degree of slope. Record also the quantity and quality of grass that grows.

2. Build a small contour model from information on a topographic map. Use cardboard of a thickness comparable to the contour interval on the map.

3. Using a sand table, the instructor can set up site situations that require alterations before landscaping can begin. Teams of students can alter the sites and discuss their reasoning with the class.

4. Order topographic maps of the local area from the U.S. Geological Survey. Find recognizable local features on the map to help students visualize how the contour lines describe terrain variations.

5. Invite a land surveyor to visit the class to demonstrate the use of surveying instruments and explain how land surveys are done.

The Landscape Process

CHAPTER 5

Objectives:

Upon completion of this chapter, you should be able to

- explain the cyclic nature of the landscape process
- list the component steps of each phase of the cycle
- distinguish between functional diagrams, preliminary designs, and final plans

Landscaping as a Process

A **process** is a sequence of steps, in the form of decisions or activities, that result in the accomplishment of a goal. A process may also be a sequence of stages during which something is created, modified, or transformed. Both interpretations of the term apply to the development of a landscape. The sequencing of decisions is what happens as a landscape architect or designer works to sort out the client's needs and desires as well as the assorted characteristics of a specific site and somehow fit one to the other. The sequencing of stages is what happens as a landscape project, beginning as the need or desire of a client, moves to the design stage, then to the construction stage, and finally into continuing maintenance and management.

Unlike a manufacturing process, landscaping is frequently a cyclic process. When an automobile rolls off the assembly line, the process that led to its construction is completed. However, the completion of a landscape project at one level of client satisfaction often sows the seeds of desire for a continuation or upgrading of the same project to a higher level of satisfaction, Figure 5-1. For example, consider two scenarios, one public and one private. A community that has no public parks desires to create a park. Open green space is the first priority of the community. Their sequential process would include such stages as finding a central location where the greatest number of residents could be served, acquiring rights to the land, and clearing and planting the area. In like manner, a homeowner who has no patio may desire one. Longing for a place to sit and cool off on a warm day, the client has little else in mind except a place where a few lawn chairs can be set out to enjoy conversation with friends. The sequence of stages would be short. A size would be determined, a paving material selected, and the construction completed. In both cases the story would end here;

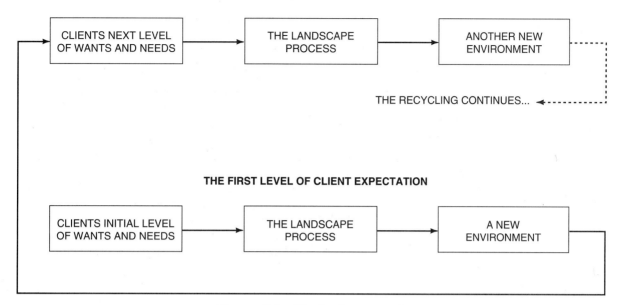

A HIGHER LEVEL OF CLIENT EXPECTATION

Figure 5-1 The landscape cycle

but it is highly probable that it would not. The community, which formerly had no park but now does, may begin to compare its park to those of other communities. Seeing that the parks of other communities have features that their park does not have, such as a swimming pool or tennis courts, will stir new desires. Once a new wish list is generated, the process begins again, but with a higher level of satisfaction as the goal. The homeowner who had no patio but now does may be very contented sitting in his lawnchair conversing with friends, until one of the friends describes the fishpond or cooking center that is part of her home patio. Suddenly the homeowner begins to compare his patio to hers, and his satisfaction level drops. Next week he may call back the landscape company to begin the development of a bigger and better patio. Landscaping cycles and recycles.

Project Development as a Process

The development of a landscape project has many stages. A typical sequence is as follows.

1. The client formulates a need or series of objectives to be accomplished by a particular time.

2. A designer is contacted and the design is developed through a sequence of steps known as "the design process."

3. The client accepts the design, and the project is assigned to a landscape contractor through competitive bidding or noncompetitive negotiation.

4. The landscape contractor arranges for suppliers and subcontractors to provide the materials and/or work needed to accomplish the project.

5. The project is built, often with the client and/or the designer monitoring the contractor to assure that it is being built as anticipated.

6. Acceptance is given by the client. Final billing and payment follow.

7. A separate maintenance contract may be negotiated with a landscape management firm.

Project Maintenance as a Process

In similar fashion, the desire for continuing maintenance of a property by a landscape management firm also sets in motion a necessary sequence of actions and activities.

1. The client desires a level of maintenance quality and/or an array of services that he or she is unable or unwilling to accomplish personally.

2. Through competitive or noncompetitive means a landscape management company is selected.

3. The company assesses the needs of the client and presents a proposal, which may be accepted or modified to suit more closely the desires and budget of the client.

4. The work to be done, the crews, equipment, and materials required, and the frequency of performance are all scheduled by the company. Work is carried out on schedule, and the client is billed at regular intervals.

5. Regular meetings with the client allow for changes and upgrading of the contract as desired by the client.

Design as a Process

Discovering that creativity is a process rather than a sudden flash of inspiration may disappoint some who prefer to regard it all somewhat mystically. However, it should reassure others that good designing can come from nearly anyone who takes the time to maximize the contribution of each stage of the design process. Accepting that design is a sequential accumulation of data, followed by sequential analysis of the data, and the equally sequential organization of ideas and solutions, permits the designer to fit client needs with site capabilities in a logical manner.

The design process occurs in two stages, which are ongoing simultaneously, but which are necessarily independent of each other for awhile. The two phases, **site analysis** and **program analysis**, are diagrammed in Figure 5-2. During the site

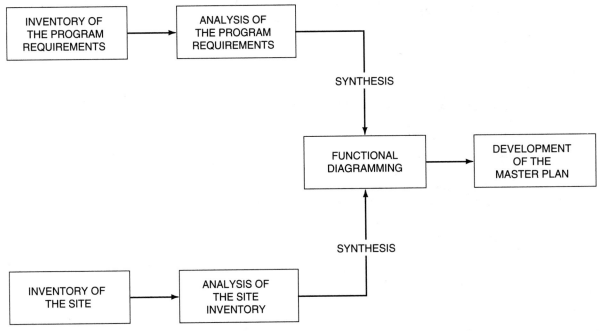

Figure 5-2 The two-track design process: program analysis and site analysis

analysis, the designer seeks a compilation of the natural, man-made, cultural, physical, and visual characteristics of the site. The compilation should be a thorough inventory of the site's attributes made without regard to whether they are positive or negative features. They will later be analyzed to determine what options and opportunities they offer for development. At the same time, but separate from the site analysis, the program analysis begins with a detailed compilation of the clients' needs and desires. At this stage, the designer should make no attempt to judge the propriety, possibility, or logic of the clients' needs. It is first necessary to make the inventory. Once that is done, then the analysis of the needs can be undertaken and a program written. The program organizes the clients' disparate wish lists into related groupings, which will serve to guide the development of the design.

Up to this point, it is important that the designer not let one analysis influence the other. Allowing the client's program requirements to enter into the designer's thought process too soon can cause the designer to overlook potentially important uses of the site that might not have occurred to the client and that in turn may never occur to the designer if he or she sees the site through the client's eyes.

Eventually the client's program and the site's potential for development must be meshed, if possible. That is shown in Figure 5-2 as **synthesis**. At this stage reality sets in. The program may seek features or uses that the site cannot provide. Conversely, the site may have been found suited to activities that the clients had not even considered. In either case, the client's program may be modified at this point to bring a closer match between the client's needs and the site's potential for development. Once the fitting of the program to site has been accomplished, the designer proceeds to develop specifics of the design, first in preliminary form for review by the client, and later in final form. Once the design is in final form and has been approved by the client, the designer may proceed to the development of specifications, construction drawings, and other support materials needed to get the project constructed.

Diagrams and Drawings Within the Design Process

The diagrams and drawings of the design process are also sequential. Each is increasingly more specific and detailed than the one that preceded it.

Functional diagrams begin the arrangement of the client's program on the site, Figures 5-3 and 5-4. Some designers refer to these and "bubble diagrams," because they use loosely drawn freeform shapes to represent the use areas or spaces that will accommodate the client's program features. More concerned with relationships between the spaces than with the specifics of how those spaces will be developed, functional diagrams help the designer make important logical decisions concerning layout of the site, size requirements of each use area, circulation patterns between use areas and throughout the total site, potential conflicts of use or circulation, and the relationship of off-site features to on-site areas.

Preliminary designs break the bubbles to reveal the designer's first draft vision of how each area of the landscape will be shaped. Development of the outdoor rooms of the landscape (as described in the next chapter) begins at this stage. Using the functional diagram as the basis for location and intended uses of each room, the designer gives the landscape its form. Decisions are made regarding the types of materials to be used to create the screenings, ceilings, and surfacings. The principles of design (as discussed in Chapter 8) are applied to the use areas. The designer begins to think in three-dimensional terms, planning not only the horizontal relationships between objects and areas, but the vertical relationships as well, Figure 5-5. While the graphic delineation of the preliminary design is simplistic and detailed specifics are lacking, the preliminary design encapsulated the concepts intended by the designer. It is a suitable drawing to present to the client for review. To carry the design any further without getting feedback from the client would be foolish, since few preliminary designs pass the all-important client test without modification. With difficult projects (or difficult clients) there may be a number of preliminary designs created before the project advances to the final stage.

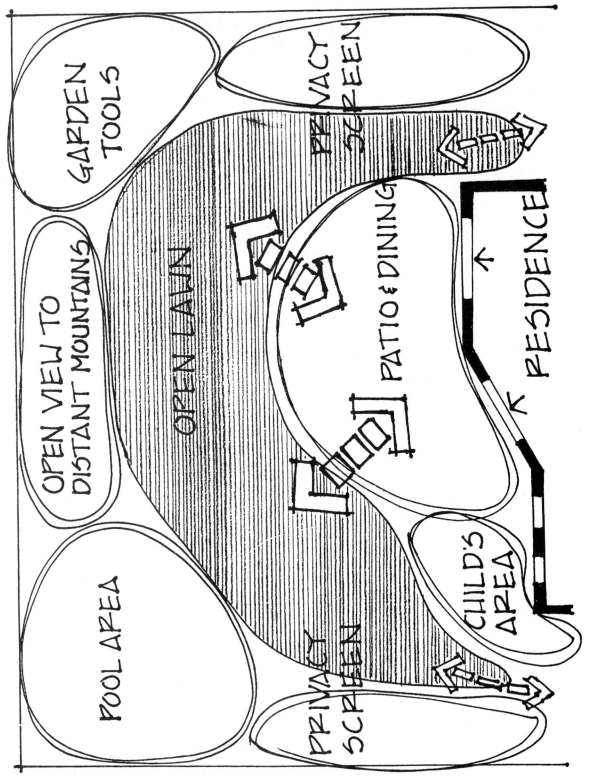

Figure 5-3 A preliminary functional diagram

Figure 5-4 A refinement of the functional diagram

Figure 5-5 A preliminary design plan

Final plans incorporate all of the suggestions and reactions of the client into a master drawing that is graphically detailed and completely specific in its intent for the landscape. Plant and hardscape materials are precisely identified. Paving patterns and enrichment features are explained and diagrammed. Graphic styling is designed to impress and to aid the client's visualization of how the completed landscape will appear. Accompanying the final plan may be a series of construction drawings or interpretive drawings to aid the client and/or the contractor in understanding how the design is to be built.

By properly and patiently working through the design process, a landscape architect or landscape designer can create an imaginative and workable solution to almost any project challenge.

Achievement Review

A. Respond to the following statement of opinion. "I would not want to be a landscape architect or designer because there are only a limited number of properties to develop, and once they are designed there will be no more work. I would eventually put myself out of a job!"

B. List sequentially the steps typical to the development of a landscape project.

C. List sequentially the steps involved in the landscape maintenance process.

D. Explain why site analysis and program analysis are done separately during the initial stages of the landscape design process.

E. Indicate if the following features are most typical of

  A. functional diagrams
  B. preliminary designs
  C. final plans

1. They provide specific details of plant selections and hardscape material choices.
2. They represent general design decisions, such as where plant masses will be located, what shape the patio and paved areas will be.
3. They display the most sophisticated graphics of the three types of drawings.
4. They are most concerned with relationships between the proposed use areas of the landscape.
5. This is usually the drawing shown to the client as the first draft of the designer's proposal.

CHAPTER 6

The Outdoor Room Concept

Objectives:

Upon completion of this chapter, you should be able to

- identify indoor and outdoor use areas
- list and define the features of the outdoor room

Modern American homes vary in size from three or four rooms to several dozen. Regardless of the simplicity or complexity of the home, the rooms are usually divided into four different categories of use, Figure 6-1. The **public area** of the home is the portion that is seen by anyone coming to the house. The public area includes the entry foyer, reception room, and enclosed porch. The **family living area** includes those rooms of the house that are used for family activities and for entertaining friends. Rooms such as the living room, dining room, family room, and game room fall into this category. The **service area** includes the rooms of the house that are used to meet the family's operating needs. These include the laundry room, sewing room, kitchen, and utility room. The **private living area** of the home is used only by the members of the family for their personal activities. The bedrooms comprise the major rooms of this area. Dressing rooms, if present, are also included.

Outdoor Use Areas

Just as a house is divided into different areas of use, so is the property around it. The public area of the landscape is the front yard. That portion of the landscape is seen by everyone who drives or walks past the house. It is also the one area of the landscape through which everyone passes who enters the house. Ideally, the public area of the landscape connects with the public area of the home, Figure 6-2. The public area should be large enough to place the house in an attractive setting, but not so large that a usable family living area is sacrificed.

The family living area of the landscape should also connect with the indoor family area whenever possible. This is often difficult because of the layout of some houses. The outdoor family living area is usually located toward the rear and often toward the side of the house. In that portion of the landscape the family relaxes and entertains guests. It includes space for patio, barbecuing, swimming, or

USE AREA CODE

⧄	PUBLIC AREA	▨	PRIVATE LIVING AREA
▤	FAMILY LIVING AREA	☐	SERVICE AREA

Figure 6-1   Examples of use of area categories within the home

NORTH   ROAD

Figure 6-2   Indoor and outdoor use areas should be matched in location as closely as possible. For example, notice that in this drawing, the outdoor service area adjoins the kitchen, the indoor service area.

whatever activities interest the family members. It is the largest of the outdoor use areas. Because of its location and its purpose, this portion of the landscape is seen by fewer people than the public area. It is developed for full or partial privacy. However, the quality of the design in both the family area and public area is important. They are both major areas of the landscape serving different purposes for the family.

The service area of the landscape plays a functional role for the family. It provides space to hang clothes, store garbage, house a dog, or grow a vegetable garden. Since it is used for service rather than beauty, the service area is usually screened from view. It is located near the kitchen or other indoor service room. The outdoor service area should be big enough to accomplish its purpose, but no larger.

All of the outdoor use areas described so far are most successful if they are connected directly to the indoor rooms they serve. A visual connection through a window is almost a necessity. A physical connection through a doorway is highly desirable. To be able to walk directly from the living room or family room into the outdoor family living area is a pleasure; to have to go through the kitchen or garage first is not so pleasant and may be a nuisance.

Figure 6-3 Assigning landscape use areas is easier on some lots than on others.

The one outdoor use area that requires a direct physical link to the indoor room it serves is the private living area. This outdoor living area is developed for total privacy. The view of outsiders is screened. This is the area of the landscape where the family may sunbathe or relax in private. If the outdoor area is to serve the bedrooms of the house, it must be possible to reach the outdoor area without walking through another part of the house or yard. Therefore, a door off the bedroom is necessary if this area is being planned.

Figure 6-3 shows several different houses and lots divided into outdoor living areas. Note the sizes of the use areas and their relationship to indoor rooms, exits, and windows.

The Outdoor Room Concept

Once the use areas have been assigned to the property, the next step is to design and develop those areas into livable outdoor spaces. To accomplish this, it helps to think of the outdoor space as an outdoor room. In this way, students can apply some of what they know about indoor rooms to the outdoors.

The composition of the indoor room is the same whether it is simple or ornate, consisting of walls, a ceiling, and a floor. While the materials may vary from room to room, the basic structure remains the same.

The outdoor room also has walls, a ceiling, and a floor, Figure 6-4. The materials are usually different from those used indoors, but they accomplish the same purpose. The **outdoor wall** defines the limits or size of the outdoor room. It can also slow or prevent movement in a certain direction. The walls of the outdoor room determine the vertical sides of the room in the same manner as the walls of an indoor room. Thus, outdoor wall materials should not be placed in the middle of the lawn, where a wall would not logically be located. Materials used to form outdoor walls may be natural (shrubs, small trees, ground covers, and flowers) or man-made (fencing and masonry).

Figure 6-4 A typical outdoor room, containing wall, ceiling, and floor elements

The **outdoor floor** provides the surfacing for the outdoor room. The materials used for the outdoor floor might be natural surfacing, such as grass, ground covers, sand, gravel, or water. They might also be man-made surfacings, such as brick, concrete, patio blocks, or tile.

The **outdoor ceiling** defines the upper limits of the outdoor room. It may offer physical protection, such as an awning or aluminum covering, or merely provide shade, such as a tree. In temperate regions, a deciduous tree is an ideal ceiling material for placement near the home. It gives shade from the hot summer sun, and then drops its leaves in the winter to allow the sunlight through, which warms the house.

Figures 6-5 and 6-6 illustrate two use areas around a home landscape. Carefully analyze the development of these areas as outdoor rooms, noting the different materials used for the walls, ceilings, and floors. Also examine the corresponding landscape plan for each area.

Figure 6-5A Example of a public area

Figure 6-5B Plan view of the public area

Figure 6-6A Example of a family living area

Figure 6-6B Plan view of the family living area

Practice Exercises

A. Figure 6-7 shows three individual houses on their own lots. All windows, doors, and rooms are marked. Also provided is information on the family living in each house. Trace each house and its lot on tracing paper. Using a straightedge and pencil, divide each lot into the three or four use areas that seem appropriate for the family. With a scale instrument, determine the approximate square footage of each area you lay out. Is each area a practical size? Are the shapes such that rooms could later be developed if necessary? Is the family living area going to receive a good deal of sunshine? (It should be located on the south or west side of the property if possible.) Is the private living area receiving morning sun from the east? If it is intended as a morning use area, western afternoon sun is of little value.

Family A:
a) An elderly couple whose children have moved away
b) Pet dog
c) Enjoy outdoor living, but are not very active
d) Garbage disposal

125'

100'

DR   BR
LR
K
G   E   BR

GREENWICH CIRCLE

LOT A

- The lot is level.
- There are neighbors on all sides of the property.
- Greenwich Circle is lightly traveled. It is used mainly as an access road for the properties adjoining it.

NORTH

SCALE: 1" = 50'

Family B:
a) A young couple with three children (ages 4, 6, and 8)
b) Pet dog
c) Enjoy outdoor living very much
d) Electric clothes dryer

WOODS   WOODS   WOODS
WOODS
WOODS

119'

K
LR   FR   BR
UR   E   BR   BR
G

122'

SUNNY LANE

- The lot is level.
- The neighborhood is suburban.
- There is a house on the adjoining lot.
- Sunny Lane is a lightly traveled paved road.

LOT B

Family C:
a) Middle-aged couple with a 20-year-old daughter
b) No pets
c) Enjoy a moderate amount of outdoor living
d) Electric clothes dryer and garbage disposal

109'

128'

K   BR   BR
DR
E
G   LR   BR

CENTRAL AVENUE

- The lot is level.
- There are neighbors on both sides of the lot.
- The rear of the lot connects with the backyard of the property on the next block.
- Central Avenue is a moderately busy street.

LOT C

Figure 6-7

B. Figures 6-8 and 6-9 illustrate two partially completed landscapes. Trace each landscape onto paper. Applying your knowledge of symbolization and the outdoor room concept, complete the landscape plans. As you insert symbols for trees, shrubs, and construction materials into the plans, be certain that each feature is playing the role of a wall, ceiling, or floor element.

Figure 6-8 Partially completed landscape

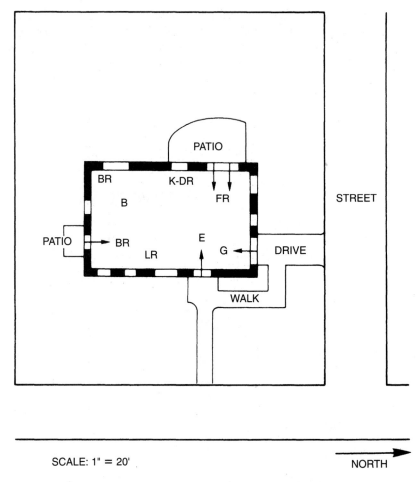

SCALE: 1" = 20'

NORTH

Figure 6-9 Partially completed landscape

Achievement Review

A. Indicate in which use area the following activities occur.

Activity	Public Area	Family Living Area	Service Area	Private Living Area
picnicking				
welcoming guests				
hanging laundry				
breakfast, coffee				
badminton				
pitch and catch				
trash can storage				

B. Indicate if the following materials are used for outdoor walls, ceilings, or floors.

Material	Outdoor Wall	Outdoor Ceiling	Outdoor Floor
brick wall			
shrubs			
crushed stone			
high-branching tree			
turf grass			
fencing			

Suggested Activities

1. Study various types of house floor plans. (These can be found in numerous magazines.) Which ones provide the best access to the outdoors through doorways? Which ones have only visual connection between the house and garden through windows?

2. Walk down a neighborhood street. Evaluate the public areas of the homes. Have the houses been placed in attractive settings? Are the wall elements where they should be or do they project into the center of the outdoor room?

3. Evaluate the property on which your school is located. Can different use areas be seen? Does the landscape appear to be a series of connecting outdoor rooms?

Plant Selection

Objectives:

Upon completion of this chapter, you should be able to

- describe the origins and forms of the plants used in landscapes
- explain plant nomenclature
- describe the factors relevant to proper plant selection

O f all the features in the designed landscape, none are likely to attract as much attention from the client as the selection of the plants that will become all or part of the walls, ceiling, and floors of the outdoor room. Logic may or may not prevail at this stage of design development. Sentimentality, pretentiousness, or economics may outweigh horticultural reality in determining what plants will be specified, purchased, installed, and maintained in the landscape. Every landscape designer and landscape architect must learn the plants of the region where he or she practices if clients are to be served most professionally. Regrettably, too few schools of landscape architecture give adequate emphasis to this important training detail, causing many graduates of these programs to believe that their site planning and engineering expertise is all that they owe to their clients. Add to this problem clients who have strong but unqualified opinions about plants, and the resulting plant specifications

for a landscape design can be limited, inappropriate, dated, or disastrous.

The Origins of Landscape Plants

Plants have found their way into American landscapes from the farthest reaches of the world. The diversity of our national flora is far greater today than it was when the Mayflower dropped anchor at Plymouth. As people from across the world have converged on America, they have brought with them seeds, plants, and propagative stock to recreate in this country's fertile soils the favored plants of their homelands. Thus, the plant palette from which modern landscape designers can select includes literally thousands of choices.

Native plants have evolved naturally within a certain geographic location over a long period of time. Examples are the eastern white pine of the northeastern United States and the Douglas-fir of the Pacific Northwest.

Exotic plants have been introduced into an area by some means other than nature. For example, many of the junipers and yews common to American landscapes actually are native to China and Japan. Brought back by plant collectors either directly or as imports via other countries, many exotics have adapted very well to growing conditions in the United States and constitute an important group of plants under production in our nurseries and greenhouses.

Naturalized plants entered the country as exotics, but have adapted so well that in many areas they have escaped cultivation and are often mistaken for native plants. They now are commonly found both in and out of landscaped settings. The bird of paradise is such a plant. Native to South America, it grows like a native plant in the desert of the southwestern United States.

Plant Forms Available

The nursery industry works closely with the landscape industry to produce the plant materials required for the country's private and public landscapes. While some plants are collected in the wild and others may be acquired from private sources, most landscape plants are produced for sale in field nurseries and greenhouses. Inventoried, described, and priced in catalogues provided by wholesale nurseries, the plants are itemized according to whatever system best categorizes the material grown. Some catalogues will distinguish the plants as coniferous (cone-bearing) or flowering; others may separate trees from shrubs and groundcovers from vines. Herbaceous (non-woody) plants may be listed as annuals or perennials, depending upon whether they live only one growing season or will survive the winter to bloom again the following year.

Trees, shrubs, vines, and many groundcovers are known collectively as **woody plants**. Other groundcovers may be **herbaceous**, as described above, a term which also applies to some vines,

grasses, and nearly all of the plants used to create floral displays. The common term for this latter group of herbaceous landscape plants is simply, flowers.

How Landscape Plants Are Sold

Woody plants grown in nurseries are sold in one of three forms: bare rooted, balled and burlapped, or containerized. Each form has certain advantages and disadvantages that landscapers should be aware of before selecting plants for installation.

A **bare-rooted plant** is one that has been dug from the nursery field, then had the soil washed away from the root system, Figure 7-1. To prevent excessive drying of the roots until the plant is purchased and installed, the root system may be wrapped in damp moss and inserted into a plastic bag, or the roots may be dipped in wax. Plants sold in bare-rooted forms are usually deciduous species that are small and are dormant at the time of harvest.

Figure 7-1 Bare-rooted plant

BARE-ROOTED PLANTS	
Advantages	**Disadvantages**
• Lightweight	• Harvesting technique reduces the root system and slows new growth
• Least expensive	• Transplanting season is limited to the early spring and late fall

Balled and burlapped (or B & B) plants are dug at the nursery with a ball of soil intact around the root system, Figure 7-2. The size of the ball is determined by the size of the plant and is further governed by standards established by The American Association of Nurserymen (the AAN). These industry standards assure that the plants will have the greatest opportunity for successful transplanting and protect the landscaper from inadvertently receiving plants with too little root system to survive the transplant. Once the root ball is dug, it is wrapped with burlap and tied or pinned to prevent the soil ball from breaking apart. A more recent harvest trend, especially suited for larger plants, replaces or supplements the burlap with a wire basket around the root ball, Figure 7-3. This gives added strength to assure that the soil ball remains intact around the roots. Plants sold in B & B form are heavier and more cumbersome to

handle, but the technique allows plants of nearly any size and type to be successfully transplanted. B & B soil balls are nearly essential for the successful transplant of all types of evergreens and large deciduous woody plants.

BALLED AND BURLAPPED PLANTS	
Advantages	**Disadvantages**
• Less disturbance to root system	• Added weight makes handling more difficult
• Allows a longer season for transplanting	• More costly
• Allows large plants to be moved and installed	

Containerized plants are grown and sold in containers such as metal cans, plastic buckets, bushel baskets, and plastic bags, Figure 7-4. Since they are sold in the same containers they have been grown in, there is no loss of roots. Therefore they transplant readily, with little or no transplant shock and no growth time lost. Containerized land-

Figure 7-3 This tree ball has been placed directly from the field into a burlap-lined wire basket, then tied at the top with twine.

Figure 7-2 Balled and burlapped plants

scape plants are young plants, seldom more than three or four years old. Although usually smaller than balled and burlapped plants, containerized plants can also be installed at almost any time of the year since their entire root system is intact.

CONTAINERIZED PLANTS

Advantages	Disadvantages
• Usually small and easy to handle	• Large plants seldom available
• Entire root system is intact. No harvest injuries to impede transplanting success	• Plants become rootbound if in container too long
• Few restrictions on time of transplant	

Figure 7-4 Containerized plant

Herbaceous plants are sold in a variety of forms, some of which are best suited to amateur gardeners, while others are more appropriate for professional landscape installations. Commonly known simply as **flowers** these plants may be started from seeds, which is often the way that home gardeners obtain their plants. It is the least expensive way to produce the colorful annuals and perennials that they desire for their gardens, Figure 7-5. However, landscape contractors and landscape management professionals are more likely to use flowers that are already growing and ready to begin blooming when they are installed. In this form, the flowers are collectively termed **bedding plants**. The plants are started in commercial greenhouses in later winter and are ready for sale to landscapers by early spring. The containers may be plastic or pressed peat moss. The peat container is biodegradable and allows the plant and its container to be installed directly into the soil without injury to the root system. The plant is able to continue growing almost without interruption. Non-biodegradable containers must be removed before the bedding plants are installed.

Bulbous perennials are available to both amateurs and professionals as propagative structures known collectively as **bulbs**. Homeowners may purchase bulbs in small quantities at their nearby garden centers, while professional landscapers usually acquire theirs in larger quantities from wholesale suppliers. Bulbous perennials may be either **tender** or **hardy**, depending upon their time of planting, time of bloom, and ability to survive the winter. Tender bulbs, such as gladiolus and begonia, will not survive the winter in most parts of the country, so they are planted in late spring, bloom in the summer, and are dug up and stored indoors until the next year. Hardy bulbs, such as tulip, crocus, and daffodil, will survive most American winters, so they are planted in the fall and flower the following spring. They do not need to be dug up and brought indoors at any time.

Other herbaceous plants, including vines and ornamental grasses, are grown by wholesale nurseries and made available to the landscape industry in assorted containers that may include any of those described above.

Figure 7-5 Packaged seeds are an easy and inexpensive way to start a flower garden.

Plant Nomenclature

Most plants have two names. One is a **common name**, by which the plant is known within a country or region of a country. Winged euonymus or grey dogwood are examples of common names. The other name is the **botanical name** of the plant, a name that is used and recognized internationally. The botanical name is expressed in Latin and is assigned to the plant by a taxonomist (a person who specializes in the classification of plants and animals). Plant taxonomy is governed by the International Rules of Botanical Nomenclature, which are established by an international botanic congress. Botanical names are recognized in all countries, regardless of native language. In the examples above, the botanical name of winged euonymus is *Euonymus alatus*, and that of grey dogwood is *Cornus racemosa*.

In *plant nomenclature* (naming of plants), each plant in the world has a unique name. Each identical plant sharing that two-part name, termed a *Latin binomial*, is referred to as a *species*. The first part of the name (such as *Cornus*) is called the

genus. The second part is called the epithet (such as *racemosa*).

NOTE: Many texts and individuals use the word *species* to mean *epithet* as well as using the term to refer to the complete binomial.

A genus (plural is *genera*) name may be applied to several closely related types of plants. For example, all pines are in the genus *Pinus*. The epithet eliminates all other types but one and creates the unique species. Thus, *Pinus sylvestris*, *Pinus resinosa*, *Pinus nigra*, and *Pinus strobus* are four distinctly different plants.

Landscapers should know both the common and botanical names of the plants with which they work. The common name is most recognizable to clients. However, to assure that the nursery provides the correct plant for installation, the landscaper must specify it by its botanical name. This is necessary because many common names are localized in their usage, and because a single plant may have numerous common names, even within one locality.

Selecting the Proper Plant

Even after the landscape designer has defined the general type of plants needed for the landscape, the selection of exact species remains. Too often the designer abbreviates this important phase of design development or simply does it incorrectly. Plants may be selected for reasons of sentimentality, easy availability, low price, or trendiness. While those reasons may have occasional validity if the chosen plants are appropriate in other more important ways, it is unlikely that plants will serve the landscape properly if selected for only those reasons.

Selection Factors

Plants have great diversity, and as such they must be matched against criteria established by the designer if they are to fill their specified roles in the landscape. While not intended to be a complete listing, here are some common factors that must be considered when selecting a specific plant for the landscape.

Role Factors. What function will it serve in the landscape?

a. Is it to be an architectural element serving as a wall, ceiling, or floor component of the outdoor room? Will it totally or partially control the views into and out from the landscape? Will it add privacy? Will it create an enframed setting for the house or other buildings, Figure 7-6?

WITH FENCING. . .
TO PROVIDE SCREENING AND SECURITY

WITH TREES. . .
TO SCREEN AN UNSIGHTLY AREA

AS A HEDGE FOR PRIVACY

Figure 7-6 Various ways to screen with shrubs

b. Is it to be an engineering element that will aid in the solution of problems such as erosion control, traffic control, diverting winds, retaining moisture, or blocking glare, Figures 7-7 and 7-8?

c. Is it to serve in climate control by filtering or blocking sunlight, or softening the impact of a prevailing breeze, Figure 7-9?

d. Is its role primarily aesthetic, contributing one or more sensory qualities such as color, fragrance, texture, or taste, as shown in the color photo section of this text?

Hardiness Factors. Will it survive the winter in the region of the country where the landscape is located? One of the most reliable measurements of a plant's potential for survival is its hardiness rating. The United States Department of Agriculture has prepared, and periodically updates, a Hardiness Zone Map, Figure 7-10, which shows the average annual minimum temperatures of the fifty states and much of Canada and Mexico. It divides the continent into eleven hardiness zones. On the most detailed maps, every county within the forty-eight adjacent mainland states is shown, along with its hardiness zone rating.

Each hardiness zone has an average annual minimum temperature variation of 10 degrees Fahrenheit. As the hardiness zone number increases, so does the temperature minimum. For example, northern Kansas temperatures (Zone 5) drop to between −10 and −20 degrees F in the winter, but southern Kansas (Zone 6) drops only to between 0 and −10 degrees F during the same months. Zones 1 and 11 represent the coldest and warmest regions of the country and are the least common zones within the United States.

Figure 7-7 This hedge, bordering the sidewalk, prevents pedestrians from cutting across the corner of the lawn.

Figure 7-8 This hedge is set back too far from the sidewalk to function as an outdoor wall. Pedestrians have cut across the corner, destroying the lawn.

Figure 7-9 Shrubs can reduce the velocity of wind, turning a gust into a breeze.

Figure 7-10 USDA Plant Hardiness Zone Map

On color coded and detailed Hardiness Zone Maps, Zones 2 through 10 are further divided into *a* and *b* zones. The *a* zone represents the colder half of the ten degree variance, and the *b* zone represents the warmer half of the ten degree variance.

NOTE: The *a* and *b* zones do not show on the map in this text.

When a plant is assigned a hardiness zone rating, it means that the plant may survive the winter in that zone or in zones with a higher numbered rating. Thus a tree or shrub rated as a Zone 5 plant may survive the winters in Zones 5 through 11, but not in Zones 4 through 1. However, there are also upper temperature limits to a plant's hardiness. Certain plants cannot survive or grow normally in climates that are too warm. That is why there are no apple trees in the Florida Keys. Still other plants can have unreliable hardiness zone ratings when grown in **microclimate** areas, regions with atypical growing conditions. For example, some plants will survive the winter because they are buried under heavy snow and escape the severe cold or drying winds that would otherwise kill them. If a winter fails to provide the protective snow cover, the plants would be unable to survive.

While hardiness zone ratings are most often associated with woody trees, shrubs, vines, and groundcovers, flowering bulbs and perennials also require hardiness considerations in their selection and use, as was described previously.

Physical Factors. Closely associated with the role of a plant are its physical features, which enable it to fulfill the role or perhaps make it unsuitable. Some of the features of plants that determine their suitability for particular uses include

a. mature size and rate of growth.
b. density of the canopy, with and without leaves.
c. branching habit.
d. thorns.
e. shape or silhouette.
f. leaf size and thickness.
g. leaf pubescence.
h. type of root system.
i. flowering characteristics.
j. seasonal variation.

Unless the landscape designer or landscape architect is also a knowledgeable horticulturist, some plants selected may grow to a mature size unsuitable for their place in the landscape. Just as puppies begin life small, but some grow to be toy poodles while others become German shepherds, so too do nursery plants grow to vastly different sizes. The shrub that matures to a height of eight feet and a width of ten feet is incorrectly used beneath a residential picture window. Yet it may be the perfect plant for creating a high, dense screen to block an unsightly view. A tall tree can seem threatening if planted too close to a low, single-story house. However, if the intent is to create a cathedral ceiling effect with a high canopy for the outdoor room, then that same tall tree may fit perfectly, Figure 7-11. Also to be considered when comparing the mature size of a possible plant choice is its rate of growth. Depending upon the species, the length of the growing season, and the growing conditions, some species will attain full size in only a few years. Others may require a century to reach maturity. The designer must match a plant's rate of growth with the expected time that the plant is to be of significance in the landscape. For example, most oaks are very large at maturity, but they grow much more slowly than ash trees, which are also large at maturity. For a residential estate, the ash tree might be a preferred choice if the design called for a large tree, simply because it would be large sooner. However, if the site is a large park or cemetery, where a stronger, longer-lived species is preferred and where there is not the same need for rapid effect as in a residential setting, then the oak might be the proper selection.

Canopy density results from a combination of leaf size and the extensiveness of the branches. Obviously, more branches with large leaves will create a denser canopy than will smaller leaves and/or more open branching. Denser canopies allow less sunlight and water to reach the surface of the ground. That may or may not be desired, so it must be considered by the landscape designer or landscape architect at the time of plant selection. It would not be appropriate for shrubs and flowers requiring dense shade to be planted beneath trees

PROPERLY SIZED TREES PROVIDE A VISUALLY
APPEALING FRAME FOR HOMES.

OVERSIZED TREES DWARF AND DEHUMANIZE A
RESIDENTIAL LANDSCAPE.

LARGE TREES PROVIDE AN EXCELLENT HIGH
CEILING FOR THE RECREATIONAL LANDSCAPE.

Figure 7-11 Size is an important factor in tree
selection.

that offered filtered sunlight. Canopy density
also affects both what can and cannot be seen
through a planting, and the amount of air that
can pass through the planting. When screening or
wind control is the role of the plants selected,
canopy density becomes an important selection
factor. While the canopy density of evergreens
is consistent year round, deciduous trees and
shrubs change with the seasons, becoming less
dense as their leaves fall. That may mean a
reduction in the amount of privacy provided by a
planting. With deciduous trees, the change in
canopy density resulting from the change of
seasons may mean that the shade provided to
the house during the hot summer becomes
the welcomed warming sunshine of winter,
Figure 7-12.

The **branching habit** of plants is a combination
of the number of branches, the average size of
the branches, the flexibility of the branches, and
the vertical or horizontal direction of their
growth. This plant characteristic becomes important when traffic control is the role of the species
being selected. Species such as the spireas,
which have small, thin branches that bend easily,
are readily trampled. They are inappropriate for
traffic control roles. However, plants such as
winged euonymus with dense, thick, inflexible
branches stand up better against invaders and
are good choices in such situations. Another
example where branching habit must be considered is when species are being used as formal
hedges. The shaping and shearing used to create
and maintain such plantings dictates the need
for species that mass well and that have some
vertically oriented branches. Horizontal branching
species may work well as informal, unsheared
hedges, but they do not do well in formal roles.

The presence of thorns makes certain species
suitable for certain roles in the landscape and
equally unsuitable for other roles. When traffic
control is the role, the presence of thorns may be
a positive supplement to the branching habit of
potential species. If safety or ease of litter cleanup
is a concern, then the thorniness of a species otherwise well suited may be a negative factor.

The shape or **silhouette** of a plant may also
determine its suitability for use in the landscape.

Figure 7-12 A deciduous tree off the southwest corner of a building provides summer shade while allowing winter warmth.

Figures 7-13 and 7-14 illustrate some of the shapes that trees and shrubs may take. Some of the more common species, their characteristics, and recommended uses are also listed. It is better to select plants that will provide the desired silhouette naturally than to select species that must be pruned and trained to do the job required in the landscape. Plants such as pin oaks and willows have low, sweeping branches as they mature, Figure 7-15. They make graceful lawn specimens where they can spread their branches fully. However, it is difficult to walk, park, or play beneath them, so they are unsuited for playgrounds or parking lots. When human activity is to occur beneath trees, a high-branching species should be selected. For example, small, vase-shaped trees (25 to 30 feet tall) make excellent outdoor ceilings for home patios. They provide shade while keeping the outdoor ceiling low and intimate. Large vase-shaped trees (50+ feet tall) function well as street tree plantings and in parking lots where cars can pass easily beneath them. Another example could be the formal approach to a commercial site where the intent of the design is to convey dignity and impress visitors. Lining both sides of a long entry drive with evenly spaced columnar forms of a single species might accomplish that design goal. With some species, the shape of the plant may suggest its best use in the design or how well it will fare during certain seasons. Many shrubs, especially needled evergreens, have rigid, geometric shapes. As such, they function well in the role of specimen or accent plants and must be used cautiously. Other shrubs are softer and less defined in their silhouettes. They mass easily and quickly lose their individual identity. Where heavy snow loads on top of plants are predictable, the most appropriate plants may be those that are more prostrate (with horizontal branches) since the branches can bend rather than break beneath the weight of the snow.

Leaf size affects the density of the canopy of trees and shrubs. It also determines the foliage texture of plants, with large leaves creating coarse textures and small leaves creating fine textures. Leaf thickness contributes to the texture of the foliage. In addition, thicker leaves tend to absorb sound and thereby can aid in noise reduction. When noise control is a function of the plants being selected for the landscape, the designer needs to

Silhouette and Examples	Characteristics	Possible Landscape Uses	Silhouette and Examples	Characteristics	Possible Landscape Uses
wide-oval Flowering crabapple Silk tree Cockspur hawthorn Flowering dogwood	• spreads to be much wider than it is tall • often a small tree • horizontal branching pattern • branches low to the ground	• focal point plant • works well to frame and screen • can be grouped with spreading shrubs beneath	**round** Shinyleaf magnolia Cornelian cherry dogwood American yellow wood Norway maple	• width and height are nearly equal at maturity • usually dense foliage • if the tree is large, a heavy shade is cast	• lawn trees • mass well to create grove effect • larger growing species may be used for street plantings • smaller growing species can be pruned and used for patio trees
vase-shaped American elm	• high, wide-spreading branches • majestic appearance • usually gives excellent shade • an uncommon tree shape	• excellent street trees • allows human activities underneath • frames structures • use above large shrubs or small trees *note*: the American elm is easily killed by Dutch elm disease; this limits its use	**columnar** Columnar Norway maple Columnar Chinese juniper Fastigiata European birch	• somewhat rigid in appearance • much taller than wide • branching strongly vertical	• useful in formal settings • accent plant • group with less formal shrubs to soften its appearance • frames views and structures
pyramidal Pines Fir Spruce Hemlock Filbert Sweetgum Pin oak Sprenger magnolia	• pyramidal evergreen trees are geometric in early years • pyramidal deciduous trees are less geometric • pyramidal shape is less noticeable as the trees mature	• accent plant • large, high-branching trees allow human activity beneath • older trees may be valued for their irregular shapes *note*: avoid planting large trees near small buildings	**weeping** Weeping willow Weeping hemlock Weeping cherry Weeping beech	• very graceful appearance • branching to the ground • easily attracts the eye • grass or other plants cannot be grown beneath them	• focal point plant • screens • attractive lawn trees *note*: avoid grouping with other plants

Figure 7-13 Typical tree silhouettes, characteristics, and landscaping uses

Shrub Silhouette and Examples	Characteristics	Recommended Landscape Uses
globular Brown's yew Globe arborvitae Burford holly Globosa red cedar	• as wide as it is tall • geometric shape • attracts attention • does not mass very well	• accent plant • use several with a single pyramidal shrub for strong eye attraction • avoid overuse
low and creeping Andorra juniper Bar-Harbor juniper Cranberry cotoneaster Prostrate holly	• low growing • much wider than it is tall • masses well • irregular shape • loose, informal shape	• use to edge walks • cascades over walls • controls erosion on banks • grown in front of taller shrubs
spreading Hetz junipers Pfitzer junipers Spreading yew Mugo pine	• wider than it is tall • medium to large shrub • masses well • usually dense foliage	• use at outcurve • place at corners of buildings • useful for screening, privacy, and traffic control
arching Forsythia Beautybush Vanhoutte spirea Large cotoneaster	• wider than it is tall • prevents the growth of other plants beneath itself • graceful silhouette • usually requires yearly thinning	• provides screening and dense enclosure • softens building corners and lines • background for flowers, statuary, fountains
pyramidal Upright yew Pyramidal junipers False cypress Arborvitae	• taller than it is wide • rigid and stiff • attracts attention • geometric shape • usually evergreen	• accent plant • focal point • use to mark entries and at incurves • group with less formal spreading shrubs
upright and loose Lilac Smoke bush Rose of Sharon Rhododendron	• taller than it is wide • loose, informal shape • usually requires pruning to prevent leggy growth	• closely spaced for privacy • use to soften building corners and lines • useful for screening and framing views
columnar Hicks yew Italian cypress Arizona cypress	• width is about half the height • geometric, flat topped, and dense	• accent plant • foundation plantings • closely spaced for hedges • mass closely when a solid wall is desired

Figure 7-14 Shrub silhouettes

Figure 7-15 The graceful silhouette of the weeping willow is attractive to the eye, but walking beneath it is difficult.

know which plants will perform most satisfactorily. As noted before, that necessitates a familiarity with the plant palette of the region.

Pubescence is the presence of fine hairs on the surface of leaves. Some species are noted for having visible, tangible amounts of pubescence. Such leaves are more likely to collect dust, pollen, and other fine particulate matter from the air than leaves without pubescence. In dusty or sooty locations, leaf pubescence might make one plant species more suitable than others if a contribution to air quality was one of the roles assigned to the selected plant.

The **root systems** of plants range from total tap root systems, with a large single root growing straight down into the soil, to full fibrous systems, which have thousands of fine, hairlike roots spreading out in all directions from the center of the plant. Fibrous root systems are useful for trapping and holding soil particles. That is an attribute that plants with tap roots do not have. If the role of the plant being selected was to reduce soil erosion due to wind or water runoff, only fibrous rooted plant species would logically be considered.

Depending upon whether the plant is woody or herbaceous, perennial or annual, cone bearing or non-cone bearing, and numerous other differences, the flowering characteristics of landscape plants vary greatly. Some plants, especially woody plants, flower once each year in the spring or summer. The flower display may be bright and showy or comparatively inconspicuous, depending upon the species. Herbaceous plants may bloom throughout the growing season, as with annual bedding plants; or they may bloom for several weeks, then wane, as with bulbs and other perennials. Some plants produce sterile flowers, incapable of producing fruit. That may or may not be important. With sterile flowers, there is no crushed fruit to stain sidewalks, but also no food to attract wildlife to the garden. Some flowers have desirable fragrances, which makes them ideal for certain settings. Others have no fragrance or even an unpleasant one, making them undesirable in certain locations at certain times. Some flowers attract bees and other insects, while others will repel certain insects. Some plants require certain soil conditions or temperatures or day lengths to induce flowering. Others will thrive under a wide range of conditions. In short, flowering characteristics are diverse and can contribute positively or negatively to the landscape's design, depending upon how a particular species matches the role assigned it by the designer.

Seasonal variation includes the presence or lack of flowers, but involves other plant attributes as well. It addresses a concern for consistency. The selected plant species must meet its obligation as a component of the outdoor room throughout all of the seasons of use, not just for part of the year. If year-round privacy screening is required, it is unacceptable for the plants to drop their leaves and permit viewing through the planting. If a flower border is to offer color enrichment throughout the growing season, then the plants used must come into bloom at different times, rather than coming and going at the same time. If the role of a plant is to be a strong focal point, a two-week color display is not good enough; it needs to catch the eye every day of the year. Conversely, if changing focal points are planned for an area, then the short-term effectiveness of individual plants is desirable.

Cultural Factors. The requirements for the growth and maintenance of plantings are their cultural requirements. Some plants establish easily, require minimal maintenance, and grow satisfactorily in a variety of locations and under assorted conditions. Other plants can only be successfully transplanted within a narrow window of time each year. Still others have very exacting requirements for growth and/or maintenance if they are to make their maximum contribution to the landscape. The more soil conditioning, spraying, pruning, thinning, weeding, fertilizing, or other requirements that are necessary to sustain a planting, the more costly it is to the client. The cost may appear at the time of installation or it may not become apparent until after a season of growth, but it is an expense in the form of materials and labor.

To be certain that the cultural requirements of the plants selected for a landscape are appropriate for the site and the client, the designer should seek answers to questions such as these:

❦ Will the plants attain and hold the size and shape desired without the need for extensive pruning and training?

❦ How frequently do the plants need pruning to prevent legginess or to induce fullness or to eliminate water sprouts and sucker growth?

❦ Are the soil and drainage characteristics of the site suitable for good plant growth?

❦ Will these species transplant easily and at the time of year that this project will be installed?

❦ Are the plants susceptible to insects and diseases common to this region?

❦ Will spraying or dusting be necessary to keep the plants pest-free?

❦ Do all of the plants used in this section of the landscape have similar cultural requirements?

❦ If replacement becomes necessary, are similar plants commonly available from nearby sources?

❦ What are the clients' attitudes about garden maintenance? Are they hobby gardeners who will enjoy giving special attention to their plants, or are they more interested in the lowest maintenance possible?

Only after the cultural requirements of the possible plant choices are matched against these questions can the proper plants be selected for the landscape.

Computer-Aided Plant Selection

Several computer software programs have been developed to aid landscape designers and landscape architects in their selection of plants for projects. These programs allow the designer to enter into the computer the specific plant features required; then the program presents a selection of plants that meet those requirements. Some of the programs even provide full-color photos of the plants, turning the computer screen into a garden catalogue. While such technology is welcomed and used increasingly by designers, there is some danger that good plant choices may be overlooked because they are not programmed into the software. There is also the potential for overlooking microclimate variations unique to sites. Computer-aided plant selection should be regarded as a tool for use by landscape architects and designers, but not a substitute for experience and knowledge of the horticultural flora of the region.

A Guide to Landscape Trees

Tree		Evergreen	Deciduous	Maximum Height		
Common Name	Botanical Name			10'–25'	25'–60'	60' and up
Almond	Prunus amygdalus		•	•		
Amur corktree	Phellodendron amurense		•	•		

Apples
  Summer
  Fall
  Winter
	Malus species		
			•
			•
			•
					
					•
					•
					•	

Apricot	Prunus armeniaca		•		•	
Arborvitae	Thuja occidentalis	•			•	

Ash
  Arizona
  Green
  White
  Flowering
	Fraxinus species
	  F. velutina
	  F. pennsylvanica
	  F. americana
	  F. ornus
			
			•
			•
			•
			•
					
					•
					•
					
					•
						
						•

Beech
  American
  European
	Fagus species
	  F. grandifolia
	  F. sylvatica
			
			•
			•
						
						•
						•

Birch
  Canoe
  European
  Sweet
	Betula species
	  B. papyrifera
	  B. pendula
	  B. lenta
			
			•
			•
			•
					
					•
					
						•
						
						•

Cherry	Prunus padus		•		•	
Chinese chestnut	Castanea mollissima		•		•	

Crabapple
  Flowering
  Fruiting
	Malus species		
			•
			•
				
				•
				•		

Crape myrtle	Lagerstroemia indica		•	•		

Cypress
  Italian
  Monterey
  Sawara false
	
	Cupressus sempervirens
	Cupressus macrocarpa
	Chamaecyparis pisifera
		
		•
		•
		•
						
						•
						•
						•

Dogwood Flowering	Cornus florida		 •	 •		
Douglas fir	Pseudotsuga menziesi	•				•
Elm	Ulmus americana		•			•
Fig	Ficus carica	•			•	

Fir
  Balsam
  White
	Abies species
	  A. balsamea
	  A. concolor
		
		•
		•
						
						•
						•

Fringe tree	Chionanthus virginicus		•	•		
Ginkgo	Ginkgo biloba		•			•
Goldenchain	Laburnum anagyroides		•	•		
Goldenrain tree	Koelreuteria paniculata		•	•		

Time of Flowering			Fruiting Time			Good Fall Color	Hardiness Zone Rating	Comment
Early Spring	Late Spring	Early Fall	Late Summer	Early Fall	Late Fall			
	•			•			8	edible fruit
	•			•			4	
•			•				4	does not fruit in warmer zones
•				•			4	
•					•		4	
•			•				5	edible fruit
							3	prunes well
	•			•		•	5	heavy seed formation
	•			•			2	
	•			•			3	
	•			•			5	
						•	3	low branching
						•	4	does not do well in city air
•						•	2	attractive but often short lived because of certain insect damage
•						•	2	
•						•	3	
•			•				4	edible fruit; attracts birds
	•			•			4	disease resistant; edible fruit
	•			•		•	3	edible fruit
	•			•		•	4	attractive flowers
		•					7	difficult to transplant
							7	pyramidal growth habit
							7	
							3	
•						•	6	good patio tree
							4	dense foliage; pyramidal shape
							2	susceptible to several diseases
•			•				6	edible fruit
							4	
							4	
	•			•			4	may also be grown as a shrub
						•	4	plant only male trees good for city conditions
	•						5	
	•				•		5	coarse textured

A Guide to Landscape Trees (continued)

Common Name	Botanical Name	Evergreen	Deciduous	10'–25'	25'–60'	60' and up
Hawthorn	Crataegus species					
Cockspur	C. Crusgalli		•		•	
English	C. oxyacantha		•	•		
Washington	C. phaenopyrum		•		•	
Hemlock	Tsuga canadensis	•				•
Holly	Ilex species					
American	I. opaca	•			•	
English	I. aquifolium	•				•
Honeylocust	Gleditsia triacanthos inermis		•			•
Hornbeam	Carpinus species					
American	C. caroliniana		•		•	
European	C. betulus		•		•	
Larch	Larix decidua		•			•
Linden	Tilia species					
American	T. americana		•			•
Little leaf	T. cordata		•			•
Silver	T. tomentosa		•			•
Magnolia	Magnolia species					
Saucer	M. soulangeana		•	•		
Southern	M. grandiflora	•				•
Star	M. stellata		•	•		
Maple	Acer species					
Amur	A. ginnala		•	•		
Hedge	A. campestre		•	•		
Japanese	A. palmatum		•	•		
Norway	A. platanoides		•			•
Red	A. rubrum		•			•
Sugar	A. saccharum		•			•
Mountain ash (European)	Sorbus aucuparia		•		•	
Oak	Quercus species					
Live	Q. virginiana	•			•	
Pin	Q. palustris		•			•
Red	Q. rubra		•			•
Scarlet	Q. coccinea		•			•
Peach	Prunus persica		•		•	
Pear	Pyrus communis		•		•	
Pecan	Carya illinoensis		•			•
Pine	Pinus species					
Austrian	P. nigra	•				•
Red	P. resinosa	•				•
Scotch	P. sylvestris	•				•
White	P. strobus	•				•

Time of Flowering			Fruiting Time			Good Fall Color	Hardiness Zone Rating	Comment
Early Spring	Late Spring	Early Fall	Late Summer	Early Fall	Late Fall			
	•			•		•	4	avoid using near children's area; plants are thorny
	•			•			4	
	•			•		•	4	
							3	grows best in full sunlight
					•		5	male and female plants are needed for fruit set
					•		6	pyramidal silhouette
				•			4	good city tree
			•			•	2	
			•			•	5	
						•	2	a deciduous, needled conifer
	•						2	good street trees
	•						3	
	•						4	
•				•			5	large, showy flowers
	•			•			7	
•				•		•	5	
•			•			•	2	has spectacular autumn foliage color where autumn days are cool and crisp
•			•				5	
•			•			•	5	
•			•				3	
•						•	3	
•			•			•	3	
	•		•			•	3	susceptible to borer insects
							7	strong trees; used widely as lawn and shade trees; good for attracting squirrels
						•	4	
						•	3	
						•	4	
•			•				5	edible fruit, but may not bear if winters are too warm
	•			•			5	edible fruit
	•			•			5	may not bear mature fruit outside of southeastern states
							4	good for use as windbreaks most effective when used in groups of three or more
							2	
							2	
							3	

A Guide to Landscape Trees *(continued)*

Tree		Evergreen	Deciduous	Maximum Height		
Common Name	Botanical Name			10'–25'	25'–60'	60' and up
Plum	Prunus species					
Fruiting	P. domestica		•	•		
Purple	P. pissardi		•	•		
Redbud	Cercis canadensis		•		•	
Russian olive	Elaeagnus angustifolia		•	•		
Sapodilla	Achras zapota	•			•	
Spruce	Picea species					
Black Hills	P. glauca densata	•				•
Blue	P. pungens glauca	•				•
Colorado	P. pungens	•				•
Norway	P. abies	•				•
White	P. glauca	•				•
Sweet gum	Liquidambar styraciflua		•			•
Sycamore	Platanus occidentalis		•			•
Tulip tree	Liriodendron tulipifera		•			•
Walnut	Juglans species					
Black	J. nigra		•			•
English	J. cinerea		•			•
Willow	Salix species					
Babylon weeping	S. babylonica		•		•	
Pussy	S. discolor		•	•		
Thurlow weeping	S. elegantissima		•		•	
Zelkova (Japanese)	Zelkova serrata		•			•

Time of Flowering			Fruiting Time			Good Fall Color	Hardiness Zone Rating	Comment
Early Spring	Late Spring	Early Fall	Late Summer	Early Fall	Late Fall			
•			•				4	
•			•				4	
•							4	attractive, delicate flowers
	•			•			2	silver foliage color
•				•			10	
							3	rigid, pyramidal trees
							2	
							2	
							2	
							2	
				•		•	5	spectacular fall color; mixed shades
	•				•		5	white, peeling bark
	•		•				4	needs room to grow and spread
				•			4	edible fruit; roots of black walnut kill most plants growing beneath the tree
				•			6	
•						•	6	grows quickly; weak wood
•							4	
•						•	4	
						•	5	good substitute for American elm

A Guide to Landscape Shrubs

Shrub		Evergreen	Deciduous	Mature Height		
Common Name	**Botanical Name**			**3'–5'**	**5'–8'**	**8' and up**
Almond, Flowering	Prunus glandulosa		•	•		
Azaelas	Rhododendron species					
Gable	R. poukanense hybrid		•	•		
Hiryu	R. obtusum 'Hiryu'		•			
Indica	R. indicum	•			•	
Kurume	R. obtusum 'Kurume'	•		•		
Mollis	R. kosterianum		•	•		
Torch	R. calendulaceum		•	•		
Barberry	Berberis species					
Japanese	B. thunbergi		•		•	
Mentor	B. mentorensis	semi			•	
Red leaved	B. thunbergi atropurpurea		•		•	
Bayberry	Myrica pennsylvanica	semi				•
Boxwood	Buxus species					
Common	B. sempervirens	•				•
Little leaf	B. microphylla	•		•		
Camellia	Camellia japonica	•				•
Coralberry	Symphoricarpos orbiculatus		•	•		
Cotoneaster	Cotoneaster species					
Cranberry	C. apiculata		•	•		
Rockspray	C. horizontalis	semi		•		
Spreading	C. divaricata		•		•	
Deutzia, slender	Deutzia gracilis		•	•		
Dogwood	Cornus species					
Cornelian cherry	C. mas		•			•
Grey	C. racemosa		•			•
Red twig	C. stolonifera		•			•
Firethorn	Pyracantha species					
Scarlet	P. coccinea	semi			•	
Formosa	P. formosana	•				•
Forsythia	Forsythia species					
Early	F. ovata		•		•	
Lynwood	F. intermedia 'Lynwood'		•			•
Showy border	F. intermedia spectabilis		•			•
Gardenia	Gardenia jasminoides	•		•		
Hibiscus	Hibiscus species					
Chinese	H. rosa-sinensis	•				•
Shrub althea	H. syriacus		•			•
Holly	Ilex species					
Chinese	I. cornuta	•				•
Inkberry	I. glabra	•				•
Japanese	I. crenata	•				•

*Semi-evergreen indicates that the plants retain their leaves all year in warmer climates, but drop them during the winter in colder areas.

Season of Bloom**			Light Tolerance			Good Fall Color	Zone of Hardiness	Comment
Early Spring	Late Spring	Early Fall	Sun	Semi-Shade	Heavy Shade			
•			•				4	very showy blooms
	•		•	•			6	requires an acidic soil condition and often iron chelate fertilizers
•	•			•			7	
				•			8	
	•			•			7	
	•		•	•			6	
	•			•		•	6	
	•		•	•		•	4	good plants for traffic control; thorny
	•		•	•			5	
	•		•	•			4	
			•	•			2	fragrant leaves and fruit good for seashore areas
			•	•			5	prunes well; good for formal hedges
			•	•			5	
•				•	•		7	blooms from late October to April
		•		•		•	2	good on banks for erosion control
	•		•	•		•	4	fall color comes from bright red fruit
	•		•	•		•	4	
	•		•	•		•	5	
	•		•				4	white flowers
•			•			•	4	red twig is very good for erosion control; all dogwoods have good fall color
•			•			•	4	
•			•			•	2	
	•		•			•	6	fall color comes from brightly colored fruit
	•		•			•	8	
•			•				4	bright yellow flowers cascading branching patterns
•			•				5	
•			•				5	
	•	•		•			8	very fragrant flowers
	•		•				9	
		•	•				5	
							7	fruit color is most attractive in the fall
							3	
							6	

**Where no rating is given, flowers are either not produced or are not of importance.

A Guide to Landscape Shrubs (continued)

Shrub		Evergreen	Deciduous	Mature Height		
Common Name	Botanical Name			3'–5'	5'–8'	8' and up
Honeysuckle	Lonicera species					
Blue leaf	L. korolkowii zabel		•			•
Morrow	L. morrowii		•		•	
Tatarian	L. tatarica		•			•
Hydrangea	Hydrangea species					
Hills of Snow	H. aborescens grandiflora		•	•		
Oak leaf	H. quercifolia		•		•	
Pee gee	H. paniculata grandiflora		•			•
Jasmine	Jasminum species					
Common white	J. officinale	semi				•
Italian		•				•
Primrose	J. primulinum	semi				•
Juniper	Juniperus species					
Andorra	J. horizontalis plumosa	•		•		
Hetz	J. chinensis hetzi	•			•	
Japanese garden	J. procumbens	•		•		
Savin	J. sabina	•			•	
Pfitzer	J. chinensis pfitzeriana	•				•
Lilac	Syringa vulgaris		•			•
Mahonia	Mahonia species					
Leatherleaf	M. bealei	•				•
Oregon grape	M. aquifolium	•		•		
Mock orange, sweet	Philadelphus coronarius		•			•
Nandina	Nandina domestica	•			•	
Ninebark	Physocarpus opulifolius		•			•
Pieris (Andromeda)	Pieris species					
Japanese	P. japonica	•			•	
Mountain	P. floribunda	•			•	
Pine, Mugo	Pinus mugo	•			•	
Poinsettia	Euphorbia pulcherrima	•				•
Pomegranate	Punica granatum		•			•
Potentilla (Cinquefoil)	Potentilla fruticosa		•	•		
Privet	Ligustrum species					
Amur	L. amurense		•			•
California	L. ovalifolium	semi				•
Glossy	L. lucidum	•				•
Regal	L. obtusifolium regelianum		•		•	
Quince, Flowering	Chaenomeles species					
Common	C. speciosa		•	•		
Japanese	C. japonica		•	•		

*Semi-evergreen indicates that the plants retain their leaves all year in warmer climates, but drop them during the winter in colder areas.

Season of Bloom**			Light Tolerance			Good Fall Color	Zone of Hardiness	Comment
Early Spring	Late Spring	Early Fall	Sun	Semi-Shade	Heavy Shade			
	•		•	•			5	
	•		•	•			4	
	•		•	•			3	
	•		•				4	
		•	•			•	5	coarse-textured shrubs
		•	•			•	4	
	•		•				7	
	•		•				8	
	•		•				8	
			•				2	grows well in hot, dry soil
			•				4	
			•				5	
			•				4	
			•				4	
	•		•				3	large, fragrant flowers
•				•			6	holly-like foliage
•				•			5	bluish, grapelike fruit
	•		•				4	creamy white fragrant flower
		•	•	•		•	7	very attractive in flower and fruit stage
	•		•			•	2	
•			•	•			5	
•			•	•			4	
			•	•			2	
		late fall	•				9	long-lasting blooms
	•		•	•		•	8	colorful in both spring and fall
	•	•	•				2	produces yellow flower all summer
	•		•	•			3	prunes well
	•		•	•			5	popular hedge plants
		•	•				7	
	•		•				3	
•			•	•			4	densely branched, thorned plants
•			•	•			4	good for traffic control

**Where no rating is given, flowers are either not produced or are not of importance.

A Guide to Landscape Shrubs (continued)

Shrub		Evergreen	Deciduous	Mature Height		
Common Name	Botanical Name			3'–5'	5'–8'	8' and up
Rhododendron	Rhododendron species					
Carolina	R. carolinianum	•			•	
Catawba	R. catawbiense	•			•	
Rosebay	R. maximum	•				•
Rose, Hybrid tea	Rosa species		•	•		
Spirea	Spiraea species					
Anthony Waterer	S. bumalda 'Anthony Waterer'		•	•		
Bridal wreath	S. prunifolia		•			•
Billiard	S. billardi		•		•	
Frobel	S. bumalda 'Froebelii'		•	•		
Thunberg	S. thunbergi		•	•		
Vanhoutte	S. vanhouttei		•		•	
Viburnum	Viburnum species					
Arrowwood	V. dentatum		•			•
Black haw	V. prunifolium		•			•
Cranberrybush	V. opulus		•			•
Doublefile	V. plicatum tomentosum		•			•
Fragrant	V. carlcephalum		•			•
Japanese snowball	V. plicatum		•			•
Leatherleaf	V. rhytidophyllum	•				•
Sandankwa	V. suspensum	•			•	
Weigela	Weigela florida		•			•
Winged Euonymus	Euonymus alatus		•		•	
Wintercreeper	Euonymus fortunei vegetus	•		•		
Yew	Taxus species					
Spreading Anglo-Japanese	T. media	•				•
Upright Anglo-Japanese	T. media hatfield	•				•
Spreading Japanese	T. cuspidata	•				•
Upright Japanese	T. cuspidata capitata	•				•
English	T. baccata	•				•
Canada	T. canadensis	•		•		

Semi-evergreen indicates that the plants retain their leaves all year in warmer climates, but drop them during the winter in colder areas.

Season of Bloom**			Light Tolerance			Good Fall Color	Zone of Hardiness	Comment
Early Spring	Late Spring	Early Fall	Sun	Semi-Shade	Heavy Shade			
	•			•			5	
	•			•			4	
	•			•			3	
	•	•	•				varies	very diversified group of plants special culture required
	•		•	•			5	very attractive when blooming most are resistant to disease and insect pests
	•		•	•		•	4	
	•		•	•			4	
•			•	•			5	
			•	•		•	4	
	•		•	•		•	4	
•			•	•		•	2	attractive spring flowers good fall foliage color many are good as wildlife food
•			•	•		•	3	
•			•	•		•	2	
•			•	•		•	4	
•			•	•		•	5	
•			•	•		•	4	
•			•	•			5	
•			•	•			9	
	•		•				5	blooms late
			•			•	3	spectacular crimson fall color
				•		•	5	
			•	•			4	excellent for foundation plantings and hedges prunes well long lived
			•	•			4	
			•	•			4	will not tolerate poorly drained soil
			•	•			4	
			•	•			6	
				•			2	

**Where no rating is given, flowers are either not produced or are not of importance.

A Guide to Groundcovers

Groundcovers									
Common Name	Botanical Name	Evergreen	Deciduous	Height	Optimum Spacing	No. Needed to Plant 100 sq. ft.	Light Tolerance	Hardiness Zone Rating	Flower or Fruit Color and Time of Effectiveness
Ajuga or bugle	Ajuga reptans		•	5"	6 inches	400	sun or shade	4	blue or white flowers in summer
Cotoneaster, creeping	Cotoneaster adpressa		•	12"	4 feet	10	sun	4	pink flowers, red fruit in summer and fall
Cotoneaster, rockspray	Cotoneaster horizontalis	semi		18" plus	4 feet	10	sun	4	pink flowers, red fruit in summer and fall
Euonymus, big leaf wintercreeper	Euonymus fortunei radicans	•		18" plus	3 feet	14	sun or shade	5	orange fruit in fall
Euonymus, purple leaf wintercreeper	Euonymus fortunei coloratus	•		18"	3 feet	14	sun or shade	5	not of significance
Honeysuckle, creeping	Lonicera prostrata		•	12"	3 feet	14	sun	5	pale yellow flowers in spring; red fruit in fall
Ivy, Baltic English	Hedera helix baltica	•		8"	18 inches	44	shade	4	none
Mondo	Ophiopogon japonicus	•		12"	10 inches	144	partial shade	8	white or pink flowers in spring
Myrtle or Periwinkle	Vinca minor	•		8"	12 inches	92	shade	4	blue flowers in spring
Oyster plant	Tragopogon porrifolius	•	•	12"	12 inches	92	sun or shade	9	not of significance
Pachysandra	Pachysandra terminalis	•		12"	12 inches	92	shade	4	white flowers in spring
Sarcococca	Sarcococca ruscifolia	•		tall—requires shearing	3 feet	14	sun or shade	7	white flowers and scarlet berries in fall
Wandering Jew	Tradescantia albiflora	•		6"	12 inches	92	shade	9	red-purple flowers in spring and summer
Weeping lantana	Lantana montevidensis	•		18" plus	24 inches	25	sun	9	lavender flowers all year
Yellowroot	Xanthorhiza simplicissima		•	18" plus	18 inches	44	sun	5	brown-purple flowers in spring

A Guide to Vines

Common Name	Botanical Name	Broad-Leaved Evergreen	Deciduous	Height	Clinging	Twining or Tendrils	Light Tolerance	Hardiness Zone Rating	Flower or Fruit Color and Time of Effectiveness
Actinidia, bower	Actinidia arguta		•	30'		•	full sun or semi-shade	4	white flowers in spring
Actinidia, Chinese	Actinidia chinensis		•	30'		•	full sun or semi-shade	7	insignificant
Akebia, fiveleaf	Akebia quinata	semi		35'		•	full sun or semi-shade	4	purple flowers in spring
Ampelopsis, porcelain	Ampelopsis brevipedunculata		•	20'		•	semi-shade	4	multicolored fruit in fall
Bignonia (or crossvine)	Bignonia capreolata	•		60'		•	full sun or semi-shade	6	orange-red flowers in spring
Bittersweet, American	Celastrus scandens		•	20'		•	sun or semi-shade	2	yellow and red fruit in fall and winter
Boston ivy	Parthenocissus tricuspidata		•	60'	•		sun or shade	4	insignificant
Bougainvillea	Bougainvillea glabra	•		20'	•		full sun	7	multicolored in summer
Clematis	Clematis species		•	3' to 25'*		•	full sun or semi-shade	4 to 7*	many colors of flowers in late spring
Euonymus, evergreen bittersweet	Euonymus fortunei vegetus	•		25'	•		sun or shade	5	yellow and red fruit in fall and winter
Fig, creeping	Ficus pumila	•		40'	•		sun or shade	9	insignificant
Honeysuckle, trumpet	Lonicera sempervirens		•	50'		•	full sun or semi-shade	3	orange flowers in summer; red fruit in fall
Hydrangea, climbing	Hydrangea anomala petiolaris		•	75'	•		full sun or semi-shade	4	white flowers in summer
Ivy, English	Hedera helix	•		70'	•		semi-shade	5	insignificant
Kudzu vine	Pueraria lobata		•	60'			sun or shade	6	insignificant
Monks hood vine	Ampelopsis aconitifolia		•	20'		•	semi-shade	4	yellow-orange fruit in fall
Rambling roses	Rosa multiflora hybrids and others		•	10' to 20'		support needed	sun	5	flowers of many colors in spring and summer
Trumpet vine	Campsis radicans		•	30'	•		sun	4	orange flowers in summer
Virginia creeper	Parthenocissus quinquefolia		•	50'	•		sun or shade	3	insignificant
Woodbine, Chinese	Lonicera tragophylla		•	50'		•	shade	5	yellow flowers in summer; red fruit in fall

*Dependent upon the actual species selected

Achievement Review

A. Define the following terms.

   1. native plant
   2. exotic plant
   3. naturalized plant
   4. bare root plant
   5. B & B plant
   6. containerized plant
   7. Latin binomial
   8. genus
   9. hardiness
   10. microclimate

B. Of the following characteristics, indicate which ones are typical of plants sold as (BR) bare rooted, (B & B) balled and burlapped, or (C) containerized.

   1. most common with very large plants
   2. the root system is totally intact
   3. may be cumbersome to handle
   4. more common with small deciduous plants
   5. limited season for planting
   6. roots may bind if kept too long in this form
   7. heavy
   8. lightweight
   9. usually the least expensive
   10. usually the most expensive

C. Explain why the use of botanical names is important when selecting plants for the landscape.

D. Identify the major categories of factors that must be evaluated when selecting specific plants for a particular project.

Suggested Activities

1. Consult the Hardiness Zone Map to determine:

   a. the hardiness zone rating of your home county.
   b. other regions of the country having winter temperatures similar to those of your home county.

2. Compare the Hardiness Zone Map with a geographic map. Determine what can influence winter temperatures in an area besides the north-south location.

3. Determine why the coast of Maine has a warmer hardiness zone rating than upstate New York, even though Maine is farther north.

4. If a light meter is available, take readings beneath several different types of trees. Be sure that the day selected for taking the readings is clear and sunny. Temperature readings can be taken as well. Rate the foliage density of the trees tested.

5. Make a collection of tree flowers, fruits, or colored leaves. Rate them on the basis of how colorful they are.

6. Collect a number of plants in the same botanical genus, but of different species (Example: different kinds of pines or dogwoods or palms). Make a listing of all of the characteristics that are common to all of them. Then look for one characteristic of each species that makes it different from all of the others in the genus. This is how taxonomists separate plant species.

   NOTE: Many of the distinctions are not easily noticed by the untrained eye.

7. Visit a nursery and compare the plants for sale based upon the form of their root systems. Compare the plant types with their methods of harvest and sale. Compare them on the basis of size, weight, price, suggested time of transplant, and any other factors that seem to illustrate differences based upon the form of the root system.

CHAPTER 8

The Principles of Design

Objectives:

Upon completion of this chapter, you should be able to

- describe the basic principles that lead to good design
- explain how these principles are applied to landscape designing
- describe features of plant materials, hardscape materials, and architecture that make them useful as design elements

The Foundation for Designing

Many texts address the subject of design principles, and each does it in a slightly different way. Since the names and number of principles vary from book to book, the reader might begin to wonder if any one source is complete or totally correct. While each author may present the subject in a slightly different way, by chapter's end most have explained the principles fully even if the terminology fails to match completely. As discussed in this chapter, **principles** refer to standards by which designs can be created, measured, discussed, and evaluated. Because design is at times very personal, it can be difficult to evaluate objectively. Whether someone likes or dislikes a certain plant selection or paving pattern does not necessarily

credit or discredit the design. Only when the design can be shown to be in compliance with or in violation of the principles that guide all design can it be judged as good, bad, or in between. Only when a judgment can be offered that the design is good or bad *because* it applies or fails to apply one or more of the principles of design does it become a judgment or critique that is understandable or justifiable to the designer.

Here then are six principles that artists have been using in the fine and applied arts for centuries. Whether the art form is oil painting or flower arranging, the same principles of design can be applied during the creation of the work and can be used to describe a viewer's reaction to the work. The reassuring thing about these principles is that they were not created centuries ago by artists. Rather, they stem from an inherent visual sense possessed by most people, whether they regard themselves as creative or not.

Balance

Balance is a state of being as well as seeing. We are physically uncomfortable when we are off balance. Whether while held hostage on top of the see-saw by a heavier playmate when we were children, or tipping over in a canoe when the contents shifted unexpectedly, we experience a lack of balance at various times in our lives, and we do not like it. It follows that we are most appreciative of and comfortable in landscape settings that are visually balanced.

There are three types of balance: symmetric, asymmetric, and proximal/distal. Symmetric balance is the balance of formal gardens. One side of the composition is a reflective mirror image of the opposite side, Figure 8-1. While it can be somewhat stiff, especially if the plants used are geometric or sheared forms, symmetric balance is acceptable to a great many people because it is comfortable and easily understood. It is not necessary that landscape composition be stiff, though, in order to be symmetric. The combination of materials used may be loose and casual, yet as long as the shapes, colors, and specific materials match on both sides of center, the balance will be symmetric. Asymmetric balance is informal balance. The visual weight on opposite sides of the composition is the same, but the materials used and their placement may vary, Figure 8-2. Asymmetric balance has the potential

Figure 8-1 Symmetric balance. One side of the outdoor room is a mirror image of the opposite side.

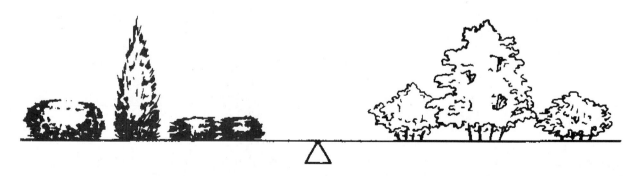

Figure 8-2 Asymmetric balance. One side of the outdoor room has as much interest as the opposite side, but does not duplicate it exactly.

to be more visually interesting to the viewer because there are two sides to observe and explore. With symmetric balance that is not true. Proximal/distal balance is asymmetric but carries it further by dealing with depth in the field of vision. In addition to balancing the left/right relationship in the composition of the landscape, there is a need to balance the near/far. This becomes a concern more frequently than might appear at first. The landscape designer is seldom able to design without concern for off-site features. Those features may be pleasant and desirable to include within the composition, but once included they must be factored into the balance of the setting. Tall buildings, distant mountains, large, off-site trees, and other such features can affect the balance of a design that looks good on paper, but would be off-balance once installed, Figure 8-3.

Focalization of Interest

Anything that is designed well has a focal point, one place within the composition where the viewer's eye is first attracted. Whether it be the enigmatic smile on the face of the Mona Lisa,

the largest, brightest flower in a table arrangement, or the jeweled pin on an evening gown, there will be something placed by the designer to command the attention of the viewer. Everything else in the composition will serve to compliment that feature. **Focalization of interest** is the principle of design that selects and positions visually strong items into the landscape composition. Focal points can be created using plants, hardscape items, architectural elements, color, movement, texture, or a combination of these and other features, Figure 8-4. Some focal points are predestined. For example, in the public area of most residential landscapes, the entry to the home is understood to be the most important focus of attention. All of the design decisions made in that area serve to support that focal point. In other areas of the landscape, the designer usually has more opportunity to select those areas to be highlighted as focal points. Beginning designers often must learn restraint in applying the principle of focalization of interest. There is often a tendency to overuse focal points, which creates complexity and visual confusion within the landscape composition.

Figure 8-3 Proximal/distal balance. The foreground tree counterbalances the distant mountain.

Figure 8-4 Focalization. Plants are arranged in an asymmetrical step-down manner to move the viewer's eye toward the entry, the focal point.

Simplicity

As with the principle of balance, **simplicity** seeks to make the viewer feel comfortable within the landscape. Few people are happy when exposed for any length of time to settings that are cluttered or fussy. Complexity is not always the opposite of simplicity, at least where landscape design is concerned. Landscapes may involve buildings with complex and intricate architecture. Other projects may be technically complex, with extensive lighting, water features, sound systems, circulation patterns, or security systems. If such things are present, they are probably important to the client or to the site; but the landscape into which they are placed can still be simple and comfortable for the user. Simplicity does not imply simplistic, boring, or lack of imagination. It does avoid the use of too many species, too many colors, too many textures, too many shapes, curves, and angles within an area or within a project.

Rhythm and Line

Rhythm and **Line** may be regarded separately or collectively as a principle of design. When something repeats enough times with a standard interval between repetitions, a rhythm is established. In landscape designing, the interval is usually a measured space. What repeats may be something structural, such as lamp posts or benches. It may be something patterned, such as a sidewalk pattern that is replicated every 50 feet from one end of a mall to the other end. As a user experiences the landscape, moving at a fairly even rate of speed, these repeating design features establish a rhythm within the user's experience. The lines of the design establish the very shape and form of the landscape. Lines are created where different materials meet, such as where turf meets pavement or where turf meets the mulched planting bed. The merger of the two edges of the materials creates the line. Lines are also created for the eye when paved areas or walls cut through lawns, or when patterns are created within paved areas. Nature too provides lines, in the crest of surrounding hills or at an edge where a lake or river meets the land. When enough lines are parallel to each other, and all can be seen by a viewer at the same time, then a rhythm of lines is established, bringing the two concepts together as one, Figure 8-5. No landscape should be without a rhythm to the layout of its major line-making features.

Proportion

Proportion is concerned with the size relationships between all of the features of the landscape. That includes both vertical and horizontal relationships as well as spatial relationships. Much of our perception of vertical proportion is influenced by the height of the viewer and in particular by the viewer's eye level, which varies between standing, seated, and reclining. Shorter people perceive the vertical space of some landscapes differently than do taller people. Children have needs for certain types of landscapes that are vastly different from those of their parents if they are to feel comfortable within the space. The concern for proper proportion extends to building size, lot size, plant size, the relationship between the areas of mass and void, and the human users of the landscape, Figure 8-6.

Unity

As a principle of design, **unity** is the easiest to measure if the other five principles have been

Figure 8-5 The repeating parallel lines of turf, pavement, and flowers add rhythm and line to the design.

A   B

Figure 8-6 Proportion. Each element of the landscape must be in the proper size relationship with all other elements.

applied properly and comprehensively to the design. A unified design is one in which all of the separate parts contribute to the creation of the total design. The color schemes and textural choices support each other rather than demanding individual attention. In a painting, the background details match the foreground features. In a flower arrangement a suggestion of each color is distributed evenly throughout the design. In a fabric pattern, the same shapes and colors repeat at regular intervals throughout the cloth. In each case, while individual components are valued and appreciated, they collectively create a single overall design. The same principle applies to landscape designing. Each component of the design, whether it is the plant materials, the shape of the planting beds, the choice and use of paving materials, the color selections, the lighting plan, or any other component of the outdoor rooms within a project, is obligated to be a part of the whole.

Applying the Principles of Design to Landscape Designing

Some references have already been made to ways that the six principles of design are used in designing and evaluating landscapes. Beginning designers may have to make a conscious effort to assure that all of the principles are applied to a specific design. However, as designers and landscape architects become more experienced, using the principles becomes part of their creative thought processes, and the application becomes nearly automatic.

It is important to understand that the principles of design must be applied at all of the levels that people will experience the landscape. While the designer may be able to see the entire project as it is spread out across the drawing board, that is not the way human users will actually experience the landscape. They will move through it one outdoor room at a time and at varying rates of speed and attentiveness. At any one time, users of the land-scape will only see what falls within their cone of vision, Figure 8-7. Thus, a focal point is only effective when it falls within that cone of vision. Balance is only apparent when the viewer is able to see both sides at the same time. As people move back and forth through a landscape, their perception of it continually changes. As a designer works to develop a plan that is responsive to those who will be using the landscape, it is important to envision the design from every logical vantage point, realizing that only birds will see the property as an overhead panorama, and birds are not the clients. The human user will experience it as an ongoing series of visual images, and no two people will link those images together in the same way.

Applying the Principle of Balance

Balance must be addressed from several ranges. At **macro-range**, the viewer sees the landscape from the most distant vantage point, such as the opposite side of the street. This extends the cone of vision and often permits the viewer to see the entire house or building plus the entire adjacent

Figure 8-7 The principles of design must apply from varying and changing vantage points.

landscape. Frequently houses are not centered within the lot, and predictably there will be a driveway connecting the street to the garage, creating a barren, no-plant space on one side of center that does not exist on the opposite side. Visible from behind the house may be large trees or utility poles that affect the balance of the setting. In addition, immediate off-site features, such as the neighbors' houses and trees, are often visible when viewed from curbside, making them a site feature to factor into the consideration of balance. It is almost mandatory that the designer take a series of photographs during the site analysis to be certain that none of these factors are overlooked when designing the public area of a landscape. It is the only way to assure that the curbside view of the landscape is balanced, Figure 8-8.

The same concern for macro-range balance must be applied to all other use areas of the landscape. For example, when viewing the family living area or the private living area from inside the house or from the patio, how much of the view is on-site and how much is off-site? What must be done

on-site to bring the view into balance? Correct design solutions may require the addition of trees or the thinning or removal of existing ones to bring the site into balance with the immediate off-site setting.

At closer range, the viewer's shortening cone of vision becomes a major concern. For example, as the person approaches the entry of a building, certain features such as the masses of trees that enframe the building move outside the viewing range and no longer contribute to the viewer's perception of balance. Instead, the placement of shrubs on either side of the entry assumes greater importance. The viewer's eye will make a judgment about the type of balance and the success or failure of the principle's application, Figure 8-9. Similarly, the viewer will pass judgment on whether the plantings on each side of an important window are or are not balanced, or if the trees lining the drive are as densely planted on one side as on the other. The viewer will decide if the shrub planting at one end of the patio has sufficient mass to counterbalance the stone grill constructed at the other end.

Figure 8-8 A balanced public area as viewed from curbside

Figure 8-9 The design is balanced when the viewer's cone of vision is limited to View A; but the balance is lost when the cone of vision widens to View B. (Drawing courtesy of James Lancaster)

The designer must envision dozens of viewing stations, looking across every section of the landscape at ground level as well as from upper story windows and decks. Further, the designer must envision a dividing line down the center of each view to determine if the design balances from each vantage point.

Applying the Principle of Focalization of Interest

As discussed earlier, focal points are places in the landscape where attention either is naturally drawn or where the designer elects to direct it. In the public area, focusing attention on the entry is just common sense, because the first question a visitor to any building asks is "How do I get into this building?" See Figure 8-4. To place anything else in the public area that will compete with the entry

area design would be a violation of the principle of focalization of interest. Both beginning and experienced designers frequently violate this principle. Among the most common mistakes are the use of flowers or night lighting in places removed from the entry. Brightly colored flowers are strong magnets for the eye. If used at all in the public area, they should be clustered at the entry and not spread across the front of the building or tucked between the shrubs bordering the lawn. Likewise, night illumination of a tree at the end of the house or in the center of the lawn will shift attention away from the entry. In the public area, night lighting should be used to reinforce the entry design so that the focal point is not lost when darkness falls.

In other use areas of the landscape, the designer often has a choice for placement of the focal points. Depending upon the size of the property and the type of design (private or public, residential or commercial), it may be suitable to have one or multiple focal points. In a typical residential family living area, a single focal point is usually appropriate, Figure 8-10, unless the area is large enough to develop additional ones without conflict. Again, the viewer's cone of vision plays a major role in deciding what works. If the property is so small that two focal points can be seen at the same time, there may be a visual conflict. As long as the focal points do not compete with each other, then more than one can be appropriate.

Many things can function as focal points in the landscape. Some of them will fulfill that role continuously and without changing. Statuary and other garden ornaments exemplify such items. Plants can also serve as focal points when they are unusually colorful or have striking growth habits. Their time of service can vary immensely. Many plants that are visually powerful when in spring flower or fall color become uninteresting during the rest of the year. By carefully selecting and positioning the focal features of the landscape, the landscape designer can introduce shifting focal points, Figure 8-11. This is an opportunity that is unique to landscape design. No other artist is able to do that.

Figure 8-10 Focalization draws the eye of the viewer to one major feature in each use area.

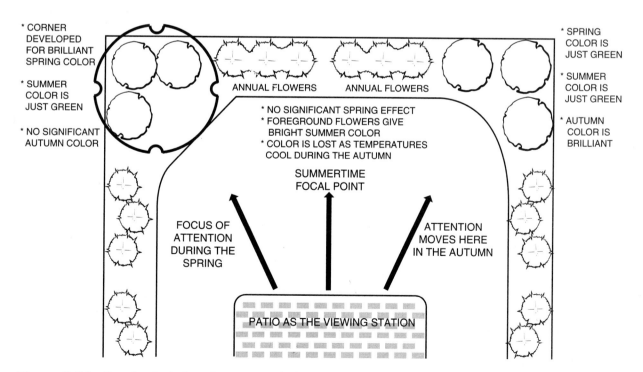

Figure 8-11 By selectively locating seasonal plants, the designer can introduce shifting focal points into the landscape composition.

The Corner Planting. One of the most natural places to position a focal point is in the corner of the outdoor room. When bedlines converge at the corner from two directions the eyes of viewers follow willingly to the point of convergence. Termed the **incurve** of the planting, the point of convergence easily accepts a focal feature such as a specimen plant (highly attracting to the eye) or an accent plant (less attracting than a specimen plant, but sufficiently different from other plants around it to be distinctive). The incurve plant is usually taller than the other plants extending from it to the outer reaches of the corner bed, the **outcurve**, Figure 8-12. The placement of plants into the corner arrangement should be done so that when the plants are mature, the viewer's eye will step up from the outcurves to the incurve, Figure 8-13.

Many variations are possible for the corner planting, Figure 8-14. A garden ornament, bench, or fountain might be used instead of a plant as the focal feature at the incurve. Flowers could be used directly at the incurve in front of the focal feature to lend some seasonal variation to the design.

Applying the Principle of Simplicity

The provision for simplicity in landscape design is the exclusion of unnecessary change. The key word here is *unnecessary*. Each time that there is a change of shape or color, complexity is introduced. Each time a bedline changes direction, complexity can result. Too many garden ornaments, too many different kinds of furniture, too many different types of wall or paving materials, can spoil a design that has everything else done correctly. It is not that change, diversity, or variety are wrong or unwarranted. A good design contains all of that. It is when the introduction of that change or variation does not advance or improve the design that it becomes a violation of the principle of simplicity. Unfortunately no magic number establishes a threshold that tells a designer when to stop. How many different colors or materials become too many? At what point does something cross the line from tasteful to garish? There are no clear answers, and there is a great deal of subjectivity in the interpretation of this principle. However, here are a few guidelines and suggestions for designing an imaginative, but simple landscape.

a. Avoid using too many angles in the layout of planting beds and paved areas. Use wide arcs in designing curvilinear features such as planting beds, and match concave arcs with convex ones of closely comparable size, Figure 8-15.

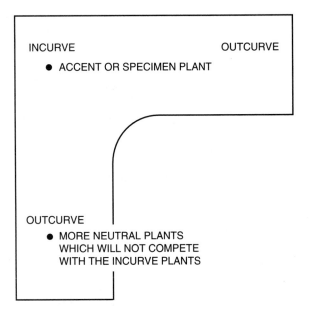

Figure 8-12 Parts of the corner planting bed: the incurve and the outcurves

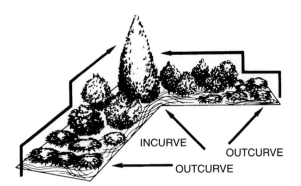

Figure 8-13 In a corner planting, attention is drawn from the outcurves to the incurve by stair-stepping plants.

Figure 8-14 Three variations of the corner planting. Notice that in each example, the eye is drawn from the ends of the planting to the center and from the front to the rear.

Figure 8-15 Bedline design: Top bedline is fussy. It has too many arcs that do not counterbalance each other. Lower bedline is well designed. It features simple curves with a good balance of concave and convex arcs.

b. Limit the number of different species of plants in any use area. While no exact number can be given because use areas can vary in size so greatly, do not create an arboretum.

c. Massing plants eliminates the need to look at each one individually.

d. Arrange plants so that they create a silhouette that flows smoothly and is not choppy, Figure 8-16.

e. Avoid paving patterns and shapes that introduce independent lines, angles, or colors into an area.

A. CHOPPY SILHOUETTE – NO MASSING

B. FLOWING SILHOUETTE – SPECIES MASSED

Figure 8-16 Simplicity in plant arrangement. Grouping B allows the viewer's eye to move smoothly along the planting.

f. When planning for color, high color contrasts are appropriate at the focal points of the design, but elsewhere it is best to use more subtle color schemes. If the architecture or nearby off-site features already contain numerous colors, the landscape should only repeat the existing colors rather than introducing new ones.

g. Avoid the overuse of geometrically shaped or sheared plants. Pyramidal forms are especially eye-attracting, so only use them where a specimen or accent plant is appropriate.

Applying the Principle of Rhythm and Line

The establishment of rhythm and line in the landscape has been compared to the creation of frozen motion. Drawing upon strong existing lines (and sometimes objects) and replicating them to the extent that the person using the landscape is consciously or subconsciously aware of them is the way this principle of design is applied. As an example, consider a typical residential landscape. The strongest and most dominant lines are the lines of the house. Architecture provides an excellent springboard from which to begin. It is seldom wrong for a landscape designer to repeat the lines and angles of the house in the lines and angles of the landscape. It's a technique that nearly always works. In other situations, the designer may choose to repeat the lines of some other visually powerful item, such as a swimming pool or a patio, Figure 8-17.

Figure 8-17 Rhythm and line: The repeating horizontal and vertical lines and repeating 45-degree angles create a rhythm of lines in this design. (Drawings courtesy of James Lancaster)

The principle of rhythm and line also can be used to evoke certain emotional responses from the users of the landscape. Strongly rectilinear designs suggest greater formality than those which are mostly curvilinear. One may seem stiffer to the viewer, while the other more casual and leisurely. One might be ideal for the design of a corporate headquarters, and the other better suited to a city park. In neither case is the use of the principle of rhythm and line incorrect. Should the lines of the landscape not be sufficiently replicated to establish at least a suggestion of rhythm, then the lines become erratic. That will also evoke an emotional reaction from the user of the landscape, but it will not be a pleasant one. Erratic lines introduce complexity and confusion.

The Line Planting. Concern for the vertical linearity of the landscape is as important as concern for the horizontal linearity. Off-site features such as distant hills or the spire of a church might offer a strong design line that could be replicated on-site. Other times, the planting design must first fill a functional role such as blocking an unsightly view or providing some degree of privacy, Figure 8-18. Still, the line created by the silhouette of the mature plants should be envisioned by the designer at the time of plant selection, and its appropriateness to the overall design should be considered.

Positioning plants in the line planting must also be done with care at the time the design is created. If not done carefully, the plants can create a vertical line, which will not match the shape intended for the outdoor wall, Figure 8-19. While the bedlines appear very prominent on the drawn plan view, the vertical linearity created by the plants and

BLOCK A VIEW

FRAME A VIEW

PROVIDE PARTIAL PRIVACY

PROVIDE TOTAL PRIVACY

Figure 8-18 Some functions of the line planting

THE REAR PLANTS ARE HIDDEN DUE TO THEIR ALIGNMENT WITH THE FOREGROUND PLANTS

THE PLANTS ARE UNNECESSARILY CROWDED

PLACEMENT OF THE PLANTS CONFLICTS WITH THE SHAPE OF THE BEDLINE

SOME PORTION OF ALL PLANTS IS VISIBLE DUE TO STAGGERED PLACEMENT

PLACEMENT OF THE PLANTS REPEATS THE SHAPE OF THE BEDLINE

Figure 8-19 Correct and incorrect placement of plants within a planting bed other materials is responsible for the creation of the shapes of landscape spaces. The plants need to be positioned so that they replicate the lines of the beds. Where smaller plants are used in front of taller plants, or where plants are staggered for shadow effects, both ranks of plants should replicate the bedlines, Figure 8-19.

The Foundation Planting. One of the most prominent line plantings is the **foundation planting**. It has been passed down from an era when houses were built on top of high, unattractive foundations. The planting was designed to screen the foundation. Even though modern houses often have no exposed foundations, the foundation planting still is used; however its functions have been broadened. The foundation planting includes the entry design, and as such must focus attention on the front door of the residence. That was discussed in detail earlier. Being so close to the building, the modern foundation planting also serves as the single most important element connecting the man-made building with the more natural landscape. In applying the principle of rhythm and line to the design of the foundation planting, a designer may use the lines and angles of the house as described previously, but with some added objectives. Where space permits, the foundation planting

should reach outward from the ends of the building, not stopping just because the building has ended. It should also reach forward from the entry rather than sticking snugly against the building. In both places, the intent is to free the planting from the constraints of the building and, by extending the lines toward the surrounding landscape, cause the house to interface more comfortably with its setting, Figure 8-20. Figures 8-21 and 8-22 illustrate contemporary foundation plantings.

Applying the Principle of Proportion

Maintaining the proper size relationships within the landscape first requires that the designer clearly understand what those relationships are to be. Anyone who has read *Alice in Wonderland* or seen the animated motion picture can understand how the scale of the landscape, in comparison to the human users, evokes predictable reactions from the users. The designer is in a position to control those reactions, assuming they are anticipated in advance. If the ceiling of the outdoor room is low, the feeling is more intimate. That is frequently desirable when developing a residential patio. If the trees loom high overhead, people feel smaller by comparison. Even small buildings appear out of scale and threatened when the nearby trees are disproportionately large. The higher the wall and ceiling elements become

Figure 8-20 The foundation plants reach outward from the corner of the house, forward from the entry, and stay below the window levels.

Figure 8-21 A modern foundation planting displaying a welcoming entry court

Figure 8-22A Before: The foundation planting not yet installed. (Courtesy of Lied's, Inc. Sussex, Wis.)

Figure 8-22B After: The completed foundation planting. (Courtesy of Lied's, Inc. Sussex, Wis.)

in the outdoor room, the more dehumanizing the landscape gets. People who walk the canyons of New York City, Chicago, or similar cities learn to appreciate the important role that street trees play in bringing the bigger-than-life cityscape back into human scale. Children enjoy play areas where the equipment and the plantings are scaled to their heights, not that of adults. Small, low-growing plants should not be planted immediately adjacent to large, tall plants. Instead, it is preferable to bring the canopy down in stages from the tallest plants to

intermediate heights, and finally down to the lowest growing forms.

Landscapes need proper proportional relationships between

❦ buildings and people.

❦ buildings and plants.

❦ plants and people.

❦ plants and plants.

❦ masses and voids.

The need to balance masses and voids changes the concern from putting things in proportion to putting spaces in proportion. If the horizontal open areas of the landscape, the **voids**, are much larger or smaller than the solid vertical areas of plantings, buildings, walls, or land forms, the **masses**, and if that size difference is apparent to the viewer, then the principle of proportion will be missing from the design. For example, if a use area such as the family living area has an unusually large central lawn (the void) and adjoins another equally large void such as a field, the planting bed (mass) separating the two voids must be sizable enough to truly separate them without looking weak and apologetic. It needs sufficient width, height, and density to match their importance.

Applying the Principle of Unity

The principle of unity must be applied within each use area of the landscape to tie together the individual components of those areas. It must also be applied to the total landscape to integrate all of the separate use areas into a single design. It can even be applied to larger areas, including residential or commercial developments or even entire geographic regions. As with the other principles, unity can be applied at both the macro and micro levels of design.

Repetition is the key to unity. As people move through the landscape, they recognize the oneness of the design when certain familiar themes are repeated often enough. Here is a partial listing of techniques often used to bring unity to a landscape design.

❦ Repeat prominent colors inside and outside the house.

- ❧ Repeat the construction materials of the house in the garden's constructed features.

- ❧ Continue the design themes of the interior rooms into the outdoor rooms.

- ❧ Use large expanses of glass to visually connect indoor and outdoor use areas.

- ❧ Raise patios, porches, decks, and landings to door level, rather than requiring users to step up or down when moving between the indoors and the outdoors.

- ❧ Repeat plant species between use areas, particularly at prominent points where the repetition will be noticed.

- ❧ Standardize the style of lighting fixtures and furniture.

- ❧ Standardize signs throughout a project or region.

- ❧ Maintain a consistent architectural style among buildings.

- ❧ If there is a strikingly prominent feature, such as a distant mountain or a city skyline, allow it to be seen from as many areas in the landscape as possible.

Material Characteristics Supportive of the Design Principles

All of the components of the landscape have characteristics that can support the application of the six design principles as long as the designer uses them correctly. If misused, certain materials may result in the weakening of one or more of the design principles.

Architecture

While there are many different styles of architecture, some of them historic and some of them contemporary, most styles can be clearly described as symmetrical or asymmetrical. Regardless of whether it is a private residence or a corporate structure, small or enormous, compact or rambling, it will be either a symmetric or asymmetric structure. Symmetric structures such as the Cape Cod frame and Federal or Georgian buildings easily accept symmetrically balanced landscape designs, Figure 8-23. Houses and other buildings that are not symmetric are often better suited for more informal, asymmetric landscape settings. However,

Figure 8-23 Symmetric architecture and symmetrically balanced landscape plantings fit together comfortably. (Drawing courtesy of James Lancaster)

it shouldn't be concluded that buildings are pre-destined for a certain type of design based solely upon the architecture. That would relegate landscape design to a paint-by-numbers status. However, some building styles do support the application of certain principles of design more obviously than do others.

Building doorways can make the focalization of interest at the entry much easier at times. Georgian doorways are traditionally ornate, making it nearly impossible for the viewer to mistake how to get into the building. Contemporary entries may feature bright colors, ornate light fixtures, double widths, or other devices to engage the viewer's eye.

Other architectural features can add to the difficulty of applying some of the design principles. For example, the wide, gaping door of an attached garage is strong competition for the viewer's eye. Bay windows, shutters, or rambling porches can all compete for attention with the front entry.

Plant Materials

Earlier reference has been made to the distinctiveness of certain plants in developing focal points within a design. To elaborate on what has already been written, specimen plants are the most satisfactory for use in focalization. Due to one or more extraordinary features, such as color or growth habit, specimen plants attract more attention than any other plants around them. If used in the public area, they need to be placed at or near the entry so that they support the intended focal point rather than competing with it. If used elsewhere, the duration of their eye appeal is important for the designer to consider. A specimen plant whose only contribution is one week of showy flowering in the early spring is not very significant as a focal point when compared to an evergreen with year-round contortion or highly glossy foliage.

While most specimen plants are naturally occurring species, it is possible to create unusual forms through special pruning or training.

Accent plants were also discussed briefly before. These are plants that can be used to counter-balance a specimen plant without competing against it for the viewer's attention. They create lesser, secondary focal points at the corners of planting beds, at secondary entries, at inter-sections, and other places where the design needs emphasis but not dominance. Examples of accent plants would be the upright forms of yew or juniper or holly. When a single upright plant is placed within a mass of spreading forms of the same genus, the upright gives emphasis to the planting. It becomes the accent plant.

Most landscape plants are neither specimen plants nor accent plants. They are the workhorses of the landscape, serving as **massing plants**. They fill large amounts of space both on the ground and in the air. As such they are most likely to be the plants used when the principle of balance is being applied.

Hardscape Materials

This is a broad category of components, and a more in-depth presentation is given in Chapter 9; therefore the suggestions made here are general and abbreviated. As a design element, hardscape materials can make no contribution until the designer decides how to use them. Through the merger of design imagination with construction fabrication, hardscape materials can be transformed into landscape features that attract the eye, add mass and weight to the composition, create themes, and add pleasure to the landscape. If sufficiently unusual or attractive, they aid in application of the principle of focalization. Their mass and visual weight make them a factor to use or contend with in applying the principle of balance. When formed into a shape or structure that repeats in all areas of the landscape, hardscape contributes to the principle of unity.

Hardscape materials, such as pavings, fencing, and walls, frequently create strong lines in the landscape. The viewer's eye will willingly follow the line to its end. Therefore it is important that the line lead somewhere, warranting the attention. When the lines created by hardscape materials need to be understated, as when they lead to the garage door rather than to the entry, the color of the material used should be restrained. A white concrete walk running to the front door of the house has a better chance of focusing attention on the entry if the driveway is paved with a less noticeable color. In turn, the drive should never be lined with flowers or other materials that would serve to outline and emphasize it.

Practice Exercises

A. Using the proper instruments and a drawing board, duplicate the corner planting bed shown in Figure 8–24, drawn to the scale of 1" = 10'. Design a planting for a viewer looking in the direction of the incurve. Select plants and dimensions from Chart A. Draw to scale and label all species used.

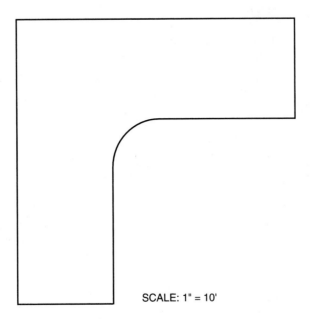

SCALE: 1" = 10'

Figure 8-24   Corner planting bed

Chart A

Species		Width	Height
Redbud tree	(D)	20 ft.	25 ft.
Viburnum	(D)	10 ft.	12 ft.
Forsythia	(D)	10 ft.	9 ft.
Cotoneaster	(D)	5 ft.	5 ft.
Spirea	(D)	5 ft.	4 ft.
Grape holly	(BLE)	3 ft.	3 ft.
Andorra juniper	(NE)	3 ft.	1½ ft.
Myrtle	(G)	Vining	1 ft.

D: Deciduous
G: Groundcover

BLE: Broad-Leaved Evergreen
NE: Needled Evergreen

B. With the proper equipment, duplicate the planting bed in Figure 8–25. Assume that the viewer is located south of the bed and that there is an attractive mountain scene north of the bed. Design a planting that frames but does not block the view. Select plants from Chart A. Draw to the scale of 1" = 10'.

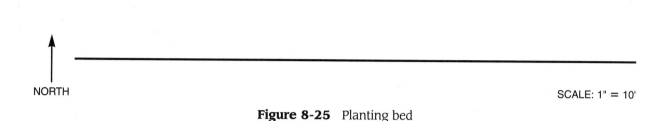

NORTH

SCALE: 1" = 10'

Figure 8-25 Planting bed

C. Figure 8-26 illustrates the entry portion of a house. The garage and driveway are also shown. With the proper instruments, design the public area portion of the foundation planting. Use the plants in Chart A. Design to the scale of 1" = 10'.

SCALE: 1" = 10'

Figure 8-26 Entry portion of house

Achievement Review

A. Indicate which principle of design is being applied in the following situations.

1. The view toward the house from the front street is framed by large masses of trees in the woodlands that border both sides of the property.

2. A dark, recessed entry door is illuminated with a bright floodlight each evening as part of the night lighting design.

3. At the point of entry into a private residential complex there is a distinctive logo affixed to a custom designed stone column. That column and logo are also featured prominently at the community pool, the clubhouse, and on all signs throughout the complex.

4. The sharp, erratic angles in the planting beds of a property being renovated are replaced with gently curvilinear beds.

5. A matching planting is placed on each side of the entry to a home.

6. A piece of garden sculpture is placed into a line planting so that it can be seen from the client's bedroom window throughout the year.

7. A house is designed to feature 90-degree and 45-degree angles prominently in its construction. The landscape designer uses those same angles in the design of the patio and pergola that adjoin the house.

8. To recreate the sense of being in a large park setting, a designer specifies that very tall tree species be selected for the canopy of a private residential family living area.

9. At regular intervals of 100 feet, a designer plans for a pedestrian mall to be lined with identical planters filled with flowers of the same color. There are a total of twenty replications of the planter from one end to the other.

10. In the above example, at the same mall, similar planters and color selections are found in the nearby parking areas, in the food court, and down the side alleys.

B. Name and explain the three types of balance.

C. Explain how the principles of design are affected by the cone of vision of the viewers.

D. Describe the parts of a corner planting.

E. What are the functions of modern foundation plantings?

F. Distinguish between the following types of plants.

1. specimen plant
2. accent plant
3. massing plant

Suggested Activities

1. Working from photographs of unlandscaped new homes or from slides of the same projected onto drawing paper, draw in different possibilities for foundation plantings. Try creating both symmetrical and asymmetrical designs for the same house. Have someone else critique the proposals using the principles of design as the basis for judgment.

2. Prepare a written critique of an existing landscape. Explain why you do or do not approve of the design. Avoid all subjective opinions and use only the principles of design to justify your statements.

3. Working from a book of house plans that show the houses in elevation and perspective views, select an array of different styles. With a pencil or marker, circle and note features of the architecture that would make it easier or more difficult to design foundation plantings that would focus attention on the entries.

Flowers

Objectives:

Upon completion of this chapter, you should be able to

- describe the uses and limitations of flowers in a landscape design
- explain the differences between annual, perennial, and biennial flowers
- list the characteristics of hardy and tender bulbs
- explain the difference between a flower bed and a flower border
- design a flower planting
- plan a naturalized bulb planting

The Uses and Limitations of Flowers in the Landscape

Flowers are one of the most desired and difficult elements of the landscape, whether the landscape is residential or commercial. While the motivation for using flowers varies, their manner of use varies little. A residential client may desire flowers to attract certain birds to the backyard, while a commercial developer may desire flowers to attract tenants to a business park. In each case, the use of flowers is driven by desire. It then falls to the designer to use the flowers in ways that complement and enhance the design rather than marring it.

Very few flowers have much impact on a viewer when used singly; however, mass several dozen, or several hundred, together and they become one of the most powerful magnets for the eye that can be placed in a landscape composition. Placed correctly, such flower plantings can reinforce a strong design, strengthen a weak one, and make unique contributions to the design. Placed incorrectly, such flower plantings can weaken or destroy an otherwise good design for all or part of the year.

In some cases, flowers are the only plants that clients know or care about. They may not know a maple from a palm tree, but they know petunias and geraniums, and how those plants are used in the landscape will be of great interest to them. A designer who knows less than clients about specific flowers and their proper use and culture will have difficulty satisfying the client's desires or guiding the client toward a proper design use of flowers.

Using Flowers Properly

In most regions of America, flowers are the most fragile of the landscape's plants in that their time of effectiveness is limited. Flowers are also the plants most likely to evoke emotional responses from people. Winter-weary residents of the northern states cheer when the first crocus breaks through the snow. Chrysanthemums evoke memories of autumns past and high school football games. Home gardeners across the country eagerly await the arrival of seed catalogues each spring to see what new varieties of flowers are being introduced. In short, flowers have many important roles to fill in the landscape:

❦ Their bright colors create strong focal points.

❦ They often possess fragrances that few other plants can match.

❦ Their colors and fragrance may attract birds and butterflies.

❦ They can soften harsh lines of the architecture and areas of the landscape such as pavement lines.

❦ Specific flowers suggest the changes of the seasons.

The Limitations of Use

Despite their many contributions, flowers can be overused or misused, so landscape architects and designers must be cautious as well.

❦ If misplaced, the focal points created by flowers can detract from the intended focal point.

❦ In addition to the birds and butterflies, flowers may attract bees, which could be troublesome for children, pets, or people with sensitivities to bees.

❦ In most regions of the country, flowers die back after a frost, and abandon their role in the landscape. If used in key areas of the landscape, the design could suffer.

❦ Flowers are expensive to purchase and time-consuming to maintain.

Different Flowers Have Different Life Spans

Annuals

An **annual** flower is one that completes its life cycle in one year. That is, it goes from seed to blossom in a single growing season and dies as winter approaches. Generally, annuals are most commonly used in summer landscapes. They bloom throughout the months of June, July, August, and September when the days are long and warm. They offer color, especially in northern regions, when many bulbous perennials are past blooming. Hundreds of annuals are commonly used throughout the country. Some examples of well-known annuals are the petunia, marigold, salvia, and zinnia.

Landscapers obtain annuals by two different methods. In one method, they are directly seeded into the ground after the danger of frost is past. Packages of annual seeds, Figure 9-1, are available at most garden centers in the spring. Seeds can also be obtained from mail-order supply houses.

Direct seeding of annuals (the placement of seeds in the ground) is the least expensive way to place flowers in the landscape. The major limitations of direct seeding are: (1) the young plants usually require thinning; (2) the plants require more time than the other method to reach blooming age; and (3) it is difficult to create definite patterns in the flower planting.

The other method used to obtain annuals is from bedding plants. These are plants that were started in a greenhouse and are already partially grown at the time they are set into the garden, Figure 9-2. Often the plants are grown in a pressed peat moss container, which can be planted directly into the ground. This creates no disturbance for the annual's root system; thus, the flower planting is well established from the beginning.

Bedding plants are more expensive than seeded annuals. However, there is no need for thinning, and definite patterns can be easily created.

Perennials

A **perennial** flower is one that does not die at the end of its first growing season. While it may

Figure 9-1 Packaged seeds are an easy and inexpensive way to start a flower garden.

Figure 9-2 Bedding plants are available from the greenhouse ready to plant in the garden. They give color faster than the direct-seeding method, but cost more.

become dormant as cold weather approaches, it lives to bloom again the following year. (When a plant is **dormant**, it is experiencing a period of rest in which it continues to live, but has little or no growth.) Most perennials live at least three or four years; many live for much longer.

Nearly all early spring flowers are perennials. Some grow from bulbs; others do not. Many special autumn flowers are also perennials. There are numerous summer perennials that act with annuals to add color to rock gardens and border plantings. Some examples of perennials are the hyacinth, iris, daffodil, tulip, poppy, phlox, gladiolus, dahlia, and mum.

Some perennials are available in seed form. However, most appear on the market as bedding plants or reproductive structures such as bulbs. Since they do not die at the end of the growing season, most perennials reproduce themselves and may eventually cover a larger area of the garden than was originally intended. This tendency to propagate may be a side benefit or a maintenance nuisance, depending upon the situation.

Bulbous Perennials. The very popular bulb comprises a large number of perennials. Bulbs survive the winter as dormant fleshy storage structures known to botanists as tubers, corms, rhizomes, tuberous roots, and true bulbs. In the landscape trade, they are usually simply called **bulbs**.

Most bulb perennials bloom only once, in the spring, summer, or fall. There are a few exceptions, however; these may bloom repeatedly. Bulbs may be classified as hardy or tender.

Hardy bulbs are perennials that are able to survive the winter outside and therefore do not require removal from the soil in the autumn. The only time they must be moved is when they are being thinned. Hardy bulbs usually bloom in the spring. Examples are the hyacinth, iris, daffodil, and tulip.

Tender bulbs are perennials that cannot survive the winter and must be taken up each fall and set out each spring after the frost is gone. These bulbs usually bloom during the summer months. Examples are the canna, gladiolus, caladium, and tuberous begonia. Refer to Figures 9-3 and 9-4 for planting instructions.

Figure 9-3 Bulbs are first spaced and then planted to assure proper color patterns and prevent crowding.

Figure 9-4 Each type of bulb has a recommended planting depth and spacing. As a general rule, bulbs should be spaced the same distance apart as their planting depth.

Biennials

Flowers known as **biennials** complete their life cycles in two years. They produce only leaves during their first year of growth and flower the second year. After they have bloomed, they die. Biennials include the English daisy, foxglove, Japanese primrose, and pansy.

Flower Beds and Flower Borders

A **flower bed** is a freestanding planting made entirely of flowers. It does not share the site with shrubs or other plants. As Figures 9-5 and 9-6 illustrate, flower beds are effective as focal points

Figure 9-5 This freestanding flower bed acts as a traffic divider while adding beauty to the area.

Figure 9-6 Walks on all sides create various points from which this flower bed may be viewed. Despite its size, only two species were used (salvia and dusty miller).

where sidewalks or streets intersect. They also work well in open lawn areas where they do not conflict with more important focal points and where the lawn is not used for activities that could be damaging to the beds.

Flower beds should never be planted in the public area of a residential landscape. When this is done, the beds attract more attention than the entry to the home, which is the most important part of any public area.

Flower beds must be designed so that they can be viewed from all sides, and must be planted accordingly. Because of this requirement, flower beds contain no woody plants to provide a backdrop for the blossoms. This has given the flower bed the reputation of being the most difficult flower planting to design. Probably for this reason, it is used much less often than the flower border.

The **flower border** is a planting that is placed in front of a larger planting of woody shrubs. The foliage of the shrubbery provides a background to set off the colors of the blossoms, Figure 9-7. Since the flower border can only be viewed from one side, it is more easily controlled by the designer.

Figure 9-8 illustrates the difference between the viewing perspectives of beds and borders.

Modern landscapers are more likely to use flowers in borders than beds for two reasons: they are easier to design, and the strong visual attraction of the flowers is more easily controlled.

Designing Flower Plantings

Flowers are high-cost, high-maintenance items, so they should be used where they will give the best return for the investment. Flowers filling a role in the landscape's design cannot be cut and taken into the house. Neither should they be wasted in side yards or stuck beneath windows where they will be looked over rather than being looked at. Regardless of whether the landscape project is large or small, commercial or residential, flowers should be used in high-profile locations. They also lend themselves to uses in novel ways, such as container plantings or hanging pots. A visit to any major theme park will display the many creative ways that flowers can be used to enrich landscapes.

Figure 9-7 The flower border has a background, as opposed to the flower bed. Here, a vine-covered wall creates a rich contrast for blossoms.

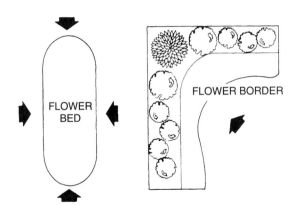

Figure 9-8 Flower beds are viewed from all sides and have no background foliage. Flower borders have a more limited viewing point and have background foliage.

Landscape designers planning the flower effects for a landscape must know the plants well. Here are some of the factors to be considered.

❦ Are the plants annuals, perennials, biennials, or a combination?

❦ Do the plants that are grouped together have similar cultural requirements? Are they compatible with the cultural requirements of the woody plants in the planting bed?

❦ Do all plants bloom at the same time? If not, what plants will be in bloom together at dif-ferent times during the growing season? Incompatible colors can coexist within the same flower planting as long as they do not bloom at the same time.

❦ How will the colors of flowers blooming simultaneously blend? Will there be subtle blends, or strong, sharp contrasts, or uncomplimentary clashes of colors or tones?

❦ Are blossom and foliage textures compatible?

❦ Dark, vibrant colors attract the eye strongly and should be used sparingly.

❦ Pale, pastel colors attract the eye less and can be used in the greatest quantities.

❦ Rigid, spike-form flowers have the strongest eye appeal and should be used in limited numbers at the center of the flower bed or at the incurve of the flower border.

❦ Flowers with round blossoms can be used in greater numbers at the outcurves of flower beds and as foreground plants in flower beds.

❦ In between the spike forms and the round forms fall the non-typical flower shapes, such as tubular blossoms, which are collectively termed the *intermediate blossom shapes*. They bridge the gap between the spike forms and the round forms of flowers. See Figure 9-9 for an example of how to arrange blossom shapes within a flower border.

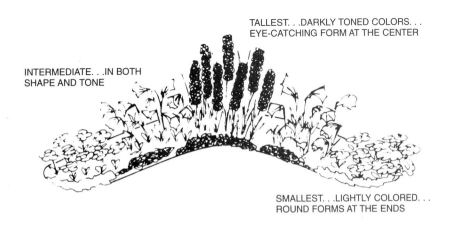

TALLEST. . .DARKLY TONED COLORS. . .
EYE-CATCHING FORM AT THE CENTER

INTERMEDIATE. . .IN BOTH
SHAPE AND TONE

SMALLEST. . .LIGHTLY COLORED. . .
ROUND FORMS AT THE ENDS

Figure 9-9 One example of how to arrange flower shapes, colors, and sizes in the flower border

If an all-season flower planting is being designed, species must be selected and arranged so that there will always be a vibrantly colored, spike-form species blooming at the center of the planting and larger quantities of shorter intermediate and round forms in complimentary colors distributed evenly throughout the bed or border.

Naturalized Plantings

In landscapes that are designed to appear more park-like or pastoral, neither the flower bed nor the flower border may be suitable. Instead, flowers may be used in a way suggestive of how nature would distribute and multiply flowers. Bulbous flowers such as daffodils are especially well suited to naturalized plantings, because once installed, they can overwinter in the ground and will reproduce freely. Thus the planting becomes more realistically natural each year.

To accomplish a naturalized bulb planting, bulbs are tossed gently onto a sunny slope at the appropriate season of the year, then planted wherever they happen to land, Figure 9-10. If enough bulbs are installed, when they bloom they will create the effect of large drifts of wildflowers sweeping across the lawn. Such plantings require no formal beds, and, once they go dormant, they disappear beneath

Figure 9-10 This iris rhizome is installed by spreading its roots evenly over a mound of soil, then backfilling to cover the roots while allowing the leaves and top of the rhizome to remain exposed.

the turf until the next spring. It is important, however, that the foliage be left uncut long enough to allow the bulb to gain strength for the next year. Naturalized plantings within lawns should not be used in landscapes where the sight of uncut grass filled with spent bulbs will upset the viewers.

Achievement Review

A. If you were a designer discussing the use of flowers with a client, what advice would you offer in response to the following client comments:

   1. "I love lots of flowers. I want them everywhere in my yard."
   2. "I think flowers spread all across the front of the house would be pretty."
   3. "I would like to use extensive flower plantings to separate the use areas of my landscape."
   4. "I hate yardwork, so let's use more flowers than trees and shrubs, since flowers don't have to be pruned and I won't have so many leaves to rake."

   5. "Flowers are so expensive and so short-lived. Wouldn't it be wiser for me to spend my money on longer-lived trees and shrubs?"

B. What are the differences between annual, biennial, and perennial flowers?

C. Indicate if the following characteristics typify

   A. hardy bulbs   C. both
   B. tender bulbs   D. neither

   1. Perennials
   2. Planted in the spring
   3. Bloom in the spring
   4. Commonly started from seed

5. Usually bloom once per year
6. Planted in the fall
7. Bloom in the summer
8. Dug and stored over winter
9. Annuals
10. Fleshy storage organs

D. Describe the technique commonly used to create naturalized plantings of bulbs within lawn areas.

E. Distinguish between flower beds and flower borders.

F. Match what is known about the eye appeal of certain flower shapes with what is known about the eye appeal of certain color tones and explain how both can contribute to the design of flower plantings.

G. If you wanted to use flowers as a means of applying the principle of simplicity to the design of the landscape, how could it be done?

Suggested Activities

1. Write to seed companies and request catalogs. The gardening section of your newspaper may have some addresses of suppliers. Posting the pictures around the room will help you become familiar with common flowers in the area.

2. With drafting tools, design a flower bed or border. It should cover an area of at least 40 square feet. Design it so that there are plants in bloom from early spring to late fall.

3. Start annual flowers from seed. The seed can be purchased at a local garden center, supermarket, or hardware store. Plant them in greenhouse flats or flowerpots and place in a greenhouse or near a sunny window. When the weather is warm enough, set them outside around the school building. As an added activity, design a planting for the building first, using the flowers being grown in class.

4. Practice blending flower colors. Cut out patches of colored paper to represent different types and quantities of flowers. Arrange them so that the brightest, most attracting colors are used where the attraction is desired. Pastel colors should be used in greater quantities and placed away from the central color. Follow the suggestions given in the text for arranging flower colors.

5. To understand better the anatomy of bulbs, select several different types for dissection. Carefully peel away each layer. Try to identify those tissues that will become the roots, stem, leaves, and flowers of each bulb. A botany text may be consulted for assistance and exact identification of all parts.

6. Study areas where wildflowers are common. Note how in the natural world flowers are grouped together in plant masses, not randomly distributed. This can help you attain natural effects when arranging landscape flowers.

CHAPTER 10

Xeriscaping

Objectives:

Upon completion of this chapter, you should be able to

- define and describe the purpose of Xeriscaping
- list the economic, environmental, and aesthetic benefits of Xeriscaping
- explain the seven principles of Xeriscaping
- apply the principles of Xeriscaping to landscape designing

Xeriscape: Water-wise Landscaping

Earth's water supply is finite. What is here now is all that will ever be available no matter how many new needs we have in the future. For generations of Americans, the distribution of our national water resources dictated where the majority of the people would live, where most of our manufacturing would be centered, and where the most pleasurable recreation would occur. In each case, the abundance of water was the key. The northeastern states grew in population and industrial prominence due largely to the generous water supplies provided by the lakes and rivers of those states. Those same lakes and rivers and the nearby ocean also gave the northeastern states a recreational advantage over other states. Eventually technology and inventiveness permitted other states to share in the bounty. Southern California learned how to tap into the snow melt of far distant mountains and divert it to the cities of the state, thereby permitting their growth and development. Manufacturers discovered that they could obtain the water they needed in the southern states by mining the fresh water of swamps and the Everglades. The landlocked midwestern states opened international ports on the Great Lakes and linked them to the ocean via the St. Lawrence Seaway. Today our national water supply is more interconnected than it has ever been. It is also experiencing a greater level of stress due to overuse and misuse than ever before, and there is no reason to believe that the stress load will ever diminish. Cities continue to sprawl and grow even though there is insufficient water to support the growth. Underground aquifers are being depleted faster than they can be recharged by natural forces. Our national hunger for new products and new synthetic materials is filling many of our water bodies with artificial waste by-products that cannot be decomposed by naturally occurring organisms.

Today, despite a generally favorable public awareness of the need for water conservation and water quality improvement, our national water supply is in jeopardy. Government at all levels is attempting to address the problem. Some of the legislation is helpful and long overdue. Some of it is short-term, knee-jerk reaction to an immediate crisis and is quickly forgotten once the crisis subsides. Landscapers are significant users of water and are frequently caught in the middle of a community's water crisis.

In the arid regions of the country the declining availability and quality of water were first apparent. Arizona, Colorado, New Mexico, Texas, and parts of California, Nevada, Oklahoma, and Utah all have regions where water quality or quantity is limited due to insufficient rainfall or other environmental factors such as drying winds or mountainous terrain. While the native plants and animals of these regions adapted long ago to their dry environments, Table 10-1, the people who have moved into these regions of the country to retire, seek recreation, or work have not adapted. Instead they have tried to bring their lifestyles with them, causing New England and Midwestern houses and gardens to spring up in areas once home to the desert. Grassy lawns grow on desert sand, and eastern shrubs bloom from planting beds filled with carefully blended soil mixtures. As the population of the Southwest has grown, the demand for water has grown disproportionately faster.

Other areas of the country are now beginning to experience water stress as well. Due to the demands of a water-greedy population and the environmental mistakes of past generations, fresh water supplies are beginning to fail in every geographic region of the nation. The resulting reactions generally take two forms. One form of action/reaction is legislative. Governments go after polluters, forcing them to stop further polluting activity and clean up their past mistakes whenever possible. The other reaction is usually to impose water use restrictions on the population, which is where landscapers get hit the hardest. Plants that can't be watered following installation

will usually die. Plants that are established will also die or be stunted if not given sufficient water to maintain them.

Xeriscaping is the term used to describe techniques of landscaping that conserve water. *Xeriscape* is a trademarked term, spelled with a capital *X*, and pronounced *zeer-escape*. It was first used in 1981 when the Denver, Colorado, Water Department and the Associated Landscape Contractors of Colorado joined together to create a program enlisting public cooperation to make landscape water use more efficient. It was an idea whose time had come. It was quickly adapted by other arid states, and in 1986 the National Xeriscape Council was created. They hold the trademark on the name *Xeriscape*. Currently cities in forty-two states have public education programs in Xeriscaping, and interest continues to grow, both nationally and internationally. Xeriscaping, if done correctly, should not be perceived as punitive or restrictive, but as an opportunity to return the designed landscape to a more natural arrangement and relationship of plants and materials. In the process, a great deal of water can be saved.

The Benefits of Xeriscaping

Economic Benefits

Simply stated, water costs money. To the community that supplies, treats, and recycles it, it is at best a break-even responsibility of local government. To those in the landscape industry, it costs to use water, and it costs to apply it. If over-applied, the water goes to waste as it goes down the drain. If under-applied and plants die, the water was of no benefit but there was still a cost. If water-deprived plants must be replaced, there is additional cost.

Comparatively, landscape designs that replace water-requiring plants and expansive lawn areas with hardscape surfacings save water costs and yield higher profit returns to the landscape contractor. Xeriscaped landscapes are also still new enough and few enough that there may be an economic marketing benefit to the homeowner at the time the property is put up for sale.

Table 10-1 A guide to selected Southwestern plants

Common Name	Botanical Name	Growth Habit	1' or less	2'–5'	6'–9'
Ash,					
Arizona	Fraxinus velutina	T			
Modesto	Fraxinus velutina var. glabra	T			
Citrus trees	Citrus sp.	T			
Coral tree	Erythrina sp.	T			
Crabapple, Flowering	Malus sp.	T			
Cypress, Arizona	Cupressus arizonica	T			
Elderberry, Desert	Sambucus arizonica	T			
Elephant tree	Bursera microphylla	T			
Elm					
Chinese	Ulmus parvifolia	T			
Siberian	Ulmus pumila	T			
Eucalyptus	Eucalyptus sp.	T			
Hackberry, Netleaf	Celtis reticulata	T			
Honeylocust					
Shademaster	Gleditsia triacanthos var. inermis 'Shademaster'	T			
Sunburst	Gleditsia triacanthos var. inermis 'Sunburst'	T			
Thornless	Gleditsia triacanthos var. inermis	T			
Ironwood, Desert	Olneya tesota	T			•
Jujube, Chinese	Ziziphus jujuba	T			
Locust					
Black	Robinia pseudoacacia	T			
Idaho	Robinia pseudoacacia 'Idahoensis'	T			
Pink flowering	Robinia pseudoacacia 'Decaisneana'	T			
Magnolia, Southern	Magnolia grandiflora	T			
Mesquite					
Honey	Prosopis glandulosa var. glandulosa	T			
Screwbean	Prosopis pubescens	T			
Mulberry, White	Morus alba	T			
Olive, European	Olea europaea	T			
Pagoda tree, Japanese	Sophora japonica	T			
Paloverde					
Blue	Cercidium floridum	T			
Little leaf	Cercidium microphyllum	T			
Mexican	Parkinsonia aculeata	T			
Pine					
Aleppo	Pinus halepensis	T			
Digger	Pinus sabiniana	T			
Italian stone	Pinus pinea	T			
Japanese black	Pinus thunbergii	T			
Pinyon	Pinus cembroides	T			

KEY: T = Trees   S = Shrubs   V = Vines   G = Groundcovers   NS = Flowers are not showy.

Mature Height				Season of Bloom					Special Use in the Landscape
10'–15'	15'–30'	30'–50'	Over 50'	Early Spring	Late Spring	Summer	Fall	Winter	
		•		NS					shade tree
		•		NS					shade tree
•				varies with the variety					excellent for containers
	•			•					brilliant flowers
	•				•				specimen plant
		•		NS					screens and windbreaks
	•			NS					screens and windbreaks
	•			NS					
		•		NS					shade tree
		•		NS					windbreak
		•		varies with the variety					many species prized for flower and/or foliage
		•		NS					shade tree
	•			NS					good in dry, desert conditions
	•			NS					
		•		NS					
	•				•				specimen tree very salt tolerant
			•		•				frequent pruning makes these attractive flowering trees
		•			•				
		•			•				
			•			•			lawn tree
	•			NS					shade trees and windbreaks
	•			NS					
		•		NS					shade tree
	•			NS					good multi-stemmed tree
	•					•			lawn tree
	•				•				specimen trees
	•				•				
	•				•				
		•		NS					grows well in poor soil
		•		NS					specimen plant
			•	NS					good in desert conditions
				NS					good in planters; prune well
•				NS					multi-stemmed effects

Table 10-1 A guide to selected Southwestern plants *(continued)*

Plant		Growth Habit	Mature Height		
Common Name	**Botanical Name**		**1' or less**	**2'–5'**	**6'–9'**
Pistache, Chinese	Pistacia chinensis	T			

Poplar
  Balm-of-Gilead
  Bolleana
  Cottonwood
  Lombardy
  White
	
	Populus balsamifera
	Populus alba var. 'Pyramidalis'
	Populus fremontii
	Populus nigra 'Italica'
	Populus alba
		
		T
		T
		T
		T
		T			

Silk tree	Albizia julibrissin	T			
Smoke tree	Dalea spinosa	T			

Sycamore
  American
  Arizona
  California
	
	Platanus occidentalis
	Platanus racemosa 'Wrightii'
	Platanus racemosa
		
		T
		T
		T			

Tamarisk
  Athel tree
  Salt cedar
	
	Tamarix aphylla
	Tamarix parviflora
		
		T
		T			

Umbrella tree, Texas	Melia azedorach 'Umbraculiformis'	T			

Willow
  Babylon
  Globe Navajo
  Wisconsin
	
	Salix babylonica
	
	Salix x blanda
		
		T
		T
		T			

Zelkova, Sawleaf	Zelkova serrata	T			
Abelia, Glossy	Abelia x grandiflora	S			•
Apache plume	Fallugia paradoxa	S		•	
Arborvitae, Oriental	Platycladus orientalis	S			

Barberry
  Darwin
  Japanese
	
	Berberis darwinii
	Berberis thunbergii
		
		S
		S
				
				•
					
					•

Bird of Paradise	Caesalpinia gilliesii	S			•
Brittlebush	Encelia farinosa	S		•	
Butterfly bush	Buddleia davidii	S			•
Cherry laurel, Carolina	Prunus caroliniana	S			
Cotoneaster, Silverleaf	Cotoneaster pannosus	S			•
Crape myrtle	Lagerstroemia indica	S			
Creosote bush	Larrea tridentata	S			•
Firethorn, Laland	Pyracantha coccinea 'Lalandei'	S			•

Hibiscus
  Chinese
  Rose of Sharon
	
	Hibiscus rosa-sinensis
	Hibiscus syriacus
		
		S
		S			

Holly
  Burford
  Wilson
  Yaupon
	
	Ilex cornuta 'Burfordii'
	Ilex wilsonii
	Ilex vomitoria
		
		S
		S
		S
					
					•
					•

KEY: T = Trees   S = Shrubs   V = Vines   G = Groundcovers   NS = Flowers are not showy.

Mature Height				Season of Bloom					Special Use in the Landscape
10'–15'	15'–30'	30'–50'	Over 50'	Early Spring	Late Spring	Summer	Fall	Winter	
			•	NS					good patio tree; good fall color
		•		NS					narrow columnar form windbreaks
		•		NS					
			•	NS					
			•	NS					
		•		NS					
		•				•			showy shade tree
•					•				
			•	NS					excellent street trees
			•	NS					
			•	NS					
		•				•			wind, drought, and salt resistant
	•					•			
		•			•				shade tree
		•		NS					
			•	NS					
		•		NS					
		•		NS					windbreak
						•			
				•					
•				NS					
					•				barrier plantings
				NS					
						•			
					•				
						•			vigorous growth
	•				•				screens and hedges
					•				wind screen
	•					•			very colorful
						•			screens and hedges
				•					espaliers well
		•				•			
•						•			
				NS					Wilson and Yaupon clip and shade well
				NS					
	•			NS					

Table 10-1 A guide to selected Southwestern plants *(continued)*

Plant		Growth Habit	Mature Height		
Common Name	Botanical Name		1' or less	2'–5'	6'–9'
Hopbush	Dodonaea cuneata	S			
Jojoba	Simmondsia chinensis	S		•	
Juniper					
Armstrong	Juniperus chinensis 'Armstrongii'	S		•	
Hollywood	Juniperus californica	S			
Pfitzer	Juniperus chinensis 'Pfitzeriana'	S			•
Lysiloma	Lysiloma sp.	S			
Myrtle	Myrtus communis	S		•	
Ocotillo	Fouquieria splendens	S			
Oleander	Neriam oleander	S			
Photina	Photina glabra	S			
Privet					
California	Ligustrum ovalifolium	S			
Glossy	Ligustrum lucidum	S			
Japanese	Ligustrum japonicum	S			
Texas	Ligustrum japonicum texanum	S			•
Rose, Floribunda	Rosa x floribunda	S		•	
Silverberry	Elaeagnus commutatus	S			
Sugar bush	Rhus ovata	S			
Bougainvillea	Bougainvillea glabra	V			•
Ivy					
Algerian	Hedera canariensis	G	•		
Boston	Parthenocissus tricuspidata	V			
Jasmine, Star	Jasminum multiflorum	V			
Lavender Cotton	Santolina chamaecyparissus	G	•		
Periwinkle	Vinca minor	G	•		
Trumpet creeper	Campsis radicans	V			
Virginia creeper	Parthenocissus quinquefolia	V			
Wisteria	Wisteria sinensis	V			

KEY: T = Trees   S = Shrubs   V = Vines   G = Groundcovers   NS = Flowers are not showy.

Environmental Benefits

Anything that reduces the demand on a region's water supply must be judged as beneficial. That is a prime objective of Xeriscaping. So many plant choices are inappropriate for their sites, surviving only because of supplemental irrigation, fertilization, and horticultural technology. By eliminating all or most of the plants that require excessive amounts of water and chemicals to sustain them, the local water supply is conserved and the remaining wastewater is less polluted and easier to clean and recycle. Also, with world hunger at record levels and insufficient quantities of fertilizer available for agricultural use in many parts of the world, it is offensive to many people that fertilizers are used to keep backyard lawns green and blemish-free. Replacing the lawn-dominant landscape with Xeriscaping can make a contribution to the survival of people throughout the world.

Mature Height				Season of Bloom					Special Use in the Landscape
10'–15'	15'–30'	30'–50'	Over 50'	Early Spring	Late Spring	Summer	Fall	Winter	
•				NS					screens
				NS					hedges
•				NS NS NS					
•					•				good for transition between garden and natural landscape
					•				prunes and shapes well
•					•				specimen plant
•						•	•		does well in heat and poor soil
	•			•					screens
•				•					all species can be pruned to lower heights
	•			•					
					•				
					•				
					•				massing effects
•				NS					good for containers
•				•					
						•			very colorful
				NS NS					
	•								
	•					•			very fragrant
						•			effective as edging
					•				
	•					•	•		
	•			NS					
•					•				may be trained as shrubs and weeping trees

Aesthetic Benefits

For many years Americans applied more horticulture than ecology to their landscapes. They so willingly accepted the trendy suggestions of the national garden magazines that much of the geographic distinctiveness of the country was lost. Xeriscaping, with its emphasis on the use of native plants and others suited to the local climatic conditions, offers the nation an opportunity to regain some of its regional distinctiveness. By combining plants with common water requirements, matching them to the climate of the region, and placing them in settings whose construction materials and design themes reflect the heritage of the region, landscape designers will begin the resurrection of our diverse national landscape. It will be possible to have a dialogue with our past while making a contribution to the nation's future.

The Seven Principles of Xeriscaping

The National Xeriscape Council has prepared a list of seven principles that serve as a guide to the development and maintenance of landscapes that use water wisely, not wastefully.

The First Principle:
Proper Planning and Design

Assuming that everything described in Section One regarding the way to analyze a site, measure a client's needs, and develop outdoor rooms has been done, this important principle of Xeriscaping is applied to the selection of use of plant materials. Plants need to be **hydrozoned**, grouped on the basis of their water needs. Three water-use zones are defined:

Low water use	Requiring little or no supplemental water after transplants become established
Moderate water use	Requiring some supplemental water during hot, dry periods
High water use	Limited areas where plants are given as much water as needed at all times

Of the three zones, high water-use zones are usually the smallest zones and are placed in the landscape in highly visible locations such as at the entry to a building or at the public entrances to commercial properties.

The Second Principle:
Proper Soil Analysis

Considering that the regions where Xeriscaping is now practiced range from the desert of Arizona to the subtropics of Florida and the mountain valleys of the Rockies, it is understandable that there is no such thing as a common soil environment. A soil analysis will give the landscaper information about the existing structure, nutrient content, water retention capacity, and drainage characteristics of the soil of a specific site. In turn, that will guide the designer in selecting plants for the site. With appropriate additives, it may be possible to improve water penetration, retention, and soil drainage. The ultimate objective is to provide the plants with a soil environment that will allow them to send their roots deep into the earth.

The Third Principle:
Appropriate Plant Selection

Plants should be chosen for their suitability and adaptability to the site. All of the factors described in the chapter on Plant Selection apply to Xeriscaping. Additionally, plants should be selected for their appropriateness for the water-use zone in which they will be located. The more low water-use zones there are within a landscape, the fewer will be the plants requiring large amounts of water to sustain them. Also, if the designer specifies small plants with wider spacing at the time of installation, less water will be required to establish the transplants, since smaller plants need less water than larger plants.

The Fourth Principle:
Practical Turf Areas

Turf areas can be the greatest users of water in the landscape. An entire industry has grown up around the irrigation requirements of turf. Too often turf has been the glue that held the other components of the landscape together. It has been used for no reason other than to fill the voids of the design. In Xeriscaping, that is no longer an affordable luxury. Turf should be used where it is the functional plant of choice and where an alternative non-living surface is inappropriate. Grasses thrive or go dormant in reaction to changes in temperature and available water. If kept separate from the other plants in the landscape, turfgrass can survive in varying hydrozones, appearing lush and green when natural water is plentiful and temperatures are optimal, and at other times becoming brown and dormant in response to the conditions of the zone.

The Fifth Principle:
Efficient Irrigation

It is a standard horticultural practice to irrigate plants in a way that encourages roots to grow deep into the soil since shallow rooting increases the plant's vulnerability to drought. In Xeriscaping it is important to select irrigation systems that match the site and the needs of the plants. Sprinkler systems waste water because they deliver too much to the foliage and too little to the root zone. Drip systems and other micro-irrigation systems conserve water by delivering low volumes of water at low pressures to precise areas. They should be the systems specified by landscape architects and designers as they practice water-efficient planning. Efficiency can also be applied to irrigation scheduling. Water applied in the early morning will benefit plants more than water applied during the heat of the day, since less will be lost through evaporation. Only enough water should be applied to replace that lost since the last application. The use of soil moisture sensors can help the landscape manager determine the amount of irrigation water needed.

The Sixth Principle:
Mulching

Mulches offer many benefits, including water retention. In Xeriscapes, the most appropriate mulches are those that are organic, fine-textured, and that do not develop a water-repelling crustiness over time. Selecting an organic mulch that is native to the local region will contribute to the natural appearance of the Xeriscape. The mulch should be applied to a thickness of 4 to 6 inches. Over time it will decompose and should be removed before adding fresh mulch.

The Seventh Principle:
Appropriate Maintenance

Any and all maintenance practices that reduce the water needs of the landscape will contribute to water conservation. Sweeping walks rather than hosing them clean will save water. Spraying plants with anti-transpirant chemicals at the time of transplanting and at other times of high water stress can reduce their water needs. Mowing turf areas often enough to prevent excessive growth, which will pull more water from the soil, will conserve water. Keeping plants disease- and insect-free will assure healthy plants, which will not need replacement. Also, if water used for other purposes, such as cooling or washing, can be used in the landscape rather than merely being poured down a drain, it can save on the consumption and expense of freshly treated municipal water.

Applying Xeriscape Principles to Designing

Landscape architects and designers should regard application of Xeriscape principles as a modern updating of their profession. It brings environmental ethics into the planning process, and this benefits everyone involved. It should not be regarded as a constraint on design creativity. Instead, Xeriscape principles offer new opportunities to freshen and recreate much of the American landscape.

Designers must know the plants of the region where they are working. In addition to knowledge of the plants' physical characteristics, they must know their water requirements if they are to fill each hydrozone correctly. There are comparatively few reference texts that consistently include reliable water-use information about the plants listed. One of the texts most often referred to is *Xeriscape Gardening*, by Connie Lockhart Ellefson, Thomas L. Stephens, and Doug Welsh, Macmillan Publishers, New York, 1992. Another helpful listing of plants and their water needs is found in a publication developed by the Georgia Water-Wise Council in Marietta, Georgia, entitled *Xeriscape: A Guide to Developing a Water-Wise Landscape*. The publication may be available through the Xeriscape council in individual states. Designers working in mountain regions may wish to consult *The Xeriscape Flower Gardener*, by Jim Knopf, Johnson Books Publisher, Boulder, Colorado, 1991. As new texts are written and current texts are revised, water requirements and tolerances should be included into the plant lists. Table 10-2 provides a rudimentary list of plants that can be used in designing a landscape that seeks to apply the principles of Xeriscaping.

Table 10-2 Plants Suitable for Xeriscaping

Trees

Common Name	Botanical Name	Hardiness Zone	Hydrozone*
Acacia, Sweet	Acacia farnesiana	9	M
Ash, Arizona	Fraxinus velutina	5	L/M
Calabash, Mexican	Crescentia alata	10	M
Cornelian cherry	Cornus mas	5	M/H
Crabwood	Alteraminus lucida	9	M
Elm, Chinese	Ulmus pumila	5	L/M
Locust, Black	Robinia pseudoacacia	3	M
Maples	Acer species	Varies	M
Oak, Live	Quercus virginiana	7	M
Olive, Black	Bucida buceras	8	M
Olive, Russian	Elaeagnus angustifolia	2	M
Poplar	Populus species	Varies	L/M
Redbud, Eastern	Cercis canadensis	4	M
Serviceberry	Amelanchier species	4	M
Silver Elaeagnus	Elaeagnus multiflora	5	L/M
Wax Myrtle	Myrica cerifera	6	L/M/H

Shrubs

Common Name	Botanical Name	Hardiness Zone	Hydrozone*
Almond, Desert	Prunus fasciculata	6	L/M
Ash, Singleleaf	Fraxinus anomala	5	L/M
Barberry, Japanese	Berberis thunbergii	4	M
Bluebeard	Caryopteris clandonensis	5	L/M
Bougainvillea	Bougainvillea species	8	L/M
Butterfly bush	Buddleia alternifolia	5	L
Cowania	Cowania mexicana	9	L
Euonymus, Dwarf winged	Euonymus alatus compacta	3	M
Firethorn, Leland Lelandei	Pyracantha coccinea	6	M
Grape Holly	Mahonia species	5/6	L/M/H
Juniper	Juniperus species	Varies	L/M
Leadplant	Amorpha canescens	2	L/M
Lilac	Syringa species	Varies	M
Pine, Pinon	Pinus edulis	6	L/M
Privet, California	Ligustrum ovalifolium	5	M
Quince, Flowering	Chaenomeles lagenaria	4	M
Rose, Floribunda	Rosa floribunda	5	L/M
Rose of Sharon	Hibiscus syriacvus	5	M
Sage, Big Western	Artemisia tridentata	5	L/M
Saw palmetto	Serenoa repens	8	M
Scotch broom	Cytisus scoparius	5	M
Spirea, Rock	Holodiscus dumosus	5	M
Sumac	Rhus species	Varies	L/M
Yucca	Yucca species	4	L/M

Turf Grasses

Common Name	Botanical Name	Warm or Cool Season	Hydrozone*
Buffalograss	Buchloe dactyloides	Warm	L
Blue grama	Boutelova gracilis	Warm	L
Crested wheatgrass	Agropyron cristatum	Cool	M
Tall fescue	Fescue species	Cool	M

*L = Low water requirements   M = Medium water requirements   H = High water requirements

In summary, here are some of the ways that landscape designers can take the lead to reduce the water consumption of landscapes:

❦ Do not create lawn areas where other options, such as hard surfacings or wildflowers, will serve the clients as well and not consume so much water. When turf is used, specify varieties that require less water, such as tall fescue and Bermudagrass.

❦ Regard planting beds as hydrozones, placing plants that share common water requirements into common hydrozones.

❦ Use as many low water-use hydrozones as possible, reserving the high water-use zones for the most important, high-profile regions of the design.

❦ Select plants that can adapt to the existing soil conditions, rather than trying to condition vast areas of the site to accept inappropriate plant species.

❦ Place trees to cast shade where it will be of greatest benefit to understory plants in reducing their water consumption.

❦ Specify that water-retention rings be placed around each tree at the time of transplanting.

❦ Specify the use of organic mulches rather than inorganic mulches. Inorganic mulches, such as marble or stone chips, can retain solar heat and bake the soil beneath.

❦ Design irrigation systems that do not waste water and that deliver the water only where and when it is needed.

❦ Use containerized plants to provide spot color and interest in areas where paving replaces turf and ground beds. While potted flowers and other plants require more frequent watering than in-ground plantings, the total water consumption should be considerably less, Figure 10-1.

Figure 10-1 When containerized plantings replace inground installations, the water savings can be substantial.

Achievement Review

A. Define *Xeriscaping*.

B. Name the seven principles of Xeriscaping.

C. Explain hydrozoning and its application in designing.

D. Describe the economic benefits of Xeriscaping to the landscaper, the local community, and the client.

E. Why does it matter what irrigation system is used as long as plants receive the water they need?

F. Why are lawn-dominant landscapes inappropriate as Xeriscapes?

G. Why is a proper soil analysis important in Xeriscape designing?

H. Why are organic mulches preferable to inorganic mulches for water retention?

Suggested Activities

1. Place into three pots with the same soil mixture, three different plants: one with low water needs, one with medium water needs, and one with high water needs. All plants should be of comparable size. Apply the same amounts of water to each. Gradually reduce the amount of water and extend the time intervals between waterings until you are able to see visible signs of the plants' differing water requirements.

2. Do a similar comparative study using local turf grasses and wildflowers. As above, gradually reduce the amount of water and increase the intervals between waterings. Compare how each appears as water is reduced.

3. Fill greenhouse flats with typical local soil and place thermometers into each at the same depth. Leave one uncovered, and cover the other flats of soil with different organic and inorganic mulches to a depth of 4 inches. Place all flats outdoors or in a greenhouse where they can receive direct summer sunshine. Apply exact amounts of water to each flat. Compare soil temperatures at different times of the day and night.

CHAPTER

11

Enclosure Materials

Objectives:

Upon completion of this chapter, you should be able to

- list five functions of landscape enclosure
- describe five types of materials used for constructed landscape enclosures
- explain the need for and means of releasing water pressure from behind solid enclosure materials

Functions of Landscape Enclosure

The walls of the outdoor room serve to enclose (surround or separate) a section of property that is to act as an outdoor living area. **Enclosure materials** form these walls.

Landscape enclosures have several functions:

- To define the shape and limits of the landscape
- To control circulation patterns within the landscape
- To provide various degrees of protection and/or privacy
- To serve other engineering needs, such as retaining steep slopes and raising planting beds

- To modify the climate by creating sheltered areas where plants may grow, or by diverting or reducing the wind's force to make the area more comfortable for occupants

As discussed previously, enclosure is sometimes accomplished with plant materials. Many other types of enclosures are man-made; these may be used alone or in combination with plant materials.

The type of constructed enclosure selected by the landscaper is usually determined by the function it is to perform. If its only purpose is aesthetic (to beautify the landscape and make it appealing to the senses), the most important consideration is how well the enclosure material coordinates with the total design. To achieve total unity of design, repeat colors and materials used in the building(s) and in the surfacing and furnishings of the garden in the enclosure materials.

Choosing Man-Made Enclosures by Function

Directing Circulation Patterns

If the purpose is to direct circulation patterns, a fence or wall may be used. Height is usually more important than density when traffic control is the concern. Therefore, open-style fencing can be used if the viewer is to be able to see through the enclosure structure, Figure 11-1. The higher the enclosure is, the more effective it is in directing movement. A combination of an open-style enclosure and plants can create an outdoor wall that is both functional and attractive, Figure 11-2.

Protection

If protection is the objective, a solidly constructed fence or wall is a necessity. Open-style fences, such as chain link, allow the viewer to see through them while discouraging passage. Walls offer the greatest protection since they combine both density and strength. The height of the fence or wall chosen is determined by the type of protection it is to offer. To prevent children or small dogs from wandering into the street may require an enclosure only 3 feet high. To prevent intruders or large animals from entering, a wall 6 feet high or higher is required.

Privacy

For privacy control, the height and density of the enclosure material are again important. Where properties are small and family living areas are closely spaced, as in many older city neighborhoods, enclosure for privacy is a necessity. The enclosure must be solid where total privacy is needed, but a cramped feeling can result when a solid enclosure is used in a small space. Using a combination of plants and constructed enclosures helps to avoid this feeling.

The height of the enclosure necessary for privacy depends upon the location of the persons desiring the privacy and the vantage point of an unwelcome viewer. If a homeowner wishes to be screened from a next-door neighbor when both are in their family living areas at ground level, the enclosure should be about 6 feet high, Figure 11-3.

Figure 11-1 Post and rail fencing can direct traffic patterns without blocking the view.

Figure 11-2 The combination of constructed and natural enclosure materials is often more attractive and effective than either type would be if used alone.

Figure 11-3 Privacy between two properties on the same level requires an enclosure of at least 6 feet in height.

If the neighbor looks out a second-floor window onto the patio, the vantage point has been elevated, giving the neighbor an unobstructed view. This requires that the height of the enclosure increase also. Constructed enclosures cannot be much higher than 8 feet without becoming too imposing or threatening in appearance. In addition, there may be local laws limiting fence height. Therefore, trees are frequently used to block views in certain cases, Figure 11-4.

When determining the enclosure height needed for privacy, remember:

❦ When both the viewer and the view are parallel to the ground, privacy can be obtained by a 6-foot enclosure at any point along the viewing line. It does not matter if the land is flat or sloped as long as the line of sight parallels it, Figure 11-5 (A).

❦ When the viewer and the view are not parallel to the ground, the enclosure height necessary for privacy increases as the enclosure approaches the viewer, Figure 11-5 (B). The height decreases as it approaches the person desiring the privacy.

Engineering Uses

There are many engineering uses of constructed enclosures. The raised planting bed, for example, is one method of providing grade variation without bulldozing ground. The raised bed also permits plant growth where the existing soil or drainage conditions are unsatisfactory, Figure 11-6. An associated benefit of raised planting beds is that they can also help control traffic or function as seatwalls (walls with flat, smooth tops suitable for use as seats).

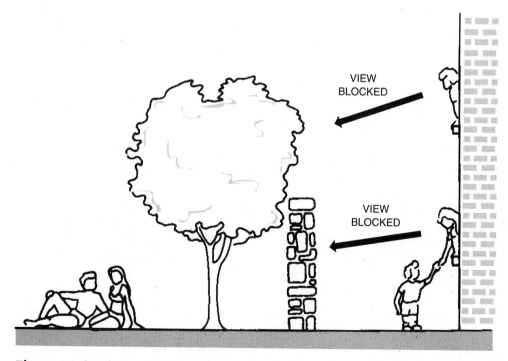

Figure 11-4 The combination of a constructed enclosure and trees gives privacy when the viewer is above ground level.

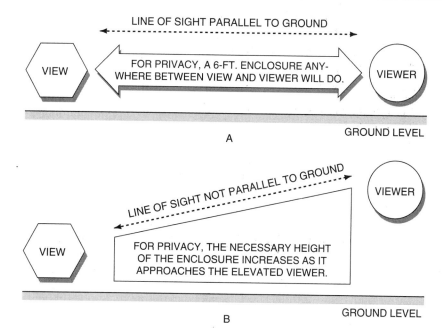

Figure 11-5 The difference in the height of the enclosure necessary for privacy when the line of sight is and is not parallel to the ground

Figure 11-6 The raised planting bed adds an interesting effect to this flat rooftop garden. It also offers protection by keeping people away from the edge of the roof.

Constructed enclosures in the form of retaining walls can transform slopes that are too steep for human use into smaller, usable areas. They may also be used to protect steep banks from erosion. For example, the bank can be turned into earthen steps, which greatly reduce the downward rush of water, Figure 11-7.

When constructed enclosures are used in engineering roles, it is important for the landscaper to choose very strong materials. In some cases, tons of soil are held back by a retaining wall. The landscaper must also carefully specify the materials to be used, such as stone, poured concrete, or brick.

Releasing Pressure Behind Retaining Walls. Each construction material has its own special requirements for installation. Some require a footing or internal steel bracing; others need wooden forms during pouring.

One concern with which all landscapers must deal when retaining walls are part of the design is the release of pressure caused by water that can build up behind the walls. Construction of retaining walls often alters the normal drainage pattern of the land. Water that once flowed easily down a bank may have no way of getting past a wall that blocks the direction of flow. If not released, the water may eventually channel under the wall and weaken or burst the wall.

Weep holes, Figure 11-8, are the means by which water is allowed to escape from behind retaining walls. Built at the bottoms of walls, they provide the necessary drain for water as it seeps down through the soil.

Climate Modification

Where climate modification is the objective of the constructed enclosure, either fences or walls may be used. It may also be desirable to combine a constructed element with plants for even better control or a more attractive appearance. A solid wall or

Figure 11-7 What was once a steep slope has been converted into a series of broad, level planting strips connected by steps. The brick retaining walls provide both strength and beauty.

Figure 11-8 Weep holes

fence built across the path of a prevailing breeze may be able to eliminate the breeze as a factor of the landscape near the wall. If a gentler breeze is desired, rather than total stillness, an open-style wall or fence is used. This slows down the wind's velocity without totally eliminating air movement.

In areas of borderline hardiness, certain species will survive the winter if they are given protected planting sites. A constructed enclosure provides a sheltered area for marginally hardy plants in Figure 11-9. Some of the most commonly used enclosure materials are compared in the following chart.

Figure 11-9 A constructed enclosure creates a sheltered area for the growth of marginally hardy plants.

A Comparison of Enclosure Styles and Materials

Style	Material	Security at 6-foot Height	Privacy at 6-foot Height	Noise Reduction at 6-foot Height	Wind Deflection at 6-foot Height	Grading Structure	Useful for Raised Beds	Comments
BASKETWEAVE FENCE	Wood	Yes	Yes	Moderate	Moderate	No	No	Available in prefabricated sections, attractive on both sides
BRICK WALL	Brick	Yes	Yes	Good	Good	Yes	Yes	An ideal material for free-standing retaining or seat walls; width will vary with height and function
CHAINLINK FENCE	Steel	Yes	No	None	None	No	No	Good for use around pet areas or to safeguard children's play areas
CONCRETE BLOCK WALL	Concrete	Yes	Yes	Good	Good	Yes	Yes	Less expensive than poured concrete, stone, or brick
GRAPE STAKE FENCING	Wood	No	Yes	Moderate	Moderate	No	No	A rustic style that weathers to an attractive gray; also comfortable in urban settings
LATTICE FENCE	Wood	No	Variable	No	Limited	No	No	Effectiveness as a screen depends upon how closely the lattice is spaced
LOUVERED FENCE	Wood	Yes	Yes	Moderate	Moderate	No	No	Louvers may be vertical or horizontal and are angled to provide privacy
PICKET FENCE	Wood or Iron	Depends on the height	No	No	No	No	No	High maintenance costs because of frequent painting needed
POST-AND-RAIL FENCE	Wood	No	No	No	No	No	No	Degree of formality varies with finish of lumber; a style valued for aesthetics more than for security or privacy
POURED CONCRETE WALL	Concrete	Yes	Yes	Good	Good	Yes	Yes	Can be smooth or textured, colored or inset with materials to add interest to the surface; requires reinforcing if it is to provide strength

A Comparison of Enclosure Styles and Materials *(continued)*

Style	Material	Security at 6-foot Height	Privacy at 6-foot Height	Noise Reduction at 6-foot Height	Wind Deflection at 6-foot Height	Grading Structure	Useful for Raised Beds	Comments
RAILROAD TIES	Wood	No	No	No	No	Yes	Yes	Ideal for rustic, natural enclosures; widely used for soil retention; care should be taken to assure that the ties have not been preserved with a phytotoxin
SLAT FENCE	Wood	Yes	Variable	Moderate	Moderate	No	No	Effectiveness as screen or wind deflector depends upon how closely slats are spaced
SOLID BOARD FENCE	Wood	Yes	Yes	Moderate	Good	No	No	Expensive but the best for security and privacy combined; maintenance easier than with other styles
SPLIT RAIL FENCE	Wood	No	No	No	No	No	No	A rustic style best used in rural settings; lumber is rough and unfinished
STOCKADE FENCE	Wood	Yes	Yes	Moderate	Good	No	No	A variation of solid board fencing
STONE WALL, ASHLAR	Stone	Yes	Yes	Good	Good	Yes	Yes	The stone is cut, usually at the quarry. Stones vary in their smoothness and finish. They are laid in a horizontal and continuous course with even joints.
STONE WALL, RUBBLE	Stone	Yes	Yes	Good	Good	Yes	Yes	The stone is not cut. No course is maintained. Small stones are avoided. Larger stones are used at the base of the wall.
WOOD RETAINING WALL	Wood	No	No	No	No	Yes	Yes	Wood must be preserved to avoid rapid decay; the preservation should not be phytotoxic; reinforcement necessary to assure strength
WROUGHT IRON GRILLS	Iron	Yes	No	No	No	No	No	Expensive and used mostly for aesthetics; grills may be continuous or used as baffles separately

Achievement Review

A. List five functions of constructed enclosures.

B. Select the best answer from the choices offered to complete each statement.

1. Enclosure can be accomplished with constructed materials or with _____ materials.

   a. design
   b. man-made
   c. plant
   d. solid

2. The _____ an enclosure element is, the more effective it is in traffic control.

   a. wider
   b. higher
   c. more colorful
   d. more solid

3. The type of constructed enclosure selected by a landscaper is usually determined by _____ .

   a. the preferences of the client
   b. the function it is to perform
   c. the money available for the job
   d. the past experience of the designer

4. Total privacy for two landscapes that are side by side and at the same level requires an enclosure of _____ in height.

   a. 4 feet
   b. 5 feet
   c. 6 feet
   d. 7 feet

5. When the vantage point for a view is highly elevated, the use of _____ helps to block the view.

   a. awnings
   b. trees
   c. a chain link fence
   d. shrubs

6. Weep holes are necessary for _____ .

   a. wall footings
   b. structural strength
   c. water pressure release
   d. appearance

Suggested Activities

1. Collect photos illustrating fencing styles to become familiar with what is available for use in landscapes. As sources, use mail order catalogs, garden center and lumber company advertisements, and house and garden magazines. Mount each style separately with its name and several examples of how it can be used in the landscape.

2. Make a similar collection of illustrations of houses. Find examples of as many different styles and construction materials as possible. Match the houses with the most appropriate fencing styles. Avoid mixing styles. Example: Split-rail fences look fine with ranch style homes, but are inappropriate with Spanish styling; picket fences go well with Cape Cod styles but not ranch styles.

3. Invite a lumber dealer to talk with the class about wood. Which woods last longest? How can less expensive woods that rot quickly be made to last longer? What prefabricated fencing styles are most popular in your town?

4. Demonstrate the different elevations of enclosure needed to provide total privacy in an area. Have two class members sit opposite each other (perhaps one can sit on a staircase or ladder). Two other class members can then separate them by holding a large blanket or other solid material between them. Measure the height needed to block the students' view of each other. Gradually elevate one student's vantage point, each time measuring the height of enclosure needed for blockage of the view. Alter the placement of the enclosure. Place it closer to the viewer first, and then closer to the person being viewed.

5. The ability of certain enclosure materials to deflect or weaken the force of the wind, as well as the patterns of protected and unprotected areas that result, can be demonstrated using a small fan and scaled replicas of enclosure materials. Build some solid and semisolid enclosures from balsa wood or cardboard. Create some shrubs with twigs in clay or Styrofoam. To determine the degree of protection offered by each enclosure, moisten a flat surface and observe patterns of drying after the fan (wind) is directed at the enclosure being compared. Note that each enclosure will have to be securely fastened to the surface before testing.

Surfacing Materials

Objectives:

Upon completion of this chapter, you should be able to

❧ select a suitable natural or constructed surface material for a specific area

❧ list the advantages and disadvantages of both hard and soft paving

❧ determine step dimensions for connecting two different levels of land

Types of Surfacing Materials

The floor of the outdoor room is formed by the application of surfacing materials. The surfacing is applied only after the soil has been prepared by grading and conditioning. Grading provides for proper water drainage away from buildings and for the movement of water across the soil without harmful erosion. Conditioning of the soil promotes water absorption necessary for the growth of natural surfacings such as grass.

The main reason surfacing is applied is to cover the exposed and disturbed soil surface with a protective material. Without this protection, erosion or compaction of the soil may occur. **Compaction** results when soil particles are pressed tightly together by heavy traffic over the area. Lack of cover can also result in puddling of the soil, which at a severe level can turn a landscape into a mud bath.

There are four basic types of surfacing:

❧ Paving, for heavy traffic areas

❧ Turf grass, for areas less subject to use

❧ Ground covers ⎫ for areas with
❧ Flowers ⎭ no pedestrians

The chart on the following page compares different types of surfacing. Numbers in parentheses in the chart are explained below it.

Paving

Whether or not paving is chosen as a surfacing material depends on how the area in question is to be used. Generally, any area with concentrated foot or vehicular traffic should be paved. Examples are driveways, patios, entries, and areas under outdoor furniture. In these locations, grass quickly wears away, with compaction and mud the result.

There are two subcategories of paving, hard and

A Comparison of Types of Surfacing

Surfacing Type	Installation Cost	Maintenance Cost	Walking Comfort	Use Intensity	Seasonal or Constant Appearance
hard paving	highest	low	lowest	highest	constant
soft paving	moderate	moderate (1)	moderate	moderate (1)	constant
turf grass	lowest	high (2)	high	moderate (6)	constant (7)
groundcover	moderate	moderate (3)	N/A (5)	low	constant to seasonal (8)
flowers	moderate	high (4)	N/A (5)	lowest	seasonal

(1) Some replacement is required each year.
(2) Fertilization, weed control, watering, and mowing are necessary.
(3) Initial cost is high due to the hand weeding that is required. Once established, costs are moderate.
(4) Much hand weeding, watering, and fertilizing are required.
(5) Should not be used in areas with pedestrians.
(6) Use intensity is greatest where people do not continuously follow the same path.
(7) Many grasses are dormant in certain seasons and may change color.
(8) Appearance of these materials depends upon whether plants are evergreen or deciduous.

soft. Hard paving includes materials such as brick, stone, poured concrete, tile, paving blocks, and wood planking, Figures 12-1, 12-2, and 12-3. The advantages of hard paving are:

Greater Durability. Once installed, hard paving provides the longest service at the least cost.

Lower Maintenance Requirements. Usually, only periodic patching is needed to keep the surfacing attractive.

Strength. Where traffic produces great pressure on the surface, hard paving is superior.

Hard paving also has certain disadvantages:

Heat Absorption. After a length of time in the hot sun, hard paving may become hot to the touch. It also releases the heat slowly at night, making patios warmer than desirable on hot summer evenings.

Hazardous When Wet. Not all pavings are slippery, but smooth concrete, tile, and even wood and stone can be hazardous after a rain. For this reason, they should not be used to surface areas near steps or ramps. Likewise, they should not be used for landscapes that are used by persons who are ill or have handicaps. ·

Glare. Smooth-surfaced hard paving reflects sunlight and can be blinding on bright, clear days. Use of a smooth concrete patio next to a glass door can fill the house with unwanted heat and reflected glare.

Expense. Hard paving is the most costly way to surface the landscape.

Figure 12-1 Patterned brick ends create an interesting and durable surface.

Figure 12-2 Brick and concrete are combined for a durable surface with an attractive pattern.

Soft paving includes asphalt and a large group of materials known as *loose aggregates*. Some loose aggregates are crushed stone, marble chips, sand, wood chips, bark chips, and tanbark. These pavings are not "soft" in the sense that they would not cause pain if an individual fell on them. While some are actually softer than the hard avings (such as sand or wood chips), the term *soft paving* indicates that the materials are less durable than hard paving and lack a solid form. They are best used in the landscape where

Figure 12-3 Wood and stone set in mortar give heavy-duty service while complementing the seaside atmosphere.

pedestrian or vehicular traffic is not intense, Figures 12-4 and 12-5. Soft paving does not perform satisfactorily if it is overused.

The advantages of soft paving are:

Lower Cost of Installation. Neither the cost of materials nor the cost of soil preparation prior to installation is as great as that for hard paving.

Faster Installation. Where the surfacing is needed quickly and/or temporarily, soft paving can be applied more rapidly.

Ease of Application in Oddly Shaped Areas. Soft paving can conform more easily to small or unusually shaped places.

Ease of Replacement. If something is spilled on the surfacing and stains it, it can be easily and inexpensively replaced.

Major disadvantages of the soft paving materials vary with the actual materials. Some of the common disadvantages are:

Greater Maintenance Requirements. Asphalt develops holes. Loose aggregates need to be frequently weeded and raked smooth.

Necessity of Replacing Materials. Loose aggregates are kicked out of place. A small amount is carried away every time someone walks on it; new material must be added annually.

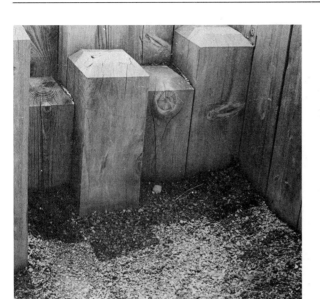

Figure 12-4 Tanbark provides the soft yet durable surface around this playground apparatus.

Figure 12-5 Crushed stone has been used to surface this planting bed. Water can reach plant roots, yet weeds are discouraged.

Tendency to Become Sticky (Asphalt) or Dusty (Aggregates). When the weather is hot and dry, soft paving may be tracked into the house, damaging carpeting or floors.

Selecting the Correct Surfacing

The final decision by a landscape designer or contractor concerning the correct surfacing to use requires an analysis of several factors:

❦ **Cost of materials and the budget of the client.** The ideal surfacing may be too expensive, requiring the substitution of a second choice which makes use of less expensive material. Landscapers must be up-to-date on cost trends in building supplies to make accurate cost decisions.

❦ **Amount of use the surfacing will receive.** If it is in a primary use area (front entry, patio, driveway), hard paving is needed. If the area receives only secondary use, soft paving or grass may suffice.

❦ **Aesthetic appearance.** The color and texture of the surfacing should harmonize with the other materials in the landscape.

❦ **Shape of the area being surfaced.** Some paving materials are fluid and can easily be molded into almost any shape (concrete, asphalt, loose aggregates). Others have definite square, rectangular, hexagonal, or other geometric shapes (bricks, patio blocks, paving stones). These are difficult to cut and are therefore most effective if used in designs that require minimal cutting.

❦ **Effect upon the building's interior.** Surfacing next to a home or other building should not create reflected glare or heat within the building, nor should it produce mud, dust, or oil that might be tracked into the building.

❦ **Maintenance required.** The amount of maintenance needed to keep the surfacing attractive should be carefully weighed against the amount of maintenance clients are willing to do themselves or pay to have done.

A Comparison of Surfacing Materials

Material Description	Hard Paving	Soft Paving	Modular	Continuous and Solid
Asphalt: A petroleum product with adhesive and water-repellant qualities. It is applied in either heated or cold states and poured or spread into place.	Semihard; allows weeds to germinate and grow through it			•
Asphalt pavers: Asphalt combined with loose aggregate and molded into square, rectangular, or hexagonal shapes. They are applied over a base of poured concrete, crushed stone, or a binder.	Semihard if not applied over concrete		•	
Brick: A material manufactured of either hard baked clay, cement, or adobe. While assorted sizes are made, the standard size of common brick is 2 1/4 × 3 3/4 × 8 inches.	•		•	
Brick Chips: A by-product of brick manufacturing. The chips are graded and sold in standardized size as aggregate material.		•	•	
Carpeting, Indoor/Outdoor: Waterproof, synthetic fabrics applied over a concrete base. They are declining in popularity. Their major contribution is to provide visual unity between indoor and outdoor living rooms.				•
Clay Tile Pavers: Similar to clay brick in comparison, but thinner and of varying dimensions (most commonly 3 × 3-inch, or 6 × 6-inch squares). They are installed over a poured concrete base and mortared into place.	•		•	
Concrete: A versatile surfacing that can be made glassy smooth or rough. It can also be patterned by insetting bricks, wood strips, or loose aggregates into it. Concrete is a mixture of sand or gravel, cement, and water. It pours into place, is held there by wood or steel forms, then hardens.	•			•

Slippery When Wet	Permeable to Water	Suitable for Vehicles	Suitable for Walks	Suitable for Patios
		•	•	Certain formulations are suitable. Others may become sticky in hot weather. *Note*: The application of a soil sterilant before applying the asphalt can eliminate the weed problem in walks, drives, and patios.
	If installed over crushed stone		•	•
	If installed in sand	•	•	•
	•		Edging needed to hold them in place	
	Provision must be made for surface water drainage or the carpeting becomes soggy			•
•			•	•
Only when smoothly finished		•	•	•

A Comparison of Surfacing Materials *(continued)*

Material Description	Hard Paving	Soft Paving	Modular	Continuous and Solid
Crushed Stone: Various types of stones are included in this umbrella term: limestone, sandstone, granite, and marble. Crushed stone is an aggregate material of assorted sizes, shapes, and durability.		•	•	
Flagstone: An expensive form of stone rather than a kind of stone. Flagstone can be any stone with horizontal layering that permits it to be split into flat slabs. It may be used as irregular shapes or cut into rectangular shapes for a more formal look. It is usually set into sand or mortared into place over a concrete slab.	•		•	
Granite Pavers: Granite is one of the most durable stones available to the landscaper. The pavers are quarried cubes of stone, 3 1/2 to 4 1/2 inches square, that are mortared into place. Various colors are available.	•		•	
Limestone: A quarried stone of gray coloration. Limestone can be cut to any size. It adapts to formal settings.	•		•	
Marble: An expensive quarried stone of varied and attractive colorations. It has a fine texture and a smooth surface that becomes slippery. Its use as surfacing is limited. It can be inset into more serviceable surfaces such as poured concrete.	•		•	
Marble Chips: A form of crushed stone, marble chips are more commonly used as a mulch than a surfacing. They are expensive compared to other loose aggregates; still they enjoy some use as pavings for secondary walks and areas that are seen more than walked upon.		•	•	
Patio Blocks: Precast concrete materials available in rectangular shapes of varied dimensions and colors. Limitless patterns can be created by combination of the sizes and colors. The blocks are set into sand or mortared over concrete.	•		•	

Slippery When Wet	Permeable to Water	Suitable for Vehicles	Suitable for Walks	Suitable for Patios
	•	•	Edging needed to hold the material in place	Limited use except beneath picnic tables where stains might spoil hard paving
Depends upon the rock used and how smooth the surface is			•	•
		•	•	Too rough
		•	•	•
•				Best used in dry climate where slipperiness will not be a frequent concern
	•		•	
			•	•

A Comparison of Surfacing Materials (continued)

Material Description	Hard Paving	Soft Paving	Modular	Continuous and Solid
Sandstone: A quarried stone composed of compacted sand and a natural cement such as silica, iron oxide, or calcium. Colors vary from reddish brown to gray and buff white. The stone may be irregular or cut to rectangular forms.	•		•	
Slate: A finely textured stone with horizontal layering that makes it a popular choice for flagstones. Black is the most common color, but others are available.	•		•	
Stone Dust: A by-product of stone quarrying. Stone dust is finely granulated stone, intermediate in size between coarse sand and pea gravel. It is spread, then packed down with a roller. The color is gray.		•		•
Tanbark: A by-product of leather tanning. The material is processed oak bark. It has a dark brown color and a spongy, soft consistency. It is ideal for children's play areas.		•	•	
Wood Chips: A by-product of saw mills, wood chips are available from both softwoods and hardwoods. The latter decompose more slowly than the former. Wood chips have a spongy, soft consistency. They are often used as mulches.		•	•	
Wood Decking: Usually cut from softwoods, the surfacing can be constructed at ground level or elevated. The deck is valuable as a means of creating level outdoor living spaces on uneven terrain. Space should be left between the boards to allow water to pass through and the wood to dry quickly. Use of a wood preservative will slow decay.	•		•	
Wood Rounds: Cross sections of wood cut from the trunks of trees resistant to decay, such as redwood, cypress, and cedar. The rounds are installed in sand. Individual rounds are replaced as they decay.	•		•	

Slippery When Wet	Permeable to Water	Suitable for Vehicles	Suitable for Walks	Suitable for Patios
		•	•	•
•			•	•
	•		•	
	•		Edging needed to hold the material in place	
	•		•	
	•		•	•
	•		•	•

Steps and Ramps

If the outdoor room has several levels, each level has a separate surfacing. The surfacing may change or remain the same between levels, depending upon the decision of the designer. Regardless, the levels must be connected with steps, ramps, or ramped steps.

The surfacing of the steps or ramps is usually selected to match that used on the levels being connected. However, there may be variations for either aesthetic or practical reasons. For example, it may be desirable to combine materials used in the surfacing and nearby enclosures in the design of the steps. One practical variation might be the necessity of constructing the steps with a more durable material than that used in the surfacing. Another variation might take into account the fact that some materials suitable for surfacing flat areas are slippery as steps or ramps.

A step is constructed of two parts: an elevating portion, or **riser**, and the part on which the foot is placed, or **tread**, Figure 12-6.

Figure 12-6 The parts of a step

Outdoor steps are usually not restricted to the same space limitations as indoor steps, Figures 12-7 and 12-8. For this reason, treads can be built wider and risers lower to create a stairway that is more comfortable than indoor steps. The design and construction of outdoor steps should be such that users of the landscape are allowed to maintain the most natural stride pattern possible. This is best accomplished by constructing steps in accordance with a widely accepted formula: the total length of two risers and one tread should equal 26 inches

Figure 12-7 Solid risers of wood are combined with treads of loose aggregates for coordination with the surrounding landscape.

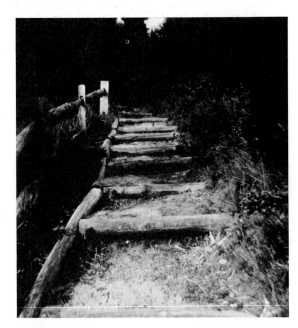

Figure 12-8 Because of the log risers and earthen treads, these steps blend comfortably with the woodland setting.

(T + 2R = 26"). Furthermore, the tread should be at least wide enough so that an entire foot can be placed upon it (12-inch minimum). Using this formula, if outdoor steps are built with a 12-inch tread, each riser will be 7 inches high (12" + (2 × 7") = 26"), Figure 12-9. As the tread dimension increases, the riser dimension decreases:

Riser	Tread
7 inches	12 inches
6 inches	14 inches
5 inches	16 inches
4 inches	18 inches

The number of steps required to connect two levels is calculated by dividing the elevation by the riser dimension desired. For example, if two levels are 48 inches apart, and a 6-inch riser is desired, eight steps are required (48" ÷ 6 = 8 steps), Figure 12-10. The amount of horizontal space required for the steps is determined by multiplying the tread dimension by the number of steps. In the example, the eight steps, with seven 14-inch treads, require

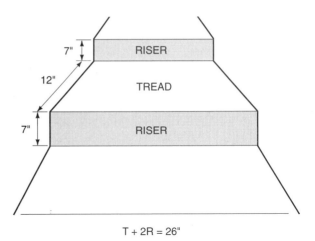

Figure 12-9 Calculating the dimensions of outdoor steps

98 inches (8.16 feet) of space in the landscape for their construction (14" × 7 treads = 98" ÷ 12" = 8.16'). It should be noted that the number of treads is always one less than the number of risers, since the top riser connects with the upper level, not another tread.

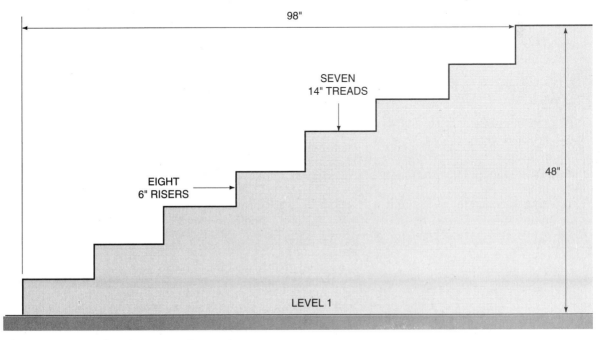

Figure 12-10 Planning space for outdoor steps

Achievement Review

A. There are four basic types of surfacing: paving (P), turf grass (T), and groundcovers and flowers (G/F). Based upon their characteristics as described in this chapter, indicate which type is most appropriate in each of the following situations.

1. This surface receives moderate foot traffic. There is no reason for fixed patterns to develop across it.
2. This surface receives heavy vehicular traffic.
3. This surface receives no traffic. It functions as a natural element of the outdoor room.
4. This surface must be durable and long lasting. It receives little or no maintenance during the year.
5. This surface offers seasonal variation in the landscape.
6. This surface is in a picnic grove where tables are moved to a different spot each week.
7. This surface is in a picnic grove where tables are never moved.
8. This surface must tolerate foot traffic while offering a high degree of walking comfort.
9. This surface must be suitable for dancing and shuffleboard.

B. Label the following paving materials as hard or soft.

1. wood rounds
2. crushed stone
3. flagstone
4. wood chips
5. brick
6. poured concrete
7. tanbark
8. asphalt
9. marble chips
10. sand

C. Indicate which paving type, hard or soft, is most suitable in the following situations.

1. The surface is in an eating area and may receive frequent food spills.
2. The surface is next to a heavily used exit from a building.
3. The surface will receive heavy and frequent auto traffic.
4. The surface is located where a dog will play in a service area.
5. The surface must allow rapid drainage of water through it.
6. The surface is to be a patio in a family living area.
7. The surfacing is to be as comfortable to the feet as possible.
8. The surfacing will receive little or no maintenance.
9. The surface must be constructed of the least expensive paving material available.

D. List the six factors that must be considered when selecting surfacing.

E. Using the formula T + 2R = 26", what size riser is best for a step with a 15-inch tread? Show how you find the answer.

F. How many steps with 5-inch risers are necessary to reach an elevation of 5 feet? Show your work.

G. How much horizontal space would the above steps require? Show your work.

H. How high above level B is level A in the following plan view?

LEVEL A

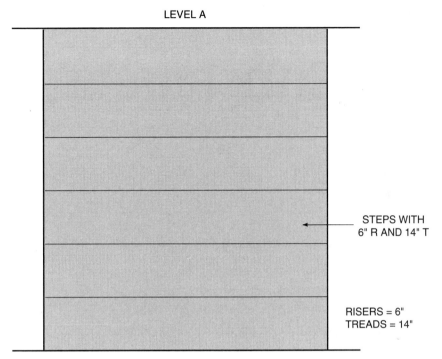

STEPS WITH
6" R AND 14" T

RISERS = 6"
TREADS = 14"

LEVEL B

CHAPTER 13

Enrichment

Objectives:

Upon completion of this chapter, you should be able to

- define landscape enrichment
- distinguish between tangible and intangible enrichment
- distinguish between natural and fabricated enrichment items
- evaluate the value of selected enrichment items

Landscape Enrichment Is Something Extra

To understand the role that enrichment items play in the design and development of a landscape, picture again an indoor room. When the basic structure of an indoor room is completed, it has walls, ceiling, and a floor, but it is not yet fully usable. To make it functional and personal, it requires furniture, lighting, pictures, music, pets, and many other things that outfit the room for use by the people for whom it is intended. Although these items are not wall, ceiling, or floor elements, their importance as components of the indoor room is easily seen.

The outdoor room has a similar need to be made usable and personal. **Enrichment items** are components of the outdoor room that are not essential to the formation of its walls, ceiling, or floors.

While they may also serve as walls, ceiling, or floor materials, enrichment items are first selected to fulfill some additional purpose.

Tangible and Intangible/ Natural and Fabricated

Enrichment items are represented by a widely diverse collection of materials. Some are **tangible**, meaning that they can be touched; while others are **intangible**, meaning that they are sensed but cannot be touched. Some are naturally occurring, while others are fabricated, meaning manufactured. Consider this partial listing to aid in understanding the contribution that enrichment items make to the landscape.

Tangible, Natural Enrichment

Stones, such as rocks, boulders, and natural outcroppings, are valuable additions to any landscape.

Figure 13-1 Large stones provide an enriching backdrop for the rock garden plants growing among them.

Where they are an existing site feature, such as rock outcroppings, the landscape designer is wise to incorporate them into the design. They can serve effectively as backdrops for border plantings or as the framework for the development of a rock garden, Figure 13-1. In other settings, naturally occurring or selected stones can serve as natural sculptures in the landscape, Figure 13-2. Where the design and size of the site permit, stones can also serve as alternatives to traditional outdoor furniture, permitting people to rest against them or children to climb safely over them, Figure 13-3.

In oriental garden design and in gardens of the American Southwest, stones have a long history of importance. The Chinese and Japanese peoples

Figure 13-3 Rock formations add natural enrichment to the landscape. (Photo courtesy of Colleen Pasquale)

Figure 13-2 Stone used as a natural sculpture

Figure 13-4 Stones play an important and traditional role in Japanese gardens.

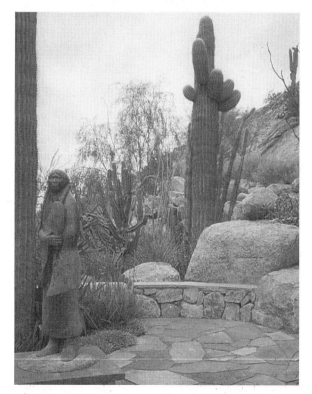

Figure 13-5 Stones are an important natural enrichment feature in this southwestern landscape.

have traditionally assigned great spiritual and cultural significance to the selection and positioning of stones within their gardens, Figure 13-4. In the Southwest, stones have been one means of interfacing designed gardens with the wide open spaces that lie outside, Figure 13-5.

Specimen plants bring something extra to the landscape's design, which may last only for a few days or which may be effective year-round. Some plants have highly attractive and unusual growth habits, making them natural outdoor sculptures, Figure 13-6. Others are striking because of their blossom, foliage, fruit, or bark colors and textures. When pruned into unusual shapes, plants can become highly styled objects of art, Figure 13-7. Such pruning of plants into unusual shapes is termed *topiary pruning*, and dates back several centuries. Because topiary pruning creates a high maintenance requirement for the garden, designers should use it sparingly as an enrichment technique.

Water is one of the most versatile enrichment materials because it can enhance the landscape's design in so many different ways. When occurring naturally on a site, as a stream, lake, or waterfall, water can enhance the recreational uses of the site.

Figure 13-6 The unusual silhouette of this spruce gives it strong eye appeal. It becomes an enriching focal feature.

Figure 13-7 The topiary pruned shrub as an enrichment item requires high maintenance and is used only in very formal settings.

It also attracts wildlife that might otherwise not approach the site. Water mirrors the sunlight and ripples in the wind, Figures 13-8 and 13-9. It makes the landscape sparkle. It is the most popular of all recreational and aesthetic enrichments.

Animals run a close second to water as people's most favored form of tangible, natural enrichment. Like water, animals bring life and movement into the landscaped setting. Because they are not stationary they add variation and surprise to the daily landscape. Whether it be the color of birds at a feeder, the scamper of a chipmunk across the patio, or a view of farm animals grazing on a distant hillside, people enjoy sharing their landscapes with animals of their choice. Designers can easily provide for the inclusion of wildlife in land-

scapes by designing outdoor rooms that serve as habitats for both humans and animals. Many plants produce berries that will attract birds and other small animals. A hedgerow can serve as an outdoor wall while simultaneously supplying nesting sites. Planning for the inclusion of feeding stations in the design will serve to assure the continued presence of wildlife through the winter as well, Figure 13-10.

Tangible, Fabricated Enrichment

Outdoor furniture has traditionally been one of two types: heavy and cumbersome or cheap and tacky. Neither type offers appropriate style or much comfort. Depending upon whether the landscape is

Figure 13-8 It is the splash and sparkle of water that make it so appealing in the landscape.

residential, public, or commercial, furnishings for the landscape may include tables and chairs, lounges, play apparatus, and trash containers. In a long overdue break with tradition, modern landscapes are finally incorporating furnishings that enhance good designs rather than detracting from them, Figure 13-11. Outdoor furniture should be selected to match the theme and style of the outdoor room in which it is placed. Some furniture may be permanent, requiring sufficient durability to withstand the winter. Other furniture may be seasonal, requiring it to be taken in when the weather becomes inclement. While the inexpensive outdoor furniture available each spring in the supermarkets and discount stores will always find buyers, the landscape designer who would be of greatest service to his or her clients should encourage those clients to select their outdoor furniture as carefully as they have selected every other component of the outdoor room, being certain that it is comfortable, color coordinated, strong, and weather resistant.

Outdoor art plays a role in the outdoor room that is similar but sometimes not identical to its role in the indoor room. In an indoor setting, pictures are selected for the walls and sculpture is placed on the

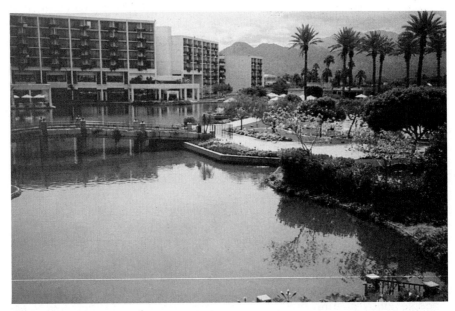

Figure 13-9 Large, still bodies of water become the reflective mirrors of the landscape.

Figure 13-10 People and wildlife can often coexist compatibly. Designers can plan for the inclusion of animals through careful selection of plants, provision of feeding stations, and development of suitable habitat areas.

Figure 13-11 Metal chairs and tables make attractive yet durable outdoor furniture. In northern areas, this type of furniture must be stored indoors during the winter.

coffee table for several reasons, not the least of which is because they fit the color scheme of the room or they are the right size to fill an empty space. In the outdoor setting the spaces are larger, and artwork must be larger as well. Whereas indoors, artwork may compliment the theme of the room, outdoors a piece of art may establish the theme for the room, Figure 13-12. In America's national landscape, public art is proving increasingly popular. The exterior walls of inner-city buildings are being brightened by the multi-story murals of local artists. Hotels, parks, campuses, and corporations are making art an important part of their landscapes, Figure 13-13. As a result, public appreciation for art and its importance to the quality of life is growing. It is predictable that most clients will be receptive to the inclusion of some type of art enrichment in their landscapes, and designers and landscape architects should be sufficiently knowledgeable to offer proper professional guidance.

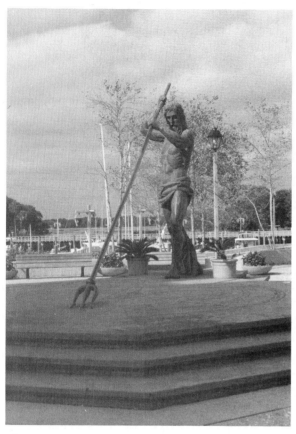

Figure 13-12 In the coastal island setting of Hilton Head, South Carolina, this looming sculpture of Neptune strikes an appropriate theme.

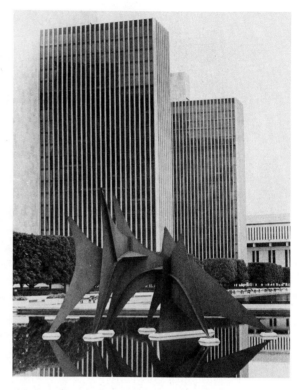

Figure 13-13 This large sculpture and reflecting pool enrich the pedestrian mall of the New York State governmental center.

The selection and use of art is very subjective, but there are some guidelines worth noting:

a. Avoid cartoonish art, such as the plastic and plaster lawn ornaments that fill the discount stores every spring.

b. Avoid anything made of pressed plastic. It is lightweight and will likely fall over or blow away (and probably should).

c. Avoid using reproductions of historic art in contemporary settings. Cherubs and gargoyles were appropriate in the gardens of Europeans 300 years ago, but they are outdated today.

d. Be sensitive to the religious significance of some artwork, and don't use it in a way that would be offensive to certain viewers.

e. Use outdoor art selectively and conservatively. Don't turn the landscape into a gallery or a gift shop.

Pools and fountains use water again, but this time in fabricated forms. Whether as a swimming pool in the backyard or as a show-stopping display fountain in a shopping mall, water in fabricated forms appeals to people. The wise designer will consider the potential for using water whenever space, weather, and the client's budget permit, Figure 13-14.

The residential landscape can easily incorporate water's enriching qualities. Small prefabricated recirculating fountains, available at a modest price, can bring new life to a patio area. Such fountains are self-contained, electrically powered units that operate with the simple flip of a switch, Figure 13-15. Also popular, and affordable, are small fish ponds such as the one illustrated in Figure 13-16.

Figure 13-14 This large splashing fountain, located in a pedestrian plaza, adds a refreshing coolness to the surroundings.

Figure 13-15 This modern sculpture, combined with the falling water, adds original and personal enrichment to the patio.

Figure 13-16 A backyard fish pond can bring enjoyment to the smallest family living area.

With a bit of imagination, landscape architects and designers can specify customized water features that will fit specific landscape sites and not rely solely upon catalogued products. Rock waterfalls are the specialty of numerous landscape contracting firms and stone masons around the country, so bringing the designers' ideas to reality is usually possible. Also, many items such as statues and tubs, that were not initially designed as water features, can be piped to spout water or lined to contain it.

Night lighting assures that the enjoyment of the landscape need not end when night falls. By including provision for the evening illumination of the garden, the landscape designer adds these and other values to the design:

❦ Increased time of use

❦ Provision of greater safety and security for users

❦ Creation of special effects such as colored lighting, silhouette lighting, shadow effects, or patterns against buildings

❦ Maintenance of the same visual relationship between indoors and outdoors that exists during daylight hours

Not too long ago most backyard lighting consisted of floodlights mounted on the roof of the house or garage. The glare of white light was unattractive and more suggestive of a prison yard than a gracious garden. Such floodlights are still excellent security lights, but lighting design has advanced to softer, more attractive forms for the outdoor room, Figure 13-17. While fixtures and manufacturers vary, there are five common techniques for lighting residential landscapes, Figure 13-18.

Walk lights offer both safety and decorative effects. They should be used wherever it is necessary to warn pedestrians that the walk is about to change direction or elevation, such as at steps or ramps.

Silhouette lighting outlines plants when placed behind them. The viewer sees a dark plant form against a background of light.

Shadow lighting places the light source in front of

Figure 13-17 Lighting in the style of an earlier era helps give this small park a restful atmosphere.

the plant and causes a shadow to be cast onto a wall or other flat surface behind the plant.

Down lighting creates patterns of light and leaf shadows on the ground. The light fixture is placed high in a tree and directed downward.

Up lighting is the reverse of the above. The light fixture is placed at the base of the object being illuminated. It is directed upward.

Two precautions are worthy of note when planning the illumination of landscapes. One is to position the lights so that they do not shine into the eyes of the landscape's users. That necessitates that the designer anticipate the probable locations of the users at night. The other precaution is to be sure that the level of brightness outside the building equals that inside. If the brightness levels are not equal, the separating glass door or window reflects like a mirror. The

SILHOUETTE LIGHTING. . .
THE LAMP IS PLACED
BEHIND THE PLANT.

UP LIGHTING GIVES HIGH-
LIGHT AND SHADOW
PATTERNS TO OBJECTS
ABOVE THE LAMP.

SHADOW LIGHTING. . .THE
LAMP IS IN FRONT AND
THE WALL IS BEHIND THE
PLANT.

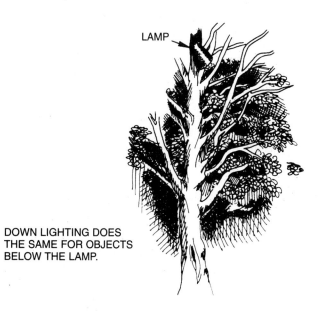

DOWN LIGHTING DOES
THE SAME FOR OBJECTS
BELOW THE LAMP.

WALK LIGHTS GIVE
SPECIAL EFFECTS, MARK
CHANGES IN DIRECTION,
AND OFFER SAFETY.

Figure 13-18 Five common outdoor lighting techniques

value of the lighting effects is then lost to the indoor viewer looking out.

Intangible Enrichment

The inability to touch certain types of enrichment does not diminish their importance to the landscape. Intangible enrichment appeals to the senses, primarily to the client's senses of hearing, smell, and sometimes taste. Consider the following contributions to the enjoyment of a landscape, and judge their importance.

❦ The sound of splashing water from a fountain or cascade

❦ The fragrance of plants in bloom

❦ The rustle of wind through a planting of trees

❦ The sounds of birds, crickets, frogs, and other wildlife

❦ The ringing of distant church bells or carillons

❦ The lapping of waves against the shore

When such enrichment occurs naturally on a site, the designer has only to take advantage of it by encouraging its continuance. When it does not occur in significant amounts, intangible enrichment can be developed by the designer. Careful selection of plant species can encourage the nesting of songbirds or provide suitable habitats for frogs or other "musical" animals. Plants known to produce fragrant flowers can be placed beneath bedroom or kitchen windows or near conversation areas of a patio where they are certain to be appreciated. Music systems can be designed into the outdoor setting with speakers hidden discreetly out of sight. Even a simple wind chime can bring enriching sounds to a private area. Additionally, the sense of taste can be enlisted by placing plants with edible and tasty fruit in areas of the landscape where people can easily sample them at certain times of the year.

For clients who are or have family members who are in some way physically impaired, intangible enrichment can play a significant role in their enjoyment of the outdoor rooms of the landscape.

Is the Enrichment Item Necessary?

It is possible and not uncommon to design a landscape with too many enrichment items. When this occurs, the effect is one of clutter, with the design principles of simplicity and unity falling victim to excess. Amateur gardeners are common culprits in the over-enriching of their home landscapes, but beginning designers are also susceptible to the temptation.

A few simple guidelines can be helpful in planning the enrichment of a landscape. First, be sure that the item being considered is truly enriching to the landscape. It must fulfill a role other than being a wall, ceiling, or surfacing material. Second, most tangible enrichment items attract the viewer's eye strongly and should be used sparingly as focal points in the landscape. Finally, when in doubt about the value of or need for an enrichment item, remove it. Then envision standing back and observing where it was. If nothing would be missed and the design would work as well without it, the enrichment item is probably unnecessary. If a visual or sensory hole would be left in the design, then the enrichment item probably belongs there.

Achievement Review

A. Indicate which of the following characteristics apply to enrichment items (E), nonenrichment items (N), or both (B).

1. Their major function is to provide shelter.
2. Their major function is to shape the outdoor room.
3. Their major function is something other than serving as a wall, ceiling, or floor element in the outdoor room.
4. They may be tangible items.
5. They may be tangible or intangible.
6. They could be used as focal points of the design.
7. They could protect people from a rainstorm.
8. They may require periodic maintenance.
9. They are sometimes created by nature.

B. Indicate if the following natural enrichment items are tangible (T) or intangible (I).

  1. a boulder
  2. a lake
  3. the sound of a bird
  4. a distorted old pine tree
  5. wind whistling through trees
  6. a waterfall
  7. the sparkle of a waterfall
  8. the sound of a waterfall
  9. berries on a shrub
  10. the taste of berries

C. Give four examples of natural, tangible enrichment items. Do not use any of those listed in question B.

D. Give four examples of natural, intangible enrichment items. Do not use any of those listed in question B.

E. Indicate whether the following enrichment items are natural (N) or man-made (MM).

  1. bird bath
  2. sun dial
  3. lake
  4. statue
  5. large boulder
  6. chaise lounge

  7. picnic table
  8. chipmunk
  9. outdoor lights
  10. wind in pine trees

F. Select the best answer from the choices offered.

  1. Where is the light for silhouette lighting located?
     a. behind the plant
     b. above the plant
     c. in front of the plant

  2. Where is the light for shadow lighting located?
     a. behind the plant
     b. above the plant
     c. in front of the plant

  3. What is a good way to test the value of an enrichment item in a landscape?
     a. Sell it and determine its worth.
     b. Remove it temporarily.
     c. Add another one like it.

  4. Permanent outdoor furniture should be attractive and _____.
     a. lightweight
     b. weather resistant
     c. upholstered

Suggested Activities

1. Demonstrate different lighting techniques in the classroom. Use a small flashlight for the light source. Use a box to represent the building or wall background. Select a small leafed branch from a shrub to represent the plant. Darken the classroom and try up lighting, down lighting, silhouette lighting, and shadow lighting the plant and building.

2. This activity demonstrates the necessity of designing outdoor lighting so that the level of brightness is close to that of the indoors. Go outside on a bright, sunny day and try to see into the school building from the road or sidewalk. The windows, acting as mirrors, will cause the inside of the building to appear dark.

3. Study the sidewalk system around the school building. Where are walk lights needed for safety?

4. Invite a sculptor to visit the class to discuss original outdoor art: its costs, common locales, time needed for production, and modern preferences. If possible, arrange a class visit to the artist's studio or to a museum of modern art.

5. Make a bulletin board display of outdoor furniture, including pictures and information about materials and costs. As sources, use newspapers, department store catalogs, landscape trade journals, and gardening magazines.

6. List all of the natural enrichment items in your home landscape. Indicate if they are tangible, intangible, or have characteristics of both.

7. Do a similar study of a park near the school. Which landscape has more natural enrichment, the park or the home?

CHAPTER 14

Computers in Design

Objectives:

Upon completion of this chapter, you should be able to

- describe how landscape firms use computers
- explain the types of computer programs currently popular in the landscape industry
- discuss the benefits and limitations of computers to the industry
- project the future uses of computers in landscaping

The Technology and Its Uses

Trying to write a chapter about the current status of computer technology is like bobbing for apples. Just when it seems that you have accomplished what you set out to do, it slips away and something else pops up in a different place. While the landscape industry was slower than others to accept the full array of computer services, even the most reluctant technophobes have now succumbed to the siren song of the microchip.

Modern computers have changed forever the way that landscape offices and drawing rooms look and the way that business is conducted. New tools and new methods of doing traditional tasks are being presented to the industry almost daily. Educational institutions have difficulty staying current with the technology and required training of graduates that the industry now expects.

The major reason for the landscape industry's full acceptance of the computer as a valuable tool was the development of specialized, industry-specific programs. Rather than waiting for the industry to come to the technology, the technology came to the industry. At present, landscape professionals are using computer assistance for:

- correspondence.
- client accounts.
- accounts payable.
- payroll.
- taxes.
- mailing lists.
- inventory records.
- accounts receivable.
- billing.

❦ project tracking.

❦ specification writing.

❦ take-offs.

❦ cost estimating.

❦ plant selection.

❦ irrigation design.

❦ technical calculations, such as cut and fill.

❦ landscape design.

❦ graphic imaging.

❦ world-wide internetting.

The Popular Programs

Computer systems require machinery known as hardware and instructions to make the machinery operate, termed *software* or *programs*. The hardware is unable to do anything until given direction by the software in an electronic language that it can understand. Fortunately, the horticulturist using the computer need not understand the electronic language in order to make the program communicate with the hardware.

The hardware requirements include a means of entering data into the computer (input devices), a central processor of the data (the computer), a place to store the data once processed (hard or floppy disks), and a means of converting the data from the computer into printed form (output device). Input and output devices vary depending upon whether the data are written or graphic in form and what quality of output is desired.

The most common software programs are of three types:

1. Word processing
2. Numerical calculations
3. Graphic visualization

Word processing is the successor to the typewriter. Letters, reports, memos, and other written documents are prepared and stored on word processors. Paragraphs can be moved or deleted, words can be changed or inserted, envelopes addressed, key phrases highlighted, lettering styles changed at will, and numerous other techniques

applied instantaneously. Gone are smudged erasures and nameless form letters. Spelling errors and incorrect word choices can be minimized by using the proper software. Many good software programs are available and usable on IBM personal computers, micro- and minicomputers, or IBM compatible computers that are now the standard for central processors.

Input for word processing is done on a keyboard that is similar in many ways to a typewriter. Output is usually from a printer that transfers the image on the computer's monitor to letterhead stationery or other paper being fed into the printer.

Numerical calculations enlist the capability of software programs that accomplish mathematical functions. Conventional business needs such as inventory control, accounts receivable and payable, payroll and taxes, billing, time records, income statements, and profit and loss statements are all commonly handled by computers.

Computer technology allows garden centers and nurseries to update their inventories when plants and other materials are purchased. One scan by a computerized register generates a printed receipt for the customer and simultaneously deducts the item from the inventory.

Landscapers are able to quantify materials from designs and convert the data into precise bids in a fraction of the time required by conventional measuring and counting methods. Last-minute changes to design specifications can be quickly input and the bid updated. Less time is required for calculations that in the past were time-consuming, tedious, and subject to human error. At the same time the accuracy of the calculations has probably improved, since the computer's insistence on certain data input has compelled companies to keep close track of labor hours, material costs, and equipment usage.

While a number of good software programs are being marketed to nursery and landscape companies, several enjoy especially widespread acceptance. One such system is the SLICE system developed by Thornton Computer Management Systems of Mainesville, Ohio. Unlike earlier programs that were directed to a generic business market, the SLICE system customized its software to the needs of the green industries. It is adaptive to a company's existing hardware or Thornton's own computer product. As part of a successful marketing campaign, backed up by a good product and customer service, the developers have made themselves highly visible to the garden center, nursery, and landscape industries and managed to implant their product's name in the minds of industry professionals as the number one computer system for their businesses. Other producers will surely continue to develop competing software products, and some landscape firms have even developed their own programs, usually as a modification of an existing generic package.

Input devices for numerical calculation programs include the keyboard, the scanner, and the digitizer, Figure 14-1. The digitizer permits direct measurement of lengths, areas, volumes, and perimeters from drawings; so it is essential for the conversion of drawn data to numerical equivalents. The output device for numerical calculations is usually the printer.

Graphic visualization has gained acceptance within the landscape industry, but more slowly than word processing or numerical calculations. Two types of graphic visualization systems are used in the green industries, mostly by landscape designers and landscape architects: computer-aided design (CAD) and video imaging.

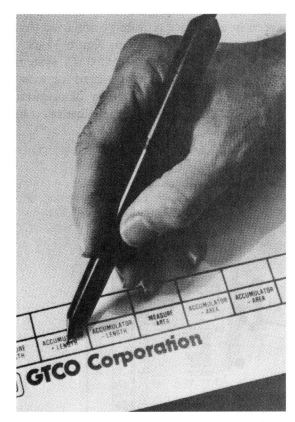

Figure 14-1   A hand-held stylus is used to select commands and symbols on a digitizer or menu tablet. (Courtesy of GTCO Corporation)

CAD Systems

CAD systems are accelerated drafting systems, designed to do what draftsmen have traditionally done with their T-square and pencils. Computer drafting has increased mainly because of a PC-based CAD software program named *AutoCAD* (a registered trademark of Autodesk, Inc.) Due to the necessity for hardware, software, and user training, AutoCAD is an expensive investment for a firm. It is not ready to use upon delivery to the office, moreover, because it is a general system, marketed to a variety of users. It is this generic nature of AutoCAD that makes it the dominant CAD system. However, to customize it for a specific type of drafting requires a third-party, discipline-specific software interface.

Landscape architects and landscape designers have most frequently selected LANDCADD as their discipline-specific software. Developed by a group of practicing professionals and released commercially in 1984, LANDCADD is marketed by LAND-CADD, Inc., of Franktown, Colorado. Presently, LANDCADD offers modular programs that can be selected by firms to fit their needs, thus avoiding the purchase of programs they will not use. At the heart of the software system is the Site Planning and Landscape Design Module. It permits the creation of two-dimensional plan views and elevations complete with contour lines, buildings, plant symbols, landscaping, and enrichment items. Surface textures, as well as lettering, can be added. As the design develops, the plant list is automatically created and tabulated. The module comes with its own menu of symbols to which users can add their own, further customizing the system. Figure 14-2 illustrates a landscape plan developed using a LANDCADD software package. Three-dimensional drawings of buildings, terrain, and grading plans are also possible using the Planning and Design Module.

An Irrigation Design Module is available as well as modules that permit material measurement, cost estimating, and plant selection, Figure 14-3. Another module provides a library of construction detail drawings that can be used directly or modified by the user before insertion into the landscape plans. Since success breeds competition, a number of competitive discipline-specific software systems have entered the market, each directed solely or in part to the landscape industry. Such competition assures a free market, innovative research and development, and reduced prices to the consumer.

The input device for computer-aided design is usually a mouse or a digitizing tablet and stylus. The mouse is held in the designer's hand and rolled around the table surface while the cursor moves about on the screen. The tablet and stylus permit the user to place drawings onto the monitor's screen in a manner somewhat similar to that of the traditional draftsman, Figure 14-4.

The output device for CAD system drawings may be the printers used for word processing, but the best quality of printout is obtained from a pen plotter, illustrated in Figure 14-5.

Video Imaging

Video imaging is a computerized technique that superimposes a landscape concept over a photographic image of the undeveloped landscape. The designer begins by taking several photographs of the building to be landscaped or filming it for several minutes using a camcorder. The photo or video image is then taken to the designer's studio, entered into the computer, and retained by the system's memory. The designer then directs the computer to apply color photo images of plants, pavings, pools, lawn, and other landscape elements over the camcorder of photographic picture. The result is a full-color suggestion of how a landscape will look before it is actually installed. It is an excellent sales tool to assist the client's visualization of the designer's ideas.

The leading video imaging system in America is Design Imaging Group, headquartered in Wood-

Figure 14-2   A LANDCADD design (Courtesy of Daniel International Corporation)

Figure 14-3 A LANDCADD irrigation design (Courtesy of Kaden Landscape)

land Hills, California. The technology is so new that there is little competition in the marketplace. Figure 14-6 illustrates before and after images of landscapes as treated by the video imaging system. In the hands of a skilled operator, its effects are realistic and impressive. The system permits the presentation of a variety of design concepts to a client for selection, before time is spent on plan view designs that may never be constructed. At present, video imaging does not create scaled drawings from which the landscape can be built. However, the Design Imaging system is designed to interact (interface) with CAD systems such as LANDCADD to apply color and realistic surface textures to the line drawings produced by the CAD system.

Figure 14-4 A drawing pad and drawing boards with cordless transducer choices (Courtesy of CalComp Digitizer Division)

Figure 14-5 An A- through E-size dual-mode plotter (Courtesy of CalComp)

Input devices for the Design Imaging system are the video camcorder, tablet and stylus, and scanner. The graphic image is composed on the monitor and can then be reproduced as a color print or as a videotape for presentation to a client audience of one or a hundred and one.

The Benefits and Limitations of Computers

Each technological advance carries with it advantages and disadvantages. If the advantages do not greatly outweigh the disadvantages, the technology is usually not accepted. Computer technology is no longer in its infancy, but neither has its development leveled off. Keyboards are being replaced by windows and icons. Projects can be transmitted in their entirety back and forth between professional offices across the city or across the country in an electronic second. Designs and cost estimates can be modified in a fraction of the time that would be required if done by hand. Yet there are two edges to the technological sword. The great benefits of time saving, reduced errors, rapid revisions, historical memory, and easy replication of tedious tasks are counterbalanced by the high costs of technological obsolescence, or the replacing or modifying of hardware or software and updating of employee training. There is also the danger of putting too much confidential or revenue-generating work into the hands of employee specialists with insufficient backup replacements.

For some landscape designers, perhaps a diminishing number, there is another limitation of computers. That is the loss of the tactile relationship with their work. Resisting the addiction of computer technology on which others have become so dependent, some landscape architects and design-

Figure 14-6 Before-and-after example of computer imaging (Image by Gary Galpin, Design Imaging Group)

ers have chosen not to discard the traditional methods of their profession. They scoff at the stiff imagery of computer-generated drawings. Traditionally, the graphic appearance of the landscape plan, including its lettering and symbol styling, has been the hallmark of a designer or a firm. Early CAD designs were simplistic and mechanical. However, improved data banks of symbols and lettering styles plus programs that accept the designer's own styles have removed most of the early objections. Recent software improvements have given CAD design systems the ability to produce perspective drawings and elevation views.

Still, to many landscape designers, the fun of designing, as well as the actual development of concepts, forms, and relationships, have been dependent upon the touch of pencil to paper. Like an electric current, the flow of creativity from these designers is dependent upon that contact. The computer removes that tactile relationship with the design. It replaces free sketching with menu selections. It offers a data bank of symbols instead of spontaneous interpretations. It asks questions instead of offering answers, and it requires exact responses from the designer if it is to serve. It is that mechanical quality of the input, not just the output, that may limit use of the current technology by the most talented designers, since many creative individuals are not proficient mechanics.

The Future

As a mechanical servant the computer's role in the landscape industry will continue to expand and, one by one, it will overcome all of the current limitations and objections. To be more than mere refinements of existing capabilities, major advances must be made in three areas:

1. The methods by which data can be acquired and input
2. A link between data as it is input and the graphic output from CAD systems
3. A greater variety of output forms

Systems that now use keyboards for data input into word processing and numerical calculation programs can be expected to be voice driven in the future. Such innovative input will permit even faster recording and processing of data. Instant spell check and grammar check as well as simultaneous translation into other languages will result from enlarged databases and more powerful computers. As computers from one company become linked nationally and internationally to all computers, communications and ordering may bypass not only the postal system and telephone, but send the FAX machine into the gadgeteering history books next to the eight-track musical tape.

Currently CAD systems have no understanding of what the lines and symbols of their graphics represent. Distinction cannot be made between a shape that represents a pool of water and one that symbolizes a brick patio. The designer must tell the estimating software how to interpret the drawing before material calculations and cost estimates can be completed. Once the computer is able to recognize from the symbols what an object is and how it will be constructed, even as it is being drawn, then construction drawings, specifications, and cost estimates can be generated, modified, and published simultaneous with the design.

Site analysis data may be entered into the CAD system directly from aerial photos or video images shot from elevated locations such as rooftops, planes, or even satellites. Optical and laser scanning systems may be the technology that permits this direct input of data. At some point in the future, voice commands may be used to direct or at least correct computer-assisted designing.

Output products can also be expected to change with time. The graphic on the monitor, which now prints out as a flat plan, elevation, or perspective view may soon appear as a video walk-through of the proposed design. The computer may recognize the differences in size and growth characteristics of the plants selected for the design and project them into the image automatically sized and correctly interpreted. It will easily provide graphic projections of the landscape at maturity or at any earlier age that interests the client or designer.

Still another computer advance will impact on areas of landscape architecture that are beyond the scope of this text. The point is that computers can be expected to move from their present status as an aid to the landscaper to the level of an assistant. Now they merely record data and do as they're directed. Soon they will be able to analyze data and offer suggestions, while providing graphic illustrations that surpass anything that can be done by conventional drafting or art media. At that point they will have moved the industry to a new level of professionalism.

Achievement Review

A. In the sequence of events following, describe when and how current computer technology could be applied.

1. initial client contact
2. site analysis
3. design development
4. cost analysis and price determination
5. billing the client
6. paying the employees

B. Define the following terms.

1. word processing
2. hardware
3. software
4. input device
5. output device
6. SLICE
7. CAD
8. AutoCAD
9. LANDCADD
10. Design Imaging Group

C. Assume the role of a computer systems salesman attempting to convince a hesitant landscape company owner to convert her design-build firm to computers. What would be your complete argument in favor of such action?

D. Compare the current status of computerization with projections for the future in the following areas:

1. word processing
2. cost estimating
3. land measurements
4. landscape designing

Suggested Activities

1. Using a personal computer and a spreadsheet software program, do the practice exercise at the end of Chapter 15 (Pricing the Proposed Design).

2. Contact local dealers for demonstrations of a CAD design program and a video imaging program.

Pricing the Proposed Design

Objectives:

Upon completion of this chapter, you should be able to

❦ explain the difference between cost and price

❦ explain the difference between an estimate and a bid

❦ describe landscape specifications

❦ prepare a design cost estimate

One of the questions that every client can be expected to ask is "What is this landscape proposal going to cost me?" The landscape contractor will ask a similar question, "What will it cost my company to build this landscape?" Although both have used the word *cost*, they do not seek the same information. The client's question really means, "What will be the total outlay of money necessary to get this landscape done from start to finish? When all is done, how much will I have spent?" The landscape contractor's question really means, "What expenses of constructing this project must I recover to not lose money?" The landscape architect or designer must know the answer to both questions. He or she must design the project to stay within the client's budget for the project. In so doing, the landscape architect or designer must have a realistic understanding of the contractor's

costs in order to anticipate the actual cost of building the project.

Cost and Price

Although the terms *cost* and *price* are often used interchangeably, they do not always mean the same thing. *Cost* refers to the recovery of expenditures. *Price* refers to an outlay of funds.

The cost of landscape services is the sum of material costs, labor costs, equipment usage costs, and overhead costs.

Cost = Materials + Labor
+ Equipment use + Overhead

These are all expenses assumed by the landscape contractor during the construction of a landscape.

If they are not recovered through payment by the client, then the landscaper will lose money.

Price is what the client pays to have the landscape built. It includes the price of the landscape architect or designer's services plus the price of the landscape contractor.

> Price = Landscape design services
> + Landscape contracting services

The price of the landscape design services will include the designer or landscape architect's fee for designing the landscape, modifying it as requested by the client, and representing the client whenever necessary. Representation of the client may include obtaining necessary permits, attending zoning meetings, conferring with individuals or groups involved in the project, selecting the landscape contractor, supervising construction of the project, and providing final inspection of the project to certify completion. The landscape architect's services may be calculated as per hour fees or as a percentage of the total value of the project. They will include cost recovery for all the time and any materials expended by the landscape architect. They will also include his or her profit on the project. The profit will be added to the costs to determine the price for landscape design services.

The landscape contractor will use a similar approach to determine the price for landscape contracting services. To the costs of materials, labor, equipment use, and overhead will be added two things; contingency and profit. The **contingency** allowance is something like insurance for the landscape contractor. Due to the unpredictable things that can disrupt a construction schedule, it is not possible to calculate precisely the costs of labor and materials. Delays caused by weather, labor problems, miscalculations of material quantities, or builders not finishing on time can wipe out the expected profit. The inclusion of a contingency fee in the price given the client can safeguard the landscape contractor's profit. However, if it is too large it may make the price too high and unacceptable to the client.

> Price = Landscape design services + Labor
> + Materials + Equipment use
> + Overhead + Contingency + Profit

NOTE: **Overhead** costs include administrative salaries, advertising costs, rent or mortgage payments, office expenses, accountant and attorney fees, subscriptions, memberships, insurance premiums, and equipment maintenance. They are operational costs of the firm. Some are general overhead costs and cannot be assigned just to a specific project. A portion of general overhead costs is charged to each project and included in the price. Other overhead costs can be assigned totally to a particular project. Costs such as portable toilet rental or temporary utility hookups can be fully assigned to the project for which they are needed. These are termed project overhead costs.

Estimates and Bids

An **estimate** is an approximation of the price that a customer will be charged for a landscape project. Estimates may be prepared by landscape architects or by landscape contractors. The time of preparation may range from minutes to hours, depending upon the size of the project and the level of accuracy needed. To keep a project within the client's budget limits, estimates are done repeatedly during the design stage. Price overruns only result in client disappointment and a waste of the designer's time.

When a price estimate is presented to a client, it is important for the client to understand that the price is not a firm quotation. Rather, it is a close approximation, subject to some change as actual costs are later determined.

A **bid** is a definite offer to provide the services and materials specified in the contract in return for the price agreed upon. Once confirmed by the signature of all parties, the price is binding upon the landscape company to do exactly what it has said it will do. If changes are made by the client or landscaper, the changes must be agreed to by everyone involved. As a result of the changes, the price may be changed, since the changes affect the original bid. Should unforeseen conditions delay the project or otherwise increase costs, the bid price remains fixed. In such cases, the landscaper's profit declines. Therefore, bids must be calculated carefully to assure that all costs are included.

When the landscape project is small, such as a residential property, there may be only one landscape firm working with the client. In such a case, the bidding process is usually informal and negotiable between the landscaper and the client. With larger projects, contracts are certain to require an exact bid, and negotiation may not be possible. On large projects, it is likely that several contractors will be competing for the same job. The firm with the lowest bid has a good chance of being selected, so a bid cannot be greatly overpriced or underpriced. These extremes will lose either jobs or profits for the landscape firm.

Specifications

Large landscape projects usually need more than drawings to explain fully what is required in the design and what quality standard is acceptable. **Specifications** are a listing of the materials, quality standards, and time schedules required to build a particular landscape. They may be prepared by the landscape architect or a professional specification writer. Copies of the specifications and the design are provided to each landscape contractor who wants to submit a bid for a project. The specifications must be carefully written to assure that the designer and client get what they are expecting from the landscape contractor. For example, specifications that describe a required species as "eight sugar maples" could result in bids ranging from a few dollars to several thousand dollars. This is because there is neither an indication of the size of the plants at the time of installation nor a specific requirement for bare-rooted or balled and burlapped plants. A client expecting a near-mature appearance in the finished landscape would be greatly disappointed by the sparseness of sapling plants. Precisely written specifications prevent client disappointment and discourage deceptive bidding.

Writing specifications can be very tedious work. Every plant in the design must be described by size at planting, root form, soil preparation, and pre- and post-transplant care. Each surfacing installation technique must be explained fully. Mulch depths, lighting fixtures, seed blends, and hundreds of other details must be fully described.

The landscape contractor uses the specifications and the drawings to prepare the bid. Each plant, every brick, every cubic yard of concrete or mulch must be counted. Then the labor and the supervisory time required for their installation are calculated. After that, charges for contingency, overhead and profit are added, and the bid is totaled. Specifications are discussed in greater detail in Chapter 19.

Preparing the Estimate

Landscape estimates and bids are prepared as spreadsheets. As the data is assembled, it is presented in columns that itemize quantities, descriptions of materials or services, unit costs of materials, unit costs of installation, and total costs for each item. The data is also grouped into categories that organize the estimate for easy reading and referral. A typical design cost estimate includes the following:

❦ Cost of site clearing and other preparation

❦ Cost of plant materials

❦ Cost of construction materials

❦ Cost of turfgrass

❦ Allowance for overhead

❦ Allowance for contingencies

❦ Fee for landscape designing

❦ Allowance for profit

❦ Name of the estimator and date of the estimate

Table 15-1 consists of two partial estimates, including an explanation of each column.

The fee for landscape design services is handled in several ways, depending upon the size of the project, the credentials of the designer, the policy of the firm, and the laws of the state. In some states, only accredited landscape architects are permitted to charge for their design services. Design-build firms often have salaried designers on the payroll. In such companies, the design costs are usually treated as overhead costs assigned to specific projects. In either case, the fee for designing, drafting, and overseeing the installation of a project to

Table 15-1

Estimate for Plant Materials

Quantity (1)	Unit (2)	Description (3)	Material Cost (4)	Installation Cost (5)	Total + 35 percent (6)
2	EA	*Acer platanoides*, 5', B & B	$ 35.00	$16.30	$138.51
1	EA	*Betula pendula*, 6', B & B	31.90	23.48	74.76
4	EA	*Cornus florida*, 4', B & B	45.00	13.80	317.52
15	EA	*Euonymus alatus*, 18", BR	8.40	11.60	405.00
200	EA	*Vinca minor*, 2" pots	1.05	0.50	418.50
			Total for plant materials (7)		$ 1,354.29

Estimate for Construction Materials

Quantity (1)	Unit (2)	Description (3)	Material Cost (4)	Installation Cost (5)	Total + 25 percent (6)
208	SY	Crushed limestone paving, 4" thick, compacted on a compacted subgrade using a motor grader	$ 7.00	$ 0.32	$1,903.20
90	SF	Timber retaining wall with deadmen, inc. excavation and backfilling	6.00	7.00	1,462.50
1	EA	Fountain, including jet, pump, drain, overflow, water linkage, underwater lighting, electric hookup, and basin	3,500.00	700.00	5,250.00
600	SF	Slate pavers, 1¼" thick, irregular fitted in a bed of sand with dry joints	6.40	4.50	8,175.00
			Total for construction materials (7)		$16,790.70

NOTES:
1. *Quantity* is the number of each separate item used in the design.
2. *Unit* is the form of measurement used to count the item. Examples include square yard (SY), square feet (SF), cubic yard (CY), cubic feet (CF), each (EA).
3. *Description* gives an explanation of what will be done or supplied. For plants it includes the species name, size at installation, and root form (BR for bare root, CT for container, and B & B for balled and burlapped). For construction materials, it includes briefly the materials to be used, means of installation, and other details needed to identify what is to be done.
4. *Material Cost* is the cost of one unit meeting the descriptions given.
5. *Installation Cost* is the charge for labor and any equipment needed to accomplish one unit of the task or service described.
6. *Total + Percent* is the sum of the material and installation cost for one unit multiplied by the quantity of units. To this is added a percentage allowance for overhead and profit. In the examples, 25 percent is used for the O & P of construction, and 35 percent is used for O & P of plant materials. The latter allows for the greater loss of plant items due to their perishability.
7. *Category Subtotals* are underlined and set off to the right for easy totaling later.

meet full client satisfaction can be estimated at 8 to 15 percent of the total cost of materials and installation. Usually a large project will use the lower percentage of the range, while a small project will use the higher percentage.

The contingency allowance may be calculated in a similar manner. Using the total cost of materials and installation as a base, a percentage may be added to cover contingencies.

The completed estimate or bid includes a client's

Price Estimate for the Design and Development of the Property of Mr. and Mrs. John Doe, 1234 Main Street, Cleveland, Ohio

I. PLANT MATERIALS

Quantity	Unit	Description	Material Cost	Installation Cost	Total + 35 percent
6	EA	*Celtis occidentalis*, 1½" cal. B & B	$ 125.00	$31.00	$1,263.60
5	EA	*Cornus florida*, 7', B & B	101.00	45.00	985.50
20	EA	*Euonymus alatus*, 15", BR	7.00	10.55	473.85
15	EA	*Philadelphus coronarius*, 3', BR	5.40	15.63	425.86
50	EA	*Vinca minor*, 2" pots	1.05	0.50	104.63
				Total for plant materials	$ 3,253.44

II. CONSTRUCTION MATERIALS

Quantity	Unit	Description	Material Cost	Installation Cost	Total + 25 percent
500	SF	Concrete pavers, 3⅛" thick, interlocking dry joints on 2" sand base, with 4" gravel subbase	$ 2.48	$ 1.53	$2,506.25
300	SF	Pressure treated timber wall, 6" × 6" timbers, gravity type, inc. excavation and backfill	6.00	7.00	4,875.00
1	EA	Flagpole, aluminum, 25' ht, cone tapered with hardware halyard and ground sleeve	1,060.00	59.02	1,398.78
				Total for construction materials	$ 8,780.03

III. TURFGRASS

Quantity	Unit	Description	Material Cost	Installation Cost	Total + 35 percent
900	SY	Kentucky bluegrass sod, on level prepared ground, rolled and watered	$ 1.00	$ 1.48	$3,013.20
				Total for turfgrass	$ 3,013.20

IV. TOTAL COST OF ALL MATERIALS AND INSTALLATION (See Note 1)		15,046.67
V. CONTINGENCY ALLOWANCE (See Note 2)		1,504.67
VI. FEE FOR LANDSCAPE DESIGN SERVICES (See Note 3)		1,805.60
VII. TOTAL COST FOR COMPLETE LANDSCAPE DEVELOPMENT (See Note 4)		**$18,356.94**

NOTES:
1. The total cost is obtained by adding the subcategory totals.
2. The contingency is taken as a percentage of total costs in IV. In the example, 10 percent is used.
3. The design fee is taken as a percentage of IV also. In the example, 12 percent is used.
4. The final figure should be the most distinctive on the spreadsheet. It should be the bottom line figure, with no others below it to create confusion.

name and address, the name of the estimator, and the date of preparation. It may also include an expiration date, after which the figures will no longer be honored because of their unreliability.

A designer will usually present the price estimate at the time the design is shown to the client. A landscape contractor will present the estimate or bid during the negotiation stage of the project. In both cases, the explicit figures shown in the examples are deleted from the proposal given to the client. Only the total figure is presented. A complete price estimate is included in this chapter. In this example, no special site preparation was required.

Practice Exercise

Prepare a complete price estimate for the property of Mr. and Mrs. Byron Lord, 1238 N. Grand Street, Harwich, Massachusetts. Use the following data and follow the format outlines in this chapter. No site preparation is required.

PLANT MATERIALS REQUIRED

    4 Chinese redbud (*Cercis chinensis*), 5' ht., B & B, cost $51.00 each and $16.25 to install

  20 Golden forsythia (*Forsythia spectabilis*), 18" ht., BR, cost $3.00 each and $11.63 to install

    2 Sugar maple (*Acer saccharum*), 1 1/4" cal, B & B, cost $58.00 each and $30.85 to install

  12 Carolina rhododendron (*Rhododendron carolinianum*), 2' ht., B & B, cost $39.00 each and $12.50 to install

  10 Japanese barberry (*Berberis thunbergi*), 15" ht., CT, cost $12.50 each and $10.55 to install

  20 Vanhoutte spirea (*Spiraea vanhouttei*), 1 gal. CT, cost $4.35 each and $10.55 to install

 300 Japanese pachysandra (*Pachysandra terminalis*), 2" pot, cost $0.83 each and $0.50 to install

    1 Babylon willow (*Salix babylonica*), 8" ht., B & B, cost $44.70 and $45.00 to install

  12 Floribunda roses (*Rosa floribunda*), 18" ht., BR, cost $6.60 each and $11.63 to install

    2 Red oak (*Quercus rubra*), 6' ht., BR, cost $21.50 each and $23.48 to install

NOTE: Profit and overhead allowance: 35 percent of the cost of materials and installation.

CONSTRUCTION REQUIRED

  90 linear feet (LF) of chain link fencing, 4' high, with galvanized posts and top rail and bonded vinyl 9 gauge fabric @ $4.37 per LF. Installation cost is $1.56 per LF.

    1 chain link gate, 4' high, 3' wide, with 2 galvanized gate posts, bonded vinyl 9 gauge fabric @ $218.00 each. Installation cost is $8.49 each.

110 square feet of brick pavers, 2 1/4" thick, set over a finished subgrade, dry joints, 2" sand base @ $2.31 per SF. Installation cost is $2.71 per SF.

    4 walk lights, 10' high, plain steel pole set into a cubic foot of concrete, with distribution system and control switch @ $2,000.00 each. Installation cost is $500.00 each.

NOTE: Profit and overhead allowance: 25 percent of the cost of materials and installation.

LAWN REQUIRED

1,000 square yards of Kentucky bluegrass sod, rolled and watered @ $1.00 per SY. Installation cost is $0.72 per SY.

NOTE: Profit and overhead allowance: 35 percent of the cost of materials and installation.

Fee for landscape design services is to be 12 percent of the total cost of all materials and installation. Contingencies will be included as 10 percent of the total cost of all materials and installation. Use your name and the current date as the name of the estimator and the date of the estimate.

Achievement Review

A. Complete the sentences below by inserting the most suitable word.

1. The price of the landscape is of concern to the client. The _____ of building it is of concern to the landscape contractor.

2. All _____ must be determined before a landscape contractor can add on his profit correctly.

3. The _____ of the landscape includes allowances for overhead, contingencies, and profit.

4. The cost of doing business is assigned as a percentage to every client's price estimate. It represents a general _____ cost.

5. The cost of temporary water and electrical hookups at a remote work site while a landscape is being built represent _____ overhead costs.

6. Cost = Materials + _____ + Equipment Use + _____

7. Price = _____ + Materials + Labor + Equipment Use + Overhead + _____ + Profit

B. Define the following terms.

1. estimate
2. bid
3. specifications

C. List the components of a typical landscape price estimate.

Landscape Contracting

SECTION

2

Landscape Calculations

Objectives:

Upon completion of this chapter, you should be able to

- ❦ describe the calculations needed for landscape take-offs
- ❦ explain the importance of organization and standardization when doing take-offs
- ❦ perform typical calculations required for take-offs

The Landscape Take-Off

As a landscape project progresses from the design stage to the point where it has received client acceptance and is ready to be built, it widens in scope and begins to involve additional landscape professionals. Creating substance from concept by turning symbols on the plan into numbers on an order form is the responsibility of the landscape contractor. The calculation of quantities from plans and specifications is termed the **take-off**.

Take-offs are done at different stages during the development of a landscape project, and each is done with different levels of refinement. Partial take-offs may be done by the landscape designer or landscape architect while the design is being developed as a means of checking to assure that the project is staying within the allotted budget. Approximate take-offs may be done by contractors who are asked to give an estimate of the cost or time required to construct a project. A final take-off is an exact calculation of quantities and time required to bring the design off the drawing board and into the ground.

Take-offs are also done by landscape management firms seeking to determine the quantities of time and materials required to maintain a landscape for a specified length of time to the standards prescribed by a client.

The calculation of material quantities and time requirements needed to install a design or maintain an existing landscape is necessary so the landscape company can (1) determine its costs, and (2) set a price for the client. Only after material quantities are determined can the company seek the best prices and sources of required materials. Only after installation or maintenance techniques are known can the time requirements be determined. Only after time requirements are calculated can the labor needs be measured and labor costs ascertained.

The Necessary Calculations

The individual usually assigned to do the landscape take-off is the **estimator**. Working with tools such as the scale, protractor, triangle, planimeter, calculator, graph paper, and/or the computer, the estimator counts, measures, and calculates every item needed for the landscape and every minute of time required as labor. Typical calculations made during a take-off include:

a. Unit counts   Examples: the number of each plant species, the number of bricks, the number of walk lights

b. Surface area calculations   Examples: the lawn area to be sodded, the size of the patio to be paved, the amount of flagstone pathway required

c. Volume calculations   Examples: the amount of soil displaced by the soil talls of the plants, the amount of concrete needed to pour the patio to desired thickness

d. Time calculations   Examples: the number of hours required to clear a woodlot, the number of hours required to build a stone wall

Sources of the Data

The information needed for the take-off comes to the landscape estimator from three sources: the drawings, the written specifications, and the landscape firm's records of crew performance capability. The drawings provide the estimator with much of the quantity data, including most of the dimensions. The written specifications may provide additional quantities as well as an explanation of desired qualities of materials, sources of supply, and information about special methods of installation or maintenance or frequency of performance. Anything that eventually translates into costs for the landscape company must be examined by the estimator and turned into a quantity that is measurable. The measurability of the data is important if cost and price figures are to be derived from it. Certain units of measurement are less precise than others and should be avoided whenever there is a more precise means of expressing the quantities required to complete a project. For example, it is less precise to list 100 bales of mulch than to list the need for 650 cubic yards. Bales vary in size and volume, so the actual quantity of material provided is imprecise, and the time required to spread it will be equally imprecise. However, if the plan calls for the mulch to be spread six inches deep, then the amount of coverage provided by 650 cubic yards is calculable.

From the calculation of quantities, the estimator must then determine the number of hours of labor required to install the materials or perform the task required. For tasks that the company's crews perform regularly, the estimator can make an accurate calculation of the number of hours required to accomplish the task. For tasks that are unfamiliar or unusual, the estimator will have no historical figures to draw upon and must calculate more cautiously.

Standardized Technique Is Important

A large landscape project, whether it is a construction project or a management project, can necessitate hundreds of calculations. To avoid mistakes, it is essential that the estimator establish a method for doing the take-off and then follow that method every time a take-off is done. In large companies, the method of doing take-offs becomes part of the corporate culture. Calculations are made the same way and in the same order every time, regardless of who does the take-off.

The reason for standardization is simple. Doing it the same way every time assures that nothing is overlooked, which could happen if the order or the method of counting and calculating was optional from project to project. Following a standard procedure also allows the estimator to establish stopping points along the way and then start again at a later time without making a mistake.

No one system for take-offs has industry-wide acceptance, nor does there need to be. Most companies need only to develop a system that works for them. Here is a typical sequence for making counts and calculations that might be used for the take-off of a landscape construction project.

Typical Sequence of Calculations for Landscape Take-Offs

I. Quantity Calculations
   A. Plant materials
      1. Make a list of all plants required
         a. List each separate species
         b. List the sizes required for each species
         c. List the quantities of each size required
      2. Make a list of all support materials required
         a. List the number of stakes
         b. List the specific guying materials
         c. List the quantities of guying materials
         d. List the quantities of tree coils and tree wraps
      3. Make a list of turf needs
         a. Describe each species or blend if more than one and whether it is to be applied as seed, sod, sprigs, or plugs
         b. Calculate the surface area(s) to be installed or renovated
   B. Materials associated with the plantings and turf
      1. Make a list of all mulches to be applied
      2. Make a list of all landscape fabrics to be used
      3. Calculate the surface area of the planting beds
      4. Calculate the volume(s) of mulch required
      5. Make a list of all soil amendments to be used
      6. Calculate the volumes of soil amendments required
      7. Calculate the volume of soil requiring disposal
      8. Make a list of all edging materials to be used
      9. Calculate the quantity of edging(s) required
   C. Hardscape materials
      1. Make a list of all hardscape materials required
         a. List and describe each separate material
         b. List and describe all associated support materials such as molds and forms, hardware, stains, etc.
      2. Calculate the quantities of each item required
      3. List and describe all prefabricated hardscape features to be provided, such as fountains, benches, lights
   D. Preliminary services
      1. List all services required to prepare the site for plant or hardscape installations
         a. List and describe each separate service, such as brush removal, excavation, or grading
         b. Calculate the areas and/or volumes affected

II. Time Calculations (These will be based upon specific crew compositions and equipment availabilities known to the estimator.)
   A. Layout times
      Calculate the time required to mark out the planting beds and other landscape features and place the plants
   B. Preparation times
      Calculate the time requirements for each preliminary service required
   C. Installation times
      1. Estimate the time required to cut out the planting beds
      2. Estimate the time required to prepare the soil for planting

3. Estimate the time required to install the plants

4. Estimate the time required for staking/guying

5. Estimate the time required for applying tree soils and wraps

6. Estimate the time required to spread landscape fabrics

7. Estimate the time required to install edgings

8. Estimate the time required to spread mulches

9. Estimate the time required for finishing details, such as pruning damaged wood and creating soil rings

10. Estimate the time required to install and/or repair lawn areas

11. Estimate the time required to clean up the site, including the time required for hauling away rubbish, excess soil

III. Maintenance Calculations

   A. Short-term requirements

     1. List all follow-up maintenance requirements such as watering, weeding, removing staking materials

     2. Estimate the materials and time required to meet the obligation

   B. Long-term requirements

   If the company obtains the contract for continuing maintenance of the property, a separate take-off will probably be done at this point

In this example, calculations of materials are grouped together, as are calculations of time. In other systems the materials, quantities, and labor times for each separate task, area, or phase of the project may be grouped together. Companies that use computerized estimating systems may have their time requirements already tied to certain tasks, so that when the task is described and measured, the calculation of labor hours required is instantaneous.

Once the take-off is completed, the estimator applies cost figures for materials, labor, and overhead to the numbers to arrive at a cost to the company. After that, the desired profit is added and a price to the client is finalized.

Typical Calculations Required for Take-Offs

Most of the calculations required for landscape take-offs use simple algebra, geometry, and old fashioned arithmetic. What follows are some examples and sample problems that illustrate how take-off calculations are made.

Linear and Surface Area Measurements

Table 16-1 contains the linear equivalents and formulas most commonly used to express lengths and measure surface areas.

Sample Problem 1:

In this example, what is the surface area of the patio? Of the lawn?

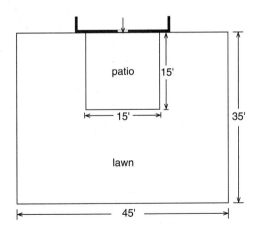

Solution:

I. The patio, since it has sides of equal length, is a square. Use the formula for the area of a square to find the area of the patio.

$$A = side \times side$$
$$A = 15\ LF \times 15\ LF$$
$$= 225\ SF$$

Table 16-1 Linear and Area Measurement

Units of Linear Measurement

12 inches	1 linear foot (LF)
3 linear feet	1 linear yard
1,760 yards	1 statute mile
5,280 feet	1 statute mile

Units of Area Measurement

144 sq. inches	1 sq. ft. (SF)
9 sq. ft.	1 sq. yd. (SY)
4,840 sq. yds.	1 sq. acre
43,560 sq. ft.	1 sq. acre
640 sq. acres	1 sq. mile

Formulas for Areas

Code

$\pi = 3.14$	$d =$ diameter	$\Theta =$ angle in degrees
$A =$ area	$c =$ circumference	$S =$ short radius of ellipse
$r =$ radius	$b =$ length of arc	$L =$ long radius of ellipse

Circumference and Area of a Circle

$$c = 2\pi r$$

$$A = \pi r^2$$

Area of a Sector

$$A = \frac{br}{2}$$

$$A = \pi r^2 \times \frac{\Theta}{360}$$

Circumference and Area of an Ellipse

$$c = 2\pi \sqrt{\frac{S^2 + L^2}{2}}$$

$$A = \pi SL$$

Area of a Square

$$A = \text{side} \times \text{side}$$

$$A = \text{side}^2$$

Area of a Rectangle

$$A = \text{length} \times \text{width}$$
$$A = \text{base} \times \text{height}$$

Area of a Triangle

$$A = \tfrac{1}{2}(\text{base} \times \text{height})$$
$$A = \tfrac{1}{2}(\text{length} \times \text{width})$$

[Note that area is always expressed in square units and that multiplication of any units is only possible when the units match. In this example, it would not be possible to multiply linear feet by linear yards or any other dissimilar unit without first converting one unit to match the other.]

II. The lawn and patio complex, since it has unequal sides, is a rectangle. (1) Use the formula for the area of a rectangle to find the area. (2) Then subtract the area of the patio to determine the remaining lawn area.

(1) A = 45 LF × 35 LF

  = 1,575 SF [the lawn/patio complex]

(2) 1,575 SF
  − 225 SF [the patio]
  1,350 SF [the area of the lawn]

Sample Problem 2:

What is the surface area of this patio?

Solution:

The patio is two connecting rectangles. One rectangle is 15 LF × 12 LF; the other is 5 LF × 3 LF. (1) Use the formula for the area of a rectangle to determine the two areas, then (2) add the areas together.

(1) 15 LF × 12 LF = 180 SF

    5 LF × 3 LF = 45 SF

(2) 180 SF + 45 SF = 225 SF [the area of the patio]

Sample Problem 3:

In this example, what is the surface area of the pool? Of the planting bed?

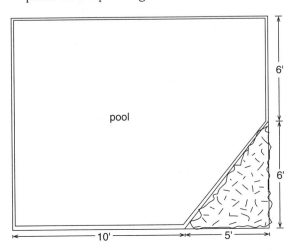

Solution:

I. If the planting bed did not exist, the pool would be a rectangle. Use the formula for the area of a rectangle to determine the area of the space before the planting bed is considered.

$$A = 15 \text{ LF} \times 12 \text{ LF}$$
$$= 180 \text{ SF}$$

II. The planting bed is a triangle. Use the formula for the area of a triangle to determine the area.

$$A = 1/2 \ (5 \text{ LF} \times 6 \text{ LF})$$
$$= 1/2 \ (30 \text{ SF})$$
$$= 15 \text{ SF [the area of the planting bed]}$$

III. Subtract the area of the planting bed from the larger area to determine the area of the pool.

$$180 \text{ SF} - 15 \text{ SF} = 165 \text{ SF [the area of the pool]}$$

Sample Problem 4:

What are the areas of flower beds A and B? How many feet of edging would be required to encircle flower beds A and B?

Flower Bed A

Flower Bed B

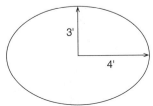

Solution:

I. Flower bed A is a circle. Use the formula for the area of a circle to determine the area. If the diameter is 8 LF, then the radius is half of that, 4 LF.

$$A = \pi r^2$$
$$= 3.14 \, (4 \, LF \times 4 \, LF)$$
$$= 3.14 \, (16 \, SF)$$
$$= 50.24 \, SF \, [\text{area of flower bed A}]$$

II. Flower bed B is an ellipse. Use the formula for the area of an ellipse to determine the area. If the short radius (S) is 3 LF and the long radius (L) is 4 LF, the formula is applied this way:

$$A = \pi SL$$
$$= 3.14 \, (3 \, LF \times 4 \, LF)$$
$$= 3.14 \, (12 \, SF)$$
$$= 37.68 \, SF \, [\text{area of flower bed B}]$$

III. Use the formula for the circumference of a circle to determine the length of edging required for flower bed A.

$$C = 2\pi r$$
$$= 2 \, (3.14 \times 4 \, LF)$$
$$= 2 \, (12.56 \, LF)$$
$$= 25.12 \, LF \, [\text{edging required for flower bed A}]$$

IV. Use the formula for the circumference of an ellipse to determine the amount of edging required for flower bed B.

$$C = 2\pi \sqrt{\frac{S^2 + L^2}{2}}$$
$$= (2 \times 3.14) \sqrt{\frac{S^2 + L^2}{2}}$$
$$= 6.28 \sqrt{\frac{9 + 16}{2}}$$
$$= 6.28 \sqrt{\frac{25}{2}}$$
$$= 6.28 \sqrt{12.5}$$
$$= (6.28) \, (3.54 \, LF)$$
$$= 22.23 \, LF \, [\text{edging required for flower bed B}]$$

Sample Problem 5:

What is the area of the deck surrounding the spa?

Solution:

I. The deck is a square. Use the formula for the area of a square to determine the area of the deck if the spa was not there.

$$A = \text{side} \times \text{side}$$
$$= 18 \, LF \times 18 \, LF$$
$$= 324 \, SF \, [\text{area of the deck/spa complex}]$$

II. The spa is a circle. Use the formula for the area of a circle to determine the area of the spa.

$$A = \pi r^2$$
$$= 3.14 \, (2 \, LF \times 2 \, LF)$$
$$= 3.14 \, (4 \, SF)$$
$$= 12.56 \, SF \, [\text{area of the spa}]$$

III. Subtract the area of the spa from the area of the deck/spa complex to determine the surface area of the deck surrounding the spa.

$$
\begin{array}{r}
324.00 \text{ SF} \\
- \quad 12.56 \text{ SF} \\
\hline
311.44 \text{ SF} \, [\text{area of the deck around the spa}]
\end{array}
$$

Sample Problem 6:

In this example, what are the surface areas of the brick patio, the deck, the planter, the red flowers, and the white flowers?

Answers:

For this problem, the answers are provided, but not the explanation of the solutions. See if you can attain these answers using the formulas from Table 16-1 and the methods explained in Sample Problems 1–5.

Area of the brick, including that beneath the deck:	116.86 SF
Area of the deck:	96.00 SF
Area of the planter:	25.00 SF
Area of the red flowers:	24.00 SF
Area of the white flowers:	24.00 SF

Volume Measurement

Table 16-2 contains the units of measurement and conversion equivalents most commonly used to calculate volumes of landscape materials.

The calculation of volume requires three dimen-

sions, two horizontal and one vertical, such as length, width, and thickness (or depth). A formula for volume calculation could be

$$\text{Volume} = \text{Length} \times \text{Width} \times \text{Depth}$$

Another way of expressing the same calculation would be

$$\text{Volume} = \text{Surface Area} \times \text{Depth}$$

This is because surface area has been previously calculated by multiplying the two sides or horizontal dimensions.

It has been noted before and is appropriate to restate here that these formulas will only work when all dimensions or measurements are expressed in the same units. Unlike units cannot be multiplied. It is not possible to multiply feet and inches or inches and yards until one has been converted to the equivalent of the other. In the sample problems that follow, some conversion equivalencies are illustrated. These are followed by several sample problems that require the calculation of typical landscape volumes.

Table 16-2 Volume Measurement

1,728 cubic inches	1 cubic foot (CF)
46,656 cubic inches	1 cubic yard (CY)
27 cubic feet	1 cubic yard (CY)
1,000 cubic yards	1 MCY*

*NOTE: M is an abbreviation for 1,000. Thus 3,000 CY could be expressed as 3 MCY.

Sample Problem 1:

Convert 4 CY to the equivalent number of CF.

Solution:

$$4 \times 27 = 108 \text{ CF}$$

Sample Problem 2:

Convert 405 CF to the equivalent number of CY.

Solution:

$$405 \div 27 = 15 \text{ CY}$$

Sample Problem 3:

Convert 8,550 CY to the equivalent number of MCY.

Solution:

$$8,550 \div 1,000 = 8.55 \text{ MCY}$$

Sample Problem 4:

A patio is 20 LF by 18 LF. It is to be paved with concrete 4 inches thick. What volume of concrete is required?

Solution:

I. The first calculation must be the conversion of unlike units to a common unit. Since two of the dimensions are already in linear feet and since the volume of concrete is more commonly expressed in cubic feet than cubic inches, it is logical to convert the 4-inch thickness of the concrete to an equivalent number of linear feet (See Table 16-1).

$$4 \text{ inches} \div 12 \text{ inches} = 0.33 \text{ LF}$$

II. With all three dimensions now in the same units of measurement, the volume can be calculated.

$$\begin{aligned} \text{Volume} &= 20 \text{ LF} \times 18 \text{ LF} \times 0.33 \text{ LF} \\ &= 118.8 \text{ CF of concrete} \end{aligned}$$

Sample Problem 5:

A lawn requires a layer of topsoil to be spread over it prior to a new seeding. The lawn is 80 LF by 74 LF. The topsoil is to be 3 inches thick. How many CY of topsoil are required?

Solution:

I. Converting from one unit to another introduces the probability of whole numbers becoming fractions. In turn, when partial numbers are multiplied or divided, the resulting answer is usually rounded off, and that introduces inaccuracies. Therefore, when making conversions during calculations that allow several options, as in this problem, it is best to select the option that results in the fewest partial numbers being entered into the calculations.

Begin by converting the three inches to linear feet so that all three dimensions have the same units.

$$3 \text{ inches} \div 12 \text{ inches} = 0.25 \text{ LF}$$

II. Next, determine the volume of topsoil in cubic feet.

$$\begin{aligned} \text{Volume} &= 80 \text{ LF} \times 74 \text{ LF} \times 0.25 \text{ LF} \\ &= 1,480 \text{ CF} \end{aligned}$$

III. Finally, while the volume of topsoil is correctly calculated, it is not expressed in units of cubic yards, so the 1,480 CF must be converted to CY.

$$1,480 \text{ CF} \div 27 \text{ CF} = 54.81 \text{ CY of topsoil}$$

Sample Problem 6:

A swimming pool is 30 LF × 20 LF × 5 LF. It rests on a bed of gravel, which is 4 LF deep, and extends outside the area of the pool by 3 LF in all directions. How many cubic yards of gravel will be required?

Answer:

The answer is 138.67 CY of gravel required. Using Table 16-2 and the methods illustrated in the previous sample problems, work through this problem until you arrive at the same answer.

Time Calculations

The unit of measurement for time is hours. Labor costs are calculated on the basis of how many units of material or service can be accomplished per hour by a person or crew of known composition and capability, using prescribed equipment and techniques. In the sample problems that follow, various methods of calculating time are illustrated.

Sample Problem 1:

If a landscape crew can install 400 SY of sod in an 8-hour day, how many hours will be required to install 280 SY?

Solution:

The solution requires an algebraic comparison, with X representing the number of hours required to install the sod.

$$\text{I.} \quad \frac{400 \text{ SY}}{8 \text{ hrs.}} = \frac{280 \text{ SY}}{X \text{ hrs.}}$$

$$\text{II.} \quad 400\,X = (280)\,(8)$$
$$= 2{,}240$$

$$\text{III.} \quad X = 2{,}240 \div 400$$
$$= 5.6 \text{ hrs. required}$$
$$\text{to install the sod}$$

Sample Problem 2:

If the crew can clear by hand 620 SY of light density brush in an 8-hour day, how many hours will be required to clear 1 square acre of such brush?

Solution:

Although the numbers are larger, the solution is the same as in Problem 1. Establish a comparison between what is known and what is sought. Consult Table 16-1 for the equivalency between SY and SA.

$$\text{I.} \quad \frac{620 \text{ SY}}{8 \text{ hrs.}} = \frac{4{,}840 \text{ SY}}{X \text{ hrs.}}$$

$$\text{II.} \quad 620\,X = (8)\,(4{,}840)$$
$$= 38{,}720$$

$$\text{III.} \quad X = 38{,}720 \div 620$$
$$= 62.45 \text{ hrs. required}$$
$$\text{to clear the brush}$$

Sample Problem 3:

Flower plantings of 6 MSF can be installed at the rate of 75 plants per hour when spaced 1 foot apart. If the contract requires the flower plantings to be rotated three times per season, how many total hours will be spent installing flowers?

Solution:

I. If spaced 1 foot apart, one installation of flowers will require that 6,000 plants be installed.

$$6 \text{ MSF} = 6{,}000 \text{ SF}$$
$$6{,}000 \text{ SF @ 1 plant/SF} = 6{,}000 \text{ plants}$$

II. If there are three rotations per season, then the total number of flowers to be installed for the year needs to be calculated.

$$6{,}000 \text{ plants per rotation} \times 3 \text{ rotations} =$$
$$18{,}000 \text{ plants}$$

III. Finally, establish the algebraic comparison of what is known to what is sought. As before, X represents the unknown.

$$\text{A.} \quad \frac{75 \text{ plants}}{1 \text{ hr.}} = \frac{18{,}000 \text{ plants}}{X \text{ hrs.}}$$

$$\text{B.} \quad 75\,X = (1)\,(18{,}000)$$
$$= 18{,}000$$

$$\text{C.} \quad X = 18{,}000 \div 75$$
$$= 240 \text{ hrs. required}$$
$$\text{to install flowers}$$

Sample Problem 4:

A lawn of 30 MSF is mowed with two different pieces of equipment. 26 MSF of the lawn is cut using a 60-inch wide riding mower that travels at 8 mph. The remainder of the lawn is cut with a 36-inch wide walking mower that averages 2 mph. How much time is required for one mowing of the lawn?

Solution:

I. To begin, it is necessary to envision the lawn being reshaped into two pieces. One piece will be 5 LF (60 inches) wide, and the other will be 3 LF (36 inches) wide. Both pieces are of indeterminate length, which must be calculated.

II. To calculate the length of the first piece, divide the known area [26,000 SF] by the known width [5 LF] to determine the unknown length.

$$26{,}000 \text{ SF} \div 5 \text{ LF} = 5{,}200 \text{ LF}$$
$$\text{to be cut by the riding mower}$$

III. Calculate the length of the second piece by dividing the known area [4,000 SF] by the known width [3 LF] to determine the unknown length.

$$4,000 \text{ SF} \div 3 \text{ LF} = 1,333 \text{ LF to be cut by}$$
the walking mower

IV. Next, convert each length to miles. See Table 16-1.

$$5,200 \text{ LF} \div 5,280 \text{ LF} = .98 \text{ miles to be}$$
cut by the riding
mower

$$1,333 \text{ LF} \div 5,280 \text{ LF} = .25 \text{ miles to be}$$
cut by the walking
mower

V. Next, set up the algebraic comparisons of what the mowers can cut per hour to what must be cut in each area. That permits a determination of the number of hours required by each machine to cut its section of the lawn.

A. The riding mower
$$\frac{8 \text{ miles}}{1 \text{ hr.}} = \frac{.98 \text{ miles}}{X \text{ hrs.}}$$

$$8X = (1)(.98)$$
$$= .98$$
$$X = .98 \div 8$$
$$= .12 \text{ hr. to cut 26 MSF once}$$

B. The walking mower
$$\frac{2 \text{ miles}}{1 \text{ hr.}} = \frac{.25 \text{ miles}}{X \text{ hrs.}}$$

$$2X = (1)(.25)$$
$$= .25$$
$$X = .25 \div 2$$
$$= .13 \text{ hr. to cut 4 MSF once}$$

VI. Sum the time requirements for the two sections of the lawn to determine the total time requirements for mowing the lawn once.

$$.12 \text{ hr.} + .13 \text{ hr.} = .25 \text{ hr. required}$$
per mowing

Sample Problem 5:

If in the previous problem the lawn was so cluttered with trees, poles, and other obstacles that the mowers have only 70 percent efficiency, how much time is required for a single cutting?

Answer:

The reduced efficiency applies to the rate of speed of the mowers. It does not affect the amount of lawn requiring mowing. Although capable of operating at 8 mph on perfectly flat, open land, the riding mower will instead average 5.6 mph (8 mph × 0.70), and the walking mower will average 1.4 mph (2 mph × .70). Using these figures, redo problem 4 and see if you can derive these answers:

The riding mower will require 0.18 hr. to cut 26 MSF of lawn.

The walking mower will require 0.18 hr. to cut 4 MSF of lawn.

The total time required for one complete mowing is 0.36 hr.

Quantity Calculations and Conversions

Other calculations often required during landscape take-offs include the determination of quantities needed to meet the specifications of a contract. The sample problems and solutions that follow illustrate how quantity calculations are made.

Sample Problem 1:

If the recommended rate of application for grass seed is 2 lbs./MSF, what quantity of seed is required to establish a new lawn of 15,000 SF?

Solution:

I. Convert 15,000 SF to its equivalent in MSF units. (See Table 16-1.)

$$15,000 \text{ SF} \div 1,000 \text{ SF} = 15 \text{ MSF}$$

II. Multiply the rate of application for 1 MSF by the total number of units (15).

$$2 \text{ lbs./MSF} \times 15 = 30 \text{ pounds of seed}$$
required

Sample Problem 2:

If the rate of application for lawn fertilizer is 4 lbs. of nitrogen per thousand square feet, how many pounds of fertilizer with a 10-6-4 analysis are required to fertilize 20,000 SF?

Solution:

I. Calculate the number of MSF units needing fertilization.

20,000 SF ÷ 1,000 SF = 20 MSF of lawn to be fertilized

II. Calculate the amount of actual nitrogen (N) in one pound of 10-6-4 fertilizer.

1 lb. of fertilizer × .10 = 0.10 lb. of actual N

III. Calculate the number of pounds of 10-6-4 fertilizer required to provide 4 lbs. of actual nitrogen/MSF.

4 lbs. (needed) ÷ 0.10 lb. = 40 lbs. of 10-6-4/MSF needed

IV. Calculate the total number of pounds of fertilizer required to apply 4 lbs. of N/MSF to 20 MSF.

40 lbs. of fertilizer/MSF × 20 = 800 lbs. of 10-6-4 fertilizer

Sample Problem 3:

In the above problem, how many pounds of 10-6-4 are required to fertilize a five-acre lawn?

Solution:

I. Changing the size of the area to be fertilized does not alter the amount of elemental nitrogen in the fertilizer. It still requires 40 lbs. of 10-6-4 to deliver 4 lbs. of elemental N to 1,000 SF of turf. What does have to be determined are the number of MSF units in five acres. (See Table 16-1.)

A. 43,560 SF = 1 square acre
B. 43,560 SF × 5 = The number of SF in five acres
= 217,800 SF

C. $\frac{217,800 \text{ SF}}{1,000 \text{ SF}}$ = The number of MSF units in five acres
= 217.8 MSF to be fertilized

II. Multiply the total number of MSF units to be fertilized by the amount of fertilizer required per MSF to determine the total requirement for the five acres.

217.8 × 40 lbs./MSF = 8,712 lbs. of 10-6-4 fertilizer

Sample Problem 4:

Concrete dye is mixed into concrete at the rate of 4 ozs. per 100 CF. How many ounces are required to color 50 CY?

Solution:

I. Determine the number of cubic feet in fifty cubic yards. (See Table 16-2.)

27 CF/CY × 50 CY = 1,350 CF

II. Make an algebraic comparison between the quantity known and the quantity being sought. As before, X represents the unknown amount.

A. $\frac{4 \text{ ozs.}}{100 \text{ CF}} = \frac{X \text{ ozs.}}{1,350 \text{ CF}}$
B. $100X = 5,400$
C. $X = 54$ ozs. of dye required

Sample Problem 5:

An 800 SF concrete patio, 4 inches thick, is to be poured over 6 inches of crushed stone subsurface. How many CY of concrete are required and how many CY of crushed stone are required?

Solution:

I. The amounts of concrete and crushed stone must be calculated from the known surface area and depth dimensions. To do that requires that the units of measurements be converted to match. The simplest conversion is to turn the 4-inch and 6-inch measurements into equivalent LF units.

A. 4 inches ÷ 12 inches = 0.33 LF of concrete

B. 6 inches ÷ 12 inches = 0.50 LF of crushed stone

II. Multiplying the surface area by the thicknesses of each material will produce the quantities required. The unit of measurement will be CF.

A. 800 SF × 0.33 LF = 264 CF of concrete required

B. 800 SF × 0.50 LF = 400 CF of crushed stone required

III. Convert the volumes from cubic feet to cubic yards.

A. 264 CF ÷ 27 CF = 9.78 CY of concrete required

B. 400 CF ÷ 27 CF = 14.81 CY of crushed stone required

Sample Problem 6:

A planter is 5 LF in length, 3 LF wide, and 4 LF deep. It is to be filled with a soil mix that is 60 percent actual soil and 40 percent conditioners. The planter will be filled to a depth of 3 1/2 LF with the conditioned soil; then 6 inches of mulch will be added. How many cubic yards of actual soil will be required to fill the planter? How many cubic yards of mulch?

Answers:

The answers are provided below, but not the method of deriving them. Based upon the calculations shown and explained in the above sample problems, solve this problem and see if your answers match these.

The amount of actual soil is 1.17 CY.

The amount of mulch is 0.77 CY.

Soil Compaction Calculations

Soil occupies different volumes of space depending upon whether it is in a natural, undisturbed state or has been loosened by digging, thereby adding more air to the soil's composition. It also changes volume when air is driven out by compaction, as when compressed by heavy traffic. The degree of expansion or compaction in comparison to the natural, undisturbed state is dependent upon the type of soil. The more clay particles there are in the soil, the greater will be the expansion and compaction potential of the soil. The more sand particles the soil contains, the less will be the difference in the soil's volume as it changes from the undisturbed state to the expanded or contracted state.

While the differences between a soil's volume as it changes from one state to another is of little concern when installing only a few plants, it becomes a greater concern on large projects. When large quantities of soil are disturbed through grading or extensive construction, the landscape contractor may need to provide for the acquisition or the removal of large amounts of soil. As with every other task, handling material translates into time and money. Table 16-3 compares the volumes of different soil types in three different states. Undisturbed field soil is said to be *banked*. The unit of measurement is BCY (Bank Cubic Yard). When disturbed and mixed with additional air, the volume is expressed as LCY (Loose Cubic Yard). If compressed and the amount of air in the soil reduced, the volume measurement is termed CCY (Compact Cubic Yard).

Table 16-3 A Comparison of Soil Expansion and Compaction			
Soil Type	**As a BCY**	**As a CCY**	**As a LCY**
Clay	1	0.85	1.25
Clay loam	1	0.91	1.24
Clay/gravel mix	1	0.95	1.18
Loam	1	0.95	1.23
Sand	1	0.98	1.12
Sandy loam	1	0.97	1.20
Sandy clay	1	0.87	1.27
Sand and gravel	1	0.99	1.11
Gravel	1	0.98	1.12

Sample Problem 1:

To what volume would 5 BCY of loam expand after digging?

Solution:

Multiply the number of BCY of loam soil by the numerical equivalent for 1 LCY of comparable soil.

5 BCY × 1.23 = 6.15 LCY of loam

Sample Problem 2:

To what volume would 8 BCY of sandy clay expand after digging?

Solution:

8 BCY × 1.27 = 10.16 LCY of sandy clay

Sample Problem 3:

To what volume would 15 BCY of clay and gravel compress after compaction?

Solution:

15 BCY × 0.95 = 14.25 CCY of clay/gravel mix

Sample Problem 4:

To what volume would 6 LCY of loam compress after compaction?

Solution:

I. Convert 6 LCY to the equivalent number of BCYs by dividing by the expansion equivalent for 1 LCY of loam.

6 LCY of loam ÷ 1.23 = 4.88 BCY of loam

II. Convert 4.88 BCY of loam to the equivalent number of CCYs.

4.88 BCY × 0.95 = 4.64 CCY of loam soil

Sample Problem 5:

To what volume could 7 CCY of sandy loam expand?

Answer:

Using the methods shown in the previous problems, see if you can arrive at the correct answer.

The answer is 8.66 LCY of sandy loam soil.

Pit and Ball Calculations

Estimators must further deal with soil volume calculations when large numbers of plant materials are installed. A certain volume of soil is removed when the planting pit is dug. A different volume of soil, contained within the new plant's root ball, is placed into the pit. That creates two additional volumes: the volume of material required to backfill the pit around the soil ball, and the volume of soil that cannot fit back into the pit and becomes waste. Table 16-4 contains the formulas that explain how these various volumes are determined.

Sample Problem 1:

A planting pit has a volume of 10.6 CF. The plant being installed has a soil ball volume of 5.2 CF. What is the volume of soil in the backfill, and what volume of waste must be disposed of?

Solution:

I. Determine the volume of backfill using the formula from Table 16-4.

PV − BV = BF

10.6 CF − 5.2 CF = 5.4 CF of backfill

II. Determine the volume of waste using the formula from Table 16-4.

BV + SA = WV

Since there are no solid soil additives in the backfill, SA has a value of 0. The calculation becomes

5.2 CF + 0 = 5.2 CF of waste

Table 16-4 Pit and Ball Volume Calculations

Key	Formulas
PV = Pit volume	PV − BV = BF
BV = Ball volume	
BF = Backfill volume	BV + SA = WV
SA = Solid soil additives	
WV = Waste volume	

When there is no replacement of the original pit soil with solid additives such as peat moss or sand, then the waste volume is equivalent to the volume of the soil ball.

Sample Problem 2:

Four shrubs are to be installed. Each plant has a ball volume of 8.9 CF. Each planting pit has a volume of 16.0 CF. What is the total volume of soil to be excavated? The total ball volume? The total backfill volume? The total waste volume?

Solution:

I. Multiplying each known volume by 4 provides necessary quantity information.

  A. Total ball volume = 8.9 × 4 = 35.6 CF

  B. Total pit volume = 16.0 × 4 = 64.0 CF

II. Applying the formulas from Table 16-4 generates the remaining volumes.

  C. Total backfill volume =
    64.0 CF − 35.6 CF = 28.4 CF

  B. Total waste volume =
    35.6 CF + 0 CF = 35.6 CF

Sample Problem 3:

Six trees, each with a soil ball volume of 14.0 CF, are being installed in pits, each with a volume of 23.8 CF. The backfill mix is to be 50 percent original soil and 50 percent sand/peat mix. What is the total volume of soil to be excavated? Total ball volume? Total backfill volume? Total volume of soil in the backfill? Total volume of solid soil additives? Total volume of waste requiring disposal?

Solution:

  A. Total pit volume = 23.8 CF × 6 = 142.8 CF

  B. Total ball volume = 14.0 CF × 6 = 84.0 CF

  C. Total backfill volume =
    142.8 CF − 84.0 CF = 58.8 CF

  D. Total volume of soil
    in the backfill = 50 percent of total BF
    58.8 CF × .50 = 29.4 CF

  E. Total volume of soil additives =
    50 percent of total BF
    58.8 CF × .50 = 29.4 CF

  F. Total volume of waste =
    84.0 CF + 29.4 CF = 113.4 CF

There are two means of determining the volume of soil removed from planting pits and contained within plant root balls. Both methods are based upon the diameter and depth of individual pits and balls. One method uses formulas, while the other method is a chart derived from the formulas, Table 16-5. When the pit and ball dimensions fall within the chart, it is time-saving to use it. When the dimensions of either the pit or the ball are not on the chart, then the formulas can be used to calculate the volumes.

$$\text{Volume of the soil ball in CF} =$$
$$\left(\text{ball diameter in LF}\right)^2 \times \left(\text{ball depth in LF} \times \frac{2}{3}\right)$$

$$\text{Volume of the pit ball in CF} =$$
$$\frac{\pi \times \left(\text{pit diameter in LF}\right)^2}{4} \times \text{pit depth in LF}$$

Sample Problem 4:

Using Table 16-5, determine the total volume of seven planting pits, each one 5 LF in diameter and 3 LF deep. Determine the total ball volume for seven plants, each with root balls with diameters of 4 1/2 LF and depths of 2 3/4 LF. Finally, calculate the total backfill volume and the total soil waste.

Solution:

I. Consulting Table 16-5 reveals that the volume of one 5' × 3' pit is 70.6 CF. The volume of one 4 1/2' × 2 3/4' soil ball is 38.1 CF.

II. Multiplying each volume by the total number of pits and balls (seven each) gives the total pit volume and total ball volume for the planting.

  A. Total pit volume = 70.6 CF × 7 pits
    = 494.2 CF

B. Total ball volume = 38.1 CF × 7 balls
= 266.7 CF

C. Total backfill = 494.2 CF − 266.7 CF
= 227.5 CF

D. Total soil waste = 266.7 CF + 0 CF
= 266.7 CF

Sample Problem 5:

Use the formulas to calculate the volume of a planting pit that is 8 LF in diameter and 4 LF in depth. Then calculate the volume of a soil ball that is 7 3/4 LF in diameter and 3 3/4 LF deep.

Answer:

The pit volume is 200.96 CF. The ball volume is 148.95.

You can test your understanding of the formulas by creating sample problems of your own using dimensions that are in the chart and seeing if you calculate the same volumes as are shown in the chart.

Achievement Review

A. List the four different types of measurements and calculations usually required for landscape take-offs.

B. Indicate which of the following units are measurements of

(A) Unit counts   (C) Volume
(B) Surface area   (D) Time

1. square yards
2. square feet
3. hours
4. each
5. cubic yards
6. miles
7. cubic inches
8. mins/MSF

C. Why is it important that the take-off be done the same way each time?

D. Do the following calculations using appropriate tables from the chapter.

1. A 12 LF × 15 LF concrete patio is to be built at grade level. One 15' side is against the house. The remaining three sides are to be edged with standard brick. Each brick is 8 inches in length. How many bricks will be required to edge the patio?

2. What is the surface area of the patio?

3. If the concrete for the patio is 5 inches thick and it is poured over a subsurface of crushed stone that is 7 inches thick, how many cubic yards of concrete and crushed stone are required?

4. If a crew can excavate soil at the rate of 26 BCY/hr., dump and spread crushed stone at the rate of 12.5 SY/hr., then pour and finish concrete at the rate of 125 SF/hr., how much time will be required to accomplish all of these tasks for this patio?

5. Twenty trees, each with a root ball 3 1/2 LF in diameter and 3 LF deep, are to be installed into pits that are 4 LF in diameter and 3 1/4 LF deep. The pits are to be backfilled with original soil. How much soil will need to be disposed of?

Suggested Activity

Obtain a set of plans and specifications from a local landscape contractor and prepare a take-off. Follow the sequence for calculations outlined in this chapter.

Table 16-5 Planting Pits and Root Balls (Cubic Feet per Plant)

Diameters	1'	1 1/4'	1 1/2'	1 3/4'	2'	2 1/4'	2 1/2'	2 3/4'	3'	3 1/4'	3 1/2'
Depths											
Pit 1'	.94	1.47	2.13	2.88	3.77	4.78	5.89	7.13	8.48	9.96	11.5
Ball	.68	1.07	1.54	2.10	2.73	3.42	4.30	5.20	6.19	7.38	8.4
Pit 1 1/4'	1.16	1.85	2.65	3.60	4.71	5.93	7.37	8.90	10.6	12.4	14.4
Ball	.85	1.35	1.93	2.63	3.42	4.29	5.36	6.49	7.7	9.2	10.5
Pit 1 1/2'	1.40	2.22	3.18	4.32	5.65	7.16	8.83	10.7	12.7	15.0	17.3
Ball	1.02	1.62	2.32	3.15	4.10	5.19	6.39	7.8	9.2	10.9	12.7
Pit 1 3/4'	1.65	2.58	3.70	5.04	6.60	8.20	10.3	12.5	14.8	17.4	20.2
Ball	1.20	1.88	2.72	3.68	4.78	6.03	7.5	9.1	10.7	12.7	14.7
Pit 2'	1.87	2.95	4.26	5.76	7.54	9.55	11.8	14.3	17.0	19.9	23.0
Ball	1.38	2.15	3.10	4.20	5.49	6.92	8.5	10.4	12.3	14.5	18.8
Pit 2 1/4'	2.10	3.32	4.78	6.48	8.48	10.7	13.2	16.0	19.1	22.4	26.0
Ball	1.55	2.43	3.48	4.73	6.19	7.8	9.6	11.7	13.8	16.4	18.9
Pit 2 1/2'	2.34	3.69	5.30	7.20	9.42	11.9	14.7	17.8	21.2	24.9	28.9
Ball	1.70	2.70	3.87	5.25	6.89	8.7	10.7	13.0	15.4	18.2	21.0
Pit 2 3/4'	2.57	4.06	5.83	7.92	10.3	13.1	16.2	19.6	23.3	27.4	31.8
Ball	1.87	2.98	4.23	5.77	7.6	9.6	11.8	14.3	17.0	20.0	23.1
Pit 3'	2.81	4.43	6.37	8.64	11.3	14.3	17.7	21.4	25.4	29.9	34.6
Ball	2.05	3.24	4.65	6.30	8.3	10.5	12.9	15.6	18.6	21.8	25.2
Pit 3 1/4'	3.00	4.80	6.90	9.36	12.2	15.5	19.2	23.2	27.5	32.4	37.5
Ball	2.21	3.50	5.03	6.83	8.9	11.4	14.0	16.9	20.1	23.6	27.3
Pit 3 1/2'	3.28	5.17	7.41	10.1	13.2	16.7	20.7	25.0	29.6	34.9	40.4
Ball	2.39	3.77	5.42	7.4	9.6	12.2	15.0	18.2	21.7	25.4	29.4
Pit 3 3/4'	3.50	5.53	7.96	10.8	14.2	17.9	22.2	26.7	31.7	37.1	43.3
Ball	2.56	4.04	5.80	7.9	10.3	13.1	16.1	19.5	23.2	27.2	31.5
Pit 4'	3.74	5.90	8.50	11.5	15.1	19.1	23.7	28.5	33.8	39.9	40.2
Ball	2.73	4.31	6.19	8.4	11.0	14.0	17.2	20.8	24.8	29.0	33.6
Pit 4 1/4'	3.97	6.28	9.01	12.3	16.0	20.3	25.1	30.3	36.0	42.3	49.1
Ball	2.90	4.58	6.58	8.9	11.8	14.8	18.3	22.1	26.3	30.8	35.7
Pit 4 1/2'	4.20	6.64	9.55	13.0	17.0	21.5	26.6	32.0	38.1	44.8	52.0
Ball	3.07	4.85	6.97	9.5	12.4	15.7	19.4	23.4	27.8	32.6	37.8
Pit 4 3/4'	4.43	7.00	10.0	13.7	17.9	22.7	28.0	33.8	40.2	47.3	54.9
Ball	3.24	5.12	7.4	10.0	13.0	16.5	20.4	24.7	29.1	34.2	40.0
Pit 5'	4.68	7.38	10.6	14.4	18.8	23.9	29.5	35.6	42.3	49.9	57.7
Ball	3.41	5.38	7.8	10.5	13.7	17.4	21.5	26.0	30.9	36.1	42.1
Pit 5 1/2'	5.14	8.12	11.7	15.8	20.7	26.3	32.4	39.2	46.7	54.8	63.5
Ball	3.76	5.93	8.5	11.6	15.1	19.1	23.7	28.6	34.0	40.0	46.5
Pit 6'	5.61	8.86	12.7	17.3	22.6	28.7	35.4	42.8	50.9	57.6	68.3
Ball	4.10	6.46	9.3	12.6	16.5	20.9	25.9	31.2	37.1	43.6	50.5

Diameters	3 3/4'	4'	4 1/4'	4 1/2'	4 3/4'	5'	5 1/4'	6'	6 1/2'	7'	7 1/2'
Depths											
Pit 1'	13.3	15.1	17.0	19.1	21.2	23.6	28.5	33.9	39.8	46.2	53.0
Ball	9.7	14.0	12.4	13.9	15.5	17.4	20.8	24.7	29.0	33.7	43.7
Pit 1 1/4'	16.6	18.8	21.3	23.8	26.5	29.5	35.4	42.4	49.8	57.7	66.3
Ball	12.2	13.8	15.5	17.4	19.4	21.7	26.0	30.9	36.3	42.1	43.4
Pit 1 1/2'	19.9	22.6	25.5	28.6	31.9	35.4	42.8	50.9	59.7	69.3	79.5
Ball	14.5	16.5	18.6	20.9	23.2	25.8	31.2	37.1	43.6	50.5	58.0
Pit 1 3/4'	23.2	28.4	29.7	33.4	37.2	41.2	49.9	59.4	69.7	80.8	92.6
Ball	16.9	19.1	21.7	24.4	27.1	30.1	36.4	43.3	50.8	58.9	67.7
Pit 2'	26.5	30.2	34.0	38.2	42.0	47.1	57.0	67.9	79.6	92.4	106.1
Ball	19.2	21.9	24.8	27.8	31.0	34.4	41.6	49.5	58.1	67.4	77.3
Pit 2 1/4'	29.8	33.9	38.3	43.0	47.8	53.0	64.1	76.4	89.6	103.8	119.3
Ball	21.6	24.6	27.9	31.3	34.9	39.7	46.8	55.7	65.3	75.8	87.0
Pit 2 1/2'	33.1	37.7	42.5	47.7	53.1	59.0	71.3	84.9	99.6	115.5	132.8
Ball	24.1	27.3	31.0	34.7	38.8	43.0	52.0	61.9	72.6	84.2	96.7
Pit 2 3/4'	36.5	41.5	46.8	52.5	58.5	64.8	78.4	93.3	109.5	127.0	145.8
Ball	26.4	30.0	34.1	38.1	42.6	47.3	57.2	68.0	79.9	92.6	106.3
Pit 3'	39.8	45.3	51.0	57.3	63.8	70.6	85.5	101.8	119.5	138.5	159.0
Ball	28.8	32.8	37.0	41.5	46.5	51.6	62.3	74.2	87.1	101.0	116.0
Pit 3 1/4'	43.1	49.0	55.3	62.0	69.1	76.5	92.6	110.3	129.4	150.1	172.3
Ball	31.4	35.7	40.1	45.0	50.4	55.9	67.5	80.4	94.4	109.5	125.6
Pit 3 1/2'	46.4	52.8	59.6	66.8	74.3	82.5	99.7	118.7	139.4	161.7	185.6
Ball	33.9	38.5	43.2	48.5	54.4	60.2	72.7	86.6	101.6	117.9	135.3
Pit 3 3/4'	49.7	56.6	63.8	71.6	79.7	88.4	106.8	127.3	149.3	173.2	198.8
Ball	36.4	41.2	46.3	52.2	58.1	64.4	77.5	92.3	108.7	126.3	145.0
Pit 4'	53.0	60.3	68.1	76.3	85.0	94.2	113.9	135.7	159.3	184.8	212.1
Ball	38.9	44.0	49.4	55.7	62.0	68.7	82.8	98.5	116.1	134.7	154.6
Pit 4 1/4'	56.3	64.1	72.3	81.1	90.3	100.1	121.0	144.2	169.2	196.3	222.3
Ball	41.4	46.8	52.5	59.2	65.9	73.0	88.1	104.8	123.0	142.4	163.4
Pit 4 1/2'	59.6	67.9	76.6	85.9	95.6	106.2	128.2	152.7	179.2	207.9	238.6
Ball	43.9	49.5	55.6	62.7	69.8	77.4	93.4	111.0	130.3	150.9	172.9
Pit 4 3/4'	63.0	71.6	80.9	90.7	101.0	112.0	135.3	161.2	189.1	219.4	251.9
Ball	46.0	52.3	58.8	66.2	73.7	81.7	98.6	117.2	137.6	159.7	182.4
Pit 5'	66.3	75.4	85.2	95.5	106.3	117.9	142.5	169.7	199.1	230.9	265.1
Ball	48.5	55.0	62.0	69.7	77.5	86.0	103.9	123.4	144.9	168.2	192.0
Pit 5 1/2'	72.9	83.0	93.6	105.2	116.9	129.6	156.8	186.7	219.0	254.0	291.6
Ball	53.2	60.5	68.3	76.6	85.2	94.6	114.2	135.9	159.6	184.6	211.0
Pit 6'	79.5	90.5	102.1	114.5	127.5	141.5	171.0	203.6	238.9	277.1	318.1
Ball	58.0	66.0	74.5	83.5	93.0	103.1	124.7	148.5	174.2	202.0	230.0

17

Understanding Contracts and Contractors

Objectives:

Upon completion of this chapter, you should be able to

- list and describe the components of a contract
- describe the different contractual associations that occur in the landscape industry
- explain the different ways that landscape contractors are selected for jobs

The Importance of Contracts

A contract is an agreement, usually between two parties, that describes certain services and/or materials that will be provided in return for an established payment or other compensation. There is nothing new or unusual about the concept of contracts or the definition given here. People have been making contracts for centuries. When a parent tells a child that he can go out and play after he eats everything on his plate, the conditions of a contract are specified. When two people wed, they do so in the presence of witnesses and simultaneously agree to love, honor and cherish ... for as long as they both shall live ... until death parts them. Thus, the conditions of the partnership are described: what each will do, for what length of time, and under what conditions the agreement will be ended. Perhaps the simplest and most traditional of all contracts has been the handshake,

often referred to as a "gentlemen's agreement." By looking each other in the eye and agreeing to honor conditions expressed orally, the handshake contract has been accepted and honored by many societies for a long time. It is an important part of business today.

Still other types of contracts that are entered into by most people at one or more times during their lives are loans, credit card agreements, labor agreements, buying a car or home, renting, and hundreds of other arrangements wherein some item or service is provided in return for payment. Take a minute to analyze the last major purchase that you made that required borrowing money. How many different parties were involved? You and a bank, parent, or friend who provided the money, and the supplier of the item purchased were all parties in the transaction. What were the conditions for payment to the supplier and for repayment to the lender? How many separate agreements were made before the transaction was complete?

Contracts are part of the way we live, play, and do business today. Our challenge is not merely to begin the study of contractual relationships, but rather to understand more fully what we are already involved with. In particular, it is important to understand the importance and uses of contracts within the landscape industry. Although the handshake and verbal commitment are still part of the landscape profession, it is most common and most wise to put the agreement between parties into written form. Thus, most modern contracts are written to assure mutual understanding of the terms of the contract. Written contracts are also less likely to result in litigation if they are carefully prepared and understood in advance by all parties.

The Components of a Contract

Contracts range in size and complexity from a single page to voluminous documents. Some are so simple they can be prepared on a standardized, pre-printed form, Figure 17-1. Others are so complex that a team of lawyers is needed to prepare and interpret them. Yet all contain certain features that are common to every contract.

a. Names and addresses of the parties involved
b. Date of contract preparation
c. Description of the work to be accomplished and the materials or services to be provided, including the location where it is to be accomplished
d. The terms of completion, including such things as inspection, the completion date, and the penalty for not completing on time
e. The terms of payment
f. The signatures of both parties or their legal representatives
g. The date of signing

Certain of the features can be extensive and complex, causing the contract to stretch across many pages and through numerous clauses. In addition, there may be drawings or other attachments to the signed documents that are also part of the contract. All of these will be described and discussed in later chapters.

Types of Contractual Associations

A landscape project may involve clients and/or their representatives, architects, engineers, landscape architects or designers, landscape construction firms, specialty construction firms, maintenance firms, and/or material suppliers. Numerous contracts must be negotiated to bind all parties together in a manner that will accomplish the project properly.

The **client** is the person(s) or organization that owns and provides financing for a project. Funding may be from the client's personal resources, or there may be another source. The **contractor** is a party in contract with the client or the client's representatives. The prime contractor or general contractor is the firm directly responsible to the client for construction of the project. Depending upon the project and the company's capabilities, the prime contractor may build the entire project. More often, the prime contractor retains full responsibility for management control and supervision of the project, but hires one or more specialty firms to construct portions of the project. Subcontractors are firms in contract with the prime contractor. They may do the electrical work, or the plumbing, or pool construction, or masonry, or countless other specialized construction tasks that the prime contractor is unable to do or could not do as economically as the subcontractor can do it. The subcontractor and prime contractor enter into a relationship known as a subcontract. The subcontractor is usually accountable to the general contractor, not the project owner. Responsibility for the satisfactory performance of the subcontractor on the owner's project is assumed by the general contractor. Figure 17-2 illustrates a typical prime contract and subcontract relationship for a residential landscape project.

In the example, Ajax Landscape Contractors could be a design-build firm. They would have prepared a landscape plan, received approval from the owners of the property, and entered into a prime contractual relationship to build the landscape. They may use their crew to do the

CONTRACT

At Your Service Landscape Company
Your Full Service Landscape Professionals
1515 West Roosevelt Parkway   Lincoln, Nebraska 12043
518 234-5646

Client Name _____ Telephone _____ Date _____

Address _____ Project Title _____

_____ Project Location _____

We agree to provide the following services and materials on behalf of the client and project identified above:

WE AGREE to furnish the material and labor, complete and in accordance with the above specifications and with the attached guarantees for the sum of

_____ dollars ($_____)

Payment is to be made as follows: _____

_____

Authorized Signature for the Company _____ Date _____

ACCEPTANCE: The above prices, descriptions of work, conditions, and terms of payment are satisfactory. You are authorized to do the work as specified.

Authorized Signature for the Client _____ Date _____

Figure 17-1   A simple one-page contract

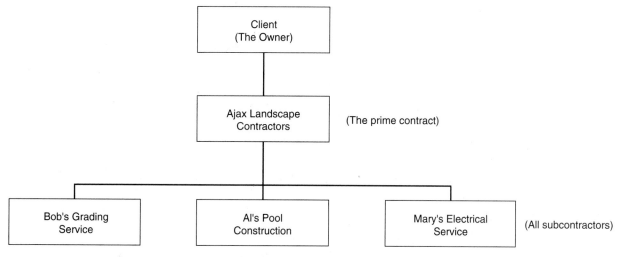

Figure 17-2 A typical prime contract and subcontractor relationship

finish grading and installation of the lawn and all plantings. All other construction work would be subcontracted to three specialty firms, who will be under contract to Ajax Landscape Contractors, not the homeowners. Therefore it will be the responsibility of Ajax, as the prime contractor, to assure that all work done by the subcontractors is coordinated to begin and end on schedule and without conflict among the various subcontractors. Ajax must supervise the work of the subcontractors to assure that the desired quality is obtained and that the client is satisfied. Ajax also bears responsibility for keeping the subcontractors on schedule so that the completion date that the client has specified is honored.

Figures 17-3 and 17-4 illustrate variations on the

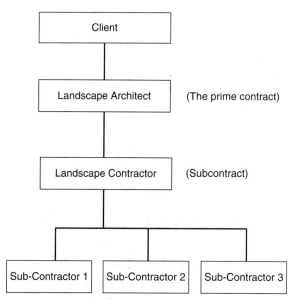

Figure 17-3 Contractual associations with a landscape architect holding the prime contract

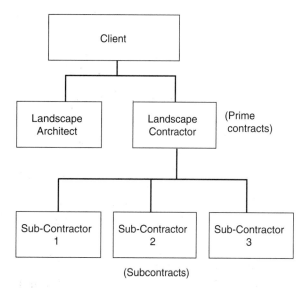

Figure 17-4 Contractual associations with a landscape architect and landscape contractor each holding contracts directly with the client

contractual relationships that are introduced by the presence of a landscape architect in the project.

The situation shown in Figure 17-3 establishes the landscape architect as the prime contract holder on the project and usually empowers the landscape architect as the owner's representative in all matters that involve the project and the subcontractors. As the principal professional on the project, the landscape architect has the responsibility for selecting the landscape contracting firm that will actually build the landscape. In addition to the drawings and renderings prepared by the landscape architect to help the client visualize the proposed design, the landscape architect must prepare detailed construction drawings and written specifications that cover every aspect of the project. Only in that way can subcontractors know exactly what is expected of them. Should questions or problems arise at any stage of the project, the subcontractor's direct line of appeal is to the landscape contractor or to the landscape architect, depending upon with which party they share their contract.

Figure 17-4 illustrates a situation typical of a larger landscape project in which the client contracts directly with both a landscape architect and a landscape contracting firm. The client may or may not have sought the assistance or advice of the landscape architect in selecting the landscape contractor. Often in this type of contractual relationship, the landscape architect may still be designated as the client's representative in dealing with the landscape contractor when the client lacks the time or expertise to deal with technical matters. However, the direct responsibility of both professionals is to the client.

On a very large construction project, the landscaping may be only one part of the work being designed and implemented. There may be an interdisciplinary team of professionals responsible for the project. Figure 17-5 diagrams such a project and the contractual relationships that they contain.

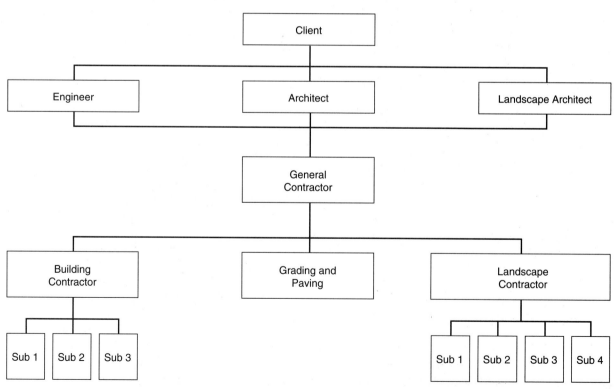

Figure 17-5 Contractual relationships for a large commercial project

As complex as the example appears, an actual construction project can be even more so. Yet as long as every participant fulfills the commitment of each contract and subcontract, most projects get built to the satisfaction of all parties.

The contracts that bind the client to the interdisciplinary team are all separate. Still, one professional is usually designated as the overall coordinator of the project to assure that the work progresses in a logical sequence with minimal confusion and congestion at the site. The general contractor may or may not be a landscape firm. On a large project, the general contractor is likely to be the coordinating overseer of the subcontractors, maintaining the schedule and quality standards established by the client's professional team. It is the responsibility of the various construction specialists and their subcontracted firms and suppliers to move on and off the site, do their work, and depart in time for the next subcontractor to begin work. The need for communication and frequent consultation among all participants in the construction project is obvious.

How Landscape Contractors Are Selected

There are three ways that landscape contractors can obtain work:

1. Direct solicitation of clients
2. Selection by the client for noncompetitive reasons
3. Selection by the client through the competitive bidding process

Direct Solicitation

Rather than waiting for the telephone to ring or for a new customer to walk through the door, a landscape contracting firm has numerous options to exercise in attracting new clients. Small firms in thin market areas, new firms, and expanding firms are the most frequent users of this means of finding new jobs. Direct solicitation is used more commonly to seek residential contracting than commercial work; however, for an expanding company that desires to become known to architects and landscape architects, the direct approach also has value.

The direct solicitation of work takes different forms. They include such techniques as:

a. Media advertising, principally the newspapers, and the telephone yellow pages; to a lesser extent radio and television.

b. Blanket mailings to all property owners within targeted neighborhoods or developments. Such mailings will emphasize the past projects of the landscape firm and their experience and quality of work.

c. Participation in community flower shows and competitions that put the company's name before a public group with a common interest in gardens.

Figures 17-6 and 17-7 show examples of media advertising and direct mail advertising. Figure 17-8 illustrates the type of garden exhibit promotion that some firms use to display their names and talents before the public.

Noncompetitive Selection

Every landscape contracting firm aspires to attain a status within its market area that will assure it a full season of production without the need to advertise or compete with other firms for the same jobs. Advertising costs money, and competitive bidding requires time and money, while not always resulting in new jobs.

Noncompetitive selection of a landscape contractor by the client or the client's representative means that the contract specifications and the price for doing the job will be negotiated between all parties until a satisfactory agreement is reached. The negotiated contract is the preferred method of doing business because it eliminates or clarifies most areas of uncertainty before construction begins.

Much private landscape construction is done by firms that were personally chosen by the owner or landscape architect. Reasons for the selection of a particular firm vary. Frequently a firm is chosen because of the professional reputation it has established within its market area or within the industry of a region. Being known as a firm that

Figure 17-6 This example of attitude advertising was run in the newspaper in the early spring. It states clearly what the firm does and suggests quality and reliability. (Courtesy of Albert Glowacki Landscape, Inc., Nantucket, Mass.)

does quality work, that begins and completes a project on schedule, that charges fairly and performs ethically, is the essence of a good reputation. This is an excellent basis for noncompetitive selection by a client.

Another reason for direct selection by a client is the proven ability to perform the work, as evidenced by past projects. The client and/or his representative is able to anticipate the performance of the landscape contractor, and that gives his confidence an important boost by removing one uncertainty from the development process. As projects are worked on, the landscape contracting firm should be clearly identified with tasteful signs and efficient workers wearing uniforms that identify the firm. Projects should be professionally photographed throughout the construction period for later use in brochures, award competitions, and for working with prospective clients.

Being familiar to the landscape architect who will actually select the contractor is another way

that many firms obtain contracts. Because the landscape architect wants to do the best job possible, he or she may select the landscape contractor from a brief list of dependable firms who have done similar projects before. It is important for the landscape contractor to maintain good relations and frequent communications with such landscape architects and similar architectural and engineering firms. Keeping them aware of past and present projects that the landscape contractor is building can cause them to call that contractor when they are involved in comparable projects.

Yet another reason why a landscape contractor may be selected to build a landscape without having to engage in competitive bidding is its proximity to the project site. Reduced travel time between the project and the contractor's base of operation can mean faster construction at less cost, at least as perceived by the client. Thus the landscape contractor who is nearby may have a decided advantage over one farther away.

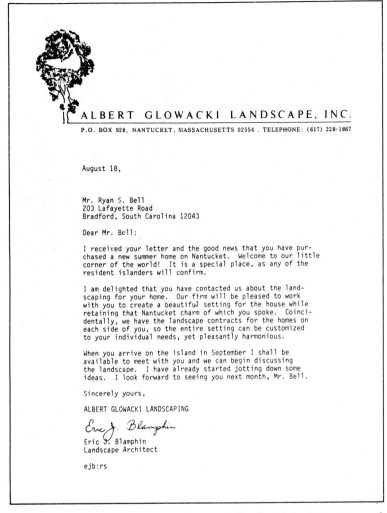

Figure 17-7 An example of direct mail advertising (Courtesy of Albert Glowacki Landscape, Inc., Nantucket, Mass.)

These four reasons, reputation of the firm, past projects, familiarity to the landscape architect, and proximity to the project site, are all common reasons why landscape contractors receive work without engaging in the bidding process.

Competitive Bidding

In the building trades, most general contracts are awarded on the basis of competitive bidding. In landscape construction, many contracts are awarded through this process. When the project is in the public sector, competitive bidding is nearly always required by law. Large private projects are also commonly awarded from among an assortment of proposals.

Competitive bidding requires that each company seeking selection as the contractor on a project submit its proposal or bid in a specified form, at a specified time and place, and usually sealed in an envelope. The sealed bids may or may not be opened in public, and then the client or clients' representative evaluates the bids.

Figure 17-8 A typical flower show display garden. Participation in such shows often generates new customers for landscape firms.

If the lowest bid does not exceed the amount budgeted by the client, then it will usually be accepted by the client. A contract will be offered to the successful bidder, and that firm will then become the contractor for the work specified.

Bids are solicited by clients in several ways. Owners of private projects may seek bids by extending "invitations to bid" to contractors who seem to have the capability of doing the work. Public projects are usually advertised in newspapers and through other media to assure that no one is denied the opportunity to bid.

The Advantages and Disadvantages

The advantages of competitive bidding are mostly on the side of the clients since they seek construction of projects at the lowest possible cost. However, bidders also benefit since partiality and other forms of unfairness are eliminated, particularly in public work. Competitive bidding easily identifies those firms that are not financially astute, since their bids are likely to be significantly higher or lower than the majority. Thus an additional advantage of competitive bidding is that it elicits a true market value for the project.

The disadvantages of competitive bidding are also borne disproportionately by the client. The requirements of the project may limit the number of firms that can qualify to do the work and submit a bid. Thus the client's options are reduced. Also, despite safeguards such as the pre-qualification of bidders, public clients cannot always be certain that the lowest bidder possesses all of the expertise or financial security needed to complete the project after the contract is signed. An additional disadvantage falls to every contracting firm that devotes time to the preparation of a bid proposal and does not get the contract. However, despite the disadvantages, competitive bidding is likely to continue as the major means of obtaining contracts for public project construction.

Achievement Review

A. List the components of a basic landscape contract.

B. Define the following terms:

    1. client
    2. contractor
    3. prime contractor
    4. subcontractor

C. Consider the following situation. Robert Jones is a homeowner desiring a sizeable residential property development. He hires an architect to build the house and a landscape architect to design the exterior areas. Both professionals have contracts directly with the client. The architect engages the services of a surveyor and an engineer. The landscape architect and the architect agree on a general contractor to build the house and prepare the site for the landscaping. The landscape contracting firm engaged to install the landscape on the prepared site will report directly to the landscape architect, not the general contractor.

Diagram the above relationships and label each relationship as a *prime* contract or a *subcontract*.

D. List the three ways that landscape contractors are selected for projects.

E. What four reasons may cause a landscape firm to be selected without competitive bidding?

F. Compare the advantages and disadvantages of competitive bidding to the client. Make the same comparison to the landscape contractor.

18 Contract Documents

Objectives:

Upon completion of this chapter, you should be able to

❦ explain the need for documents as part of a contract

❦ describe and distinguish between the numerous documents that may be part of construction contracts

The Need for Documents

A construction project typically involves many different firms. A landscape firm is normally involved in numerous projects simultaneously. It would be impossible for the work of one major project or one active company to be accomplished efficiently without all performance requirements, scheduling, and other terms of work written down for repeated referral throughout the project or season of operation. Without the commitment of the terms of the contract to writing, most projects would come to a chaotic standstill, with all parties on the telephone to their attorneys.

Some of the documents that are part of a contract are initially used during the bidding phase of a project. Once a contractor of subcontractor is selected, many of the bidding documents become part of the contract.

The Documents

While the building industry can generate scores of documents to accomplish a total construction project, the emphasis here will be on those documents that may affect a landscape contractor.

Standard Form of Agreement

Comprehension of this document necessitates a two-part understanding. First, as a form of agreement the document concisely states the relationship and obligations of the client and the major party or parties responsible for accomplishment of the project. The agreement is usually between the client and his architect, engineer, and/or landscape architect. It may also be used when direct contractual relationships are established between the client and the general contractor. It is a formal statement of the work to be accomplished, any supplementary conditions that are required, and the price agreed to. Second, it is standardized in that it follows a format recognizable to everyone working within the construction industry. That means that the document has proven fair and workable for all parties using it in the past. It further means that loopholes, weaknesses, and vague areas have been eliminated by prior users of the form, thus reducing the possibility of misunderstanding between the current parties. Standardized forms that have been used before by the parties of a contract are familiar in their content and arrangement. That offers the users the assurance that all components and implications of the agreement are understood.

There are several sources of standard forms of agreement and other standardized contract documents. The federal government and many state and local governments have prepared standard documents in an effort to treat all bidders equitably and to assure that various aspects of their laws are applied as required and without bias. Other sources include several important professional and business organizations. Foremost among them are the American Institute of Architects (AIA), the Associated General Contractors of America, the Engineers' Joint Contract Documents Committee, the American Society of Landscape Architects (ASLA), and the National Landscape Association (NLA). The standardized documents of the AIA enjoy the widest acceptance for private building construction projects. Those of the ASLA are most frequently used when a landscape architect holds the prime contract on a project. Small projects of landscape contractors commonly use one of the NLA forms.

The very nature of standardized forms suggests that there will frequently be provisions or requirements that are unique to a particular project and will not be addressed within the general wording of the form. In such cases, the standard form may be modified to include these special features, or it may refer to attached supplementary conditions that will detail them. Because the clauses that constitute these special features are not standardized, they require careful reading by all parties before the contract is signed.

Invitation to Bid

This document is an advertisement to all interested members of the construction industry that a project is seeking a builder. More common to public agency projects than to privately owned projects, the invitation to bid usually appears for a required period of time in newspapers, trade journals, or other media and is the government's way of offering equal access to the project by all qualified bidders.

Some variation in form exists for this document. Even the title varies, being called a "Pre-Solicitation Notice" by the federal government, and a "Notice to Bidders" or "Advertisement for Bids" by other agencies. Regardless of the title, the advertisement usually gives a brief description of the project; states the deadline for bidding; the place where bids will be received and (perhaps) opened publicly; the location and cost of plans, specifications, and other contract documents; requirements for bonding; and other guidelines for bidding.

Less formal invitations to bid may be used by private owners who do not choose to advertise their projects publicly. Instead, the client may screen and select a few prime contractors deemed reputable and qualified to do the work and invite them to prepare bids for the project. This method offers the client the advantage of competitive bidding while minimizing the risk of an unqualified contractor submitting a low bid and then being unable to accomplish the project. Figures 18-1 and 18-2 are examples of public and private invitations to bid.

Drawings and Specifications

At the heart of every construction contract are the illustrations and written descriptions that describe the project in finite detail. The drawings will be prepared by the architects, engineers, and/or landscape architects working on the project. For landscape contractors the drawings of concern may include planting plans; construction drawings for all landscape items such as outdoor furniture, walks, fencing, lighting, and pavings; irrigation designs; and grading plans. The specifications provide written explanations of requirements for

---

NOTICE TO BIDDERS

The Municipal Sewage Treatment Plant of the Town of Cobleskill will receive bids for the landscape installation at the new entry on Draker Road until 7:00 PM on Monday, February 24, 1997 at the Town Hall. At that time all Bids will be opened and read aloud to all interested parties.

Contract documents in the form of written specifications and a twelve page set of drawings are available for review at the Town Hall. Individual sets of the documents can be purchased for $75.00. That is refundable if all documents are returned undamaged within one week of Bid openings.

A Bid Bond is required from the Bidder or an acceptable surety agent in the amount equal to 5% of the total Bid. This will serve as a guarantee that the Bidder will accept and execute the Contract if selected. Further, it will serve as a guarantee that the Bidder will post both Performance and Labor & Material Bonds within seven days of being awarded the Contract.

The Municipal Sewage Treatment Plant of the Town of Cobleskill reserves the right to accept none of the submitted proposals or to reject specific technicalities of a proposal that are not in the best interest of the Plant.

It is agreed that the officials of the Treatment Plant may not delay longer than thirty days from the date of the Bids for the purpose of reviewing the submitted Bids and investigating the qualifications of the Bidders before awarding the Contract.

Figure 18-1 A typical public invitation to bid, usually published in newspapers

the project. They describe such things as the materials to be used in the construction, dates of beginning and completion, quality standards, and other factors that are not part of the drawings, but which are essential to the contract. The specifications are prepared by or under the supervision of the project's designers and are usually prepared in accordance with the format established by the Construction Specifications Institute (CSI).

Initially made available to all interested contractors during the bidding period, the drawings and specifications become part of the contract after the successful bidder is selected.

Bid Form

Both the AIA and ASLA have standardized forms for submission of bids. An example of a bid form is shown in Figure 18-3.

The document is most common to public bids where all bidders must be treated equally, and the lowest qualified bid is likely to be accepted.

Noncollusion Affidavit

This document may be required by owners of large projects as a means of assuring that bidders have not discussed the project prior to bidding and established a minimum bid among themselves. Such consultation is termed *collusion* and is illegal because it denies the owner the truly lowest bid, which is the purpose of competitive bidding. By signing the noncollusion affidavit, each bidder certifies that the bid submitted was calculated without consultation with other bidders. An affidavit is as legally binding upon the signer as verbally swearing before a judge in court. Figure 18-4 exemplifies a noncollusion affidavit.

THE IMOGENE SMALLWOOD HOSPITAL
LINCOLN, NEBRASKA

June 19, 1997

Mr. Rodney Ryan, VP Sales
At Your Service Landscaping, Inc.
1515 West Roosevelt Parkway
Lincoln, Nebraska 12043

Dear Mr. Ryan:

The Imogene Smallwood Association is requesting bids for the maintenance and upkeep of all surrounding grounds as specified in the attached Request for Proposal (RFP). The hospital desires to change from operating our own in-house grounds maintenance operation to having the work done by contract.

While the specifications attached represent the hospital's anticipated needs and quality standard, you (and the other vendors) are invited to take exception to any specific item and propose a better and/or more cost effective way of meeting our needs.

It is the hospital's intent to solicit meaningful proposals in order to give us a range of alternatives from which to choose. To facilitate this, we will be pleased to work with your organization in answering any questions concerning the requirements outlined in the RFP.

Certain key days should be noted. A bidders' conference is scheduled for 9:00 AM on Tuesday, June 24, 1997 in the hospital's board room. Vendor proposals must be submitted to the hospital on or before July 9, 1997. There will be no acceptance or consideration of bids submitted after that date.

Your response to this RFP, including cost data and documentation of your company's qualifications, will be our primary basis for final selection. Therefore we suggest submitting as complete a response as possible.

I look forward to the receipt of your proposal and to the development of a cordial professional relationship in the future.

Sincerely yours,

Frances A. Dickinson

IMOGENE SMALLWOOD HOSPITAL ASSOC.
Frances A. Dickinson
Administrator

Figure 18-2 A private invitation to bid, which is sent directly to specific landscape firms

BID SUBMISSION FORM

NAME OF BIDDER _____

NAME OF AUTHORIZED REPRESENTATIVE _____

PROPOSAL FOR SITEWORK PLANTING AT THE STATE UNIVERSITY OF NEW YORK AT COBLESKILL, NY

A. The work proposed herein will be completed on or before September 1, 1997. In the event that the undersigned fails to complete the specified work on or before that time or by the time to which such completion may have been extended in accordance with the Contract Documents, the undersigned agrees to pay to The University liquidated damages in the amount of five hundred dollars for each calendar day of delay in completing the work.

B. The undersigned hereby declares that all Bidding and Contract Documents have been carefully examined and that he/she has personally inspected the actual site of the work and accepts all quantities and conditions, and understands that in signing this Bid Submission Form he/she waives all future right to contest the Contract for reasons of misunderstanding.

C. The undersigned further understands and agrees to do, perform, and complete all work in full accordance with the Contract Documents and to accept as full compensation the amount of the Total Bid.

D. **TOTAL BID**

    a. Bid for supplies and materials $ _____

    b. Bid for work and labor, excluding the aforesaid supplies and materials $ _____

    c. Total Bid (Total of a + b) $ _____

Authorized Signature for the Bidder _____

Date of Signing _____

Figure 18-3 A typical bid form

Bid Bond

Consider this scenario. A client and his landscape architect have spent several months preparing their project for bidding and waiting through the bidding period, eager to get the contractor selected and construction started. They have met with dozens of would-be bidders, walked the site repeatedly, and answered hundreds of questions about the project. Finally, the bids are opened, a bid selected, the winner is notified, and the losers turn to other projects. The owner and landscape architect urge the selected contractor to sign and return the contract;

Affidavit of Noncollusion

By submission of this Bid, the person signing on behalf of the Bidder swears under penalty of perjury that to the best of knowledge and belief:

A. The prices quoted in this Bid were derived independently and without collusion, consultation, communication, or agreement with any other bidder or competitor with intent to restrict competition.

B. The prices quoted in this Bid have not been knowingly disclosed by the Bidder nor will they knowingly be disclosed by the Bidder to any other bidder or competitor prior to the time of the announced Bid Opening.

C. No attempt has been or will be made by the Bidder to discourage any other person or company from submitting a bid and thereby restrict competition.

AUTHORIZED SIGNATURE FOR THE CONTRACTOR:

In compliance with and pursuant to Section 107-A of the General Municipal Law, I hereby affirm that the above statements are true.

Signed _____

Title _____

Company Name _____

Company Address _____

_____

Date of Signing _____

Figure 18-4 A noncollusion affidavit

but the contractor hesitates and delays. Finally, the contractor announces that he has changed his mind. He never really wanted the contract. He just wanted to gain some experience at bidding. Now the client is back at square one. The bidding process must begin again; the project is delayed; the client is both angry and inconvenienced.

A bid bond is a required security deposit that is submitted by a contractor at the time a bid is submitted. It serves as a guarantee that the selected bidder will honor the bid if selected. Required on nearly all large public projects and many private

ones, the bid bond requirement is frequently stated as a percentage of the maximum bid price. The most common requirement is 5 to 10 percent, although on construction projects a larger bid bond may be required.

The bid bond is the most common form of bid security. Contractors who submit numerous bids requiring bonding will obtain the bond from a financial organization in business for that purpose. Known as a surety, the bonding organization charges the contractor a fee for the temporary loan of the money, which is like interest on a short-term loan. Companies that post bonds frequently do not want their capital tied up in security bonds, so they may pay an annual service charge to a surety bondsman to fund their bonds as needed.

Other forms of bid security may include negotiable securities, such as certified checks or cashier's checks. When a successful bidder is selected, the client retains the security until the contract is signed and required performance bonds are posted. The bid bonds of the unsuccessful bidders are returned soon after all bids are opened. Occasionally the bonds of the next lowest bidder(s) are retained as a safeguard until the contract is signed.

In the event that the chosen contractor declines to sign the contract or to meet all of the contractual requirements, the owner can recover some of the resulting damages from the defaulting contractor's bond security. Two types of bid bonds are used in the construction industry. One is known as a liquidated damages bond. With this bond, the owner claims the entire amount of the bond as the contractor's penalty for default. The other type is a difference-in-price bond. It allows the owner to withhold enough from the bond of the defaulting contractor to pay the difference between the defaulted low bid and the price of the next highest bid, up to the full value of the bond.

Performance Bond

Understanding a performance bond may be easier after considering another example. Consider the predicament of the client who has a signed contract for a project under construction, then is told that the project will not be completed because the contractor or a major subcontractor has gone bankrupt; or has accepted a larger, more profitable project; or has lost key, uniquely skilled personnel; or dozens of other reasons for default that may be the responsibility of the contractor. The damages to the client can be imagined. There may be a hole in the backyard where the swimming pool was supposed to be. The family reunion that was supposed to be held in the new backyard had to be moved to a park because the new landscape was unfinished. The grand opening of the business was less than grand with construction equipment and piles of mud greeting the guests.

The purpose of a performance bond is to provide security that can be drawn upon by the owner to recover damages resulting from contractor default after the project has started. The bond often covers any warranty period that is required by the contract, as well.

The performance bond must provide enough money to the owner from the surety to complete the contract if default occurs. Most performance bonds are for 100 percent of the contract price, and some require security of more than 100 percent to allow for increased prices should default occur. As with bid bonding, the actual cost to the contractor for providing the bond is the charge assessed by the surety source. Unlike the bid bond, which is usually held for only a short time by the client, the performance bond may be held in part or in total for a long period of time. The greater value of the bond and the length of the loan it represents makes the cost of the performance bond more expensive for the contractor.

Not all companies qualify for the same levels of bonding, any more than all individuals represent the same loan risk to their lending institutions. Only larger landscape construction firms, with reputations for stability and dependability, are likely to find surety support when large performance bonds are required. Requiring proof of bonding is one way of assuring the owner that only qualified bidders and contractors are working on the project.

Subcontract and Material Bid Invitations

These forms are similar to the invitation to bid forms described previously, except they are used by a contractor to solicit bids from subcontractors and material vendors whose services or products are needed for the project. An additional difference from the invitation to bid used by the owner or owner's representative is the less formal nature of these invitations. They may be as brief as a postcard announcement or a letter generated on an office word processor. The invitation will usually identify the project and the material or service for which a quotation is needed. It will also state the place where proposals should be submitted, and perhaps the person to whom they should be submitted. If drawings or specs are involved, there will be information about where they can be obtained. Information regarding other bidding documents may be included, and the deadline for submission of the price quote is always included.

Subcontract

This document confirms the contractual relationship between the subcontractor and the prime contractor. It may be a standard formal document or, with small projects, it may be written but less formal. Formal or informal, the subcontract must detail the agreements and obligations of both the prime contractor and the subcontractor. Since the subcontractor's agreement is with the prime contractor, not the owner, architect, engineer, or landscape contractor, the completeness of the subcontract is as important as that of the prime contract, because it alone protects the subcontractor. The subcontractor has no legal line of appeal or obligation to the owner.

There is strong support among many subcontractors who regularly work on large projects for the use of a standard subcontract form. A form that is familiar because it has been used before and checked carefully by attorneys for both parties has less chance of presenting difficulties in the business relationship. A standard subcontract form is shown in Figure 18-5.

Notice to Proceed

This document is common to large projects. It is a written notice sent by the owner to the contractor, directing the contractor to begin work on the project. The contract will frequently specify that the contractor must begin work on the project within a specified number of days after receipt of the notice to proceed. The benefit of this document is that it relieves the contractor of liability in cases where the owner may not yet hold clear title to the property where the project is located.

Labor and Material Payment Bond

Contractors and subcontractors rely on other businesses to furnish materials and/or labor to support the work they are obligated to perform on a project.

For example, a local electric company may be required to bring power to an undeveloped site to permit lighting and the use of power tools; a company that provides portable toilets or trash removal may be engaged in support of the project. While having nothing to do with general or landscape construction, their services to the companies involved are essential. In the event of default by the contractor or subcontractor to whom they were bound by contract, the supplier might not get paid. The owners could not be expected to pay because they were not a party to the contract. The labor and material payment bond provides third-party suppliers of utilities, fuels, telephones, equipment rentals, and/or labor with a surety-backed security that payment will be forthcoming soon after the supplier has furnished the last of the materials or labor that it was obligated to provide. Usually written notice of the fulfillment of the obligation is required.

Change Order

It is predictable that on many projects alterations in the original contract will be made. The owner or architect may have a new idea while the project is still under construction. The contractor may find it impossible to acquire an item believed available at the time of contract signing. Unexpected developments at the site may complicate the schedule. The possible reasons necessitating a change in some part of a contract are endless. The change order document provides for those changes after all parties involved in the contract have agreed upon the change. The change order describes the amendment of materials, methods, price, and/or time. It requires the signatures of all parties and the dates of their signing. It then becomes a part of the contract that it has changed.

In construction contracts for large projects, there is usually a clause that describes how changes will be handled. Most changes are initiated by the owner, and the contract clause describing change orders will typically allow the owner that prerogative, providing a fair price or time allowance is given to the contractor. When the owner initiates the change, he or his architect, engineer, or other professional representative must

SUB-CONTRACT AGREEMENT

This agreement is established this 25th day of October, 1997 by and between AT YOUR SERVICE LANDSCAPING, hereinafter called the **General Contractor** and GREAT LAKES POOL CONSTRUCTION, hereinafter referred to as the **Sub-contractor**.

The General Contractor and the Sub-contractor agree to the following:

SECTION 1   JOB SITE

The job site is the rear of the residence at 1234 Main Street in Lincoln, NB.

SECTION 2   DESCRIPTION OF SERVICES

a. The work to be performed is outlined in the attached specifications.

b. Sub-contractor shall provide all of the labor, materials, tools, equipment, and supervisory personnel needed to perform properly the work described in the specifications. All work will be performed in a professional and safe manner in accordance with accepted industry quality standards as set forth by the American Pool and Spa Institute.

c. Sub-contractor shall at all times keep the premises free of accumulated litter, waste materials, and other debris resulting from the work. Removal of such material shall be at the sub-contractor's expense.

d. The timely completion of the work is essential and is an important element for full compensation. The work must be started on or before April 20, 1998 and completely finished no later than May 20, 1998. The Sub-contractor may request in writing for an extension of times for delays attributable to weather or other uncontrollable events. Approval is at the discretion of the General Contractor.

SECTION 3   INSURANCE COVERAGES

a. Sub-contractor agrees to carry and maintain at its expense these minimum insurance requirements:

- Worker's compensation and Employer's Liability
- Comprehensive general liability
- Comprehensive automobile liability
- Umbrella excess liability coverage of at least $1,000,000.

b. Sub-contractor agrees to indemnify and save harmless the General Contractor against any and all liabilities, damages, losses, claims, demands, expenses, and actions including personal injury, death, or property damage that arise from any act of negligence by the Sub-contractor, its owners, employees, or agents while engaged in the performance of the services described in this sub-contract.

SECTION 4   STANDARD CONDITIONS AND PROVISIONS

a. Sub-contractor shall act as an independent contractor, employing and directing all personnel required to perform the specified services, securing all required permits or licenses, and complying with all federal, state, and local ordinances, laws, rules, and regulations.

b. Sub-contractor shall be liable for any damages to existing property on the site, property that is in its care, or property that is being installed until the General Contractor has fully accepted and paid for such work.

Figure 18-5   A subcontract agreement

c. Sub-contractor has thoroughly evaluated the site, the work to be done, and the general work conditions. No additional payment will be made for alleged or unknown contingencies or difficulties. Sub-contractor also agrees to all terms and conditions of the Contract that the General Contractor has with the Owner and the Owner's Agents.

d. No extra work or changes in the work or terms of the sub-contract will be recognized or compensated unless agreed to in writing before the work is done or the changes made.

e. Sub-contractor shall pay and discharge all costs and expenses including reasonable attorney fees that may be incurred or expended by the General Contractor to enforce the provisions of this Sub-contract.

f. Should the Sub-contractor fail to provide the contracted services herein described, the General Contractor may contract with another company to complete the work and the additional cost to the General Contractor shall be deducted from any sums due the Sub-contractor.

g. Sub-contractor shall guarantee all work performed for a period of one year.

h. Sub-contractor shall not contact or deal direct with the owner of the project where the work is being done nor with any of the owner's family or agents. All communication and correspondence shall be directed to the General Contractor who will solely deal with the Owner.

i. Sub-contractor shall immediately, within twenty-four hours, give written notice to the General Contractor of any accidents, injuries, damages, or additional costs or charges incurred during and in performance of this Agreement.

j. Nothing appearing herein shall represent a waiver of any right that the General Contractor may have.

SECTION 5   COMPENSATION

Payment is to be made net 30 days after completion and acceptance by AT YOUR SERVICE LANDSCAPING and after receipt by same of a completed full waiver of lien and a detailed invoice.

SIGNATURES TO THE AGREEMENT

For the Sub-contractor: For the General Contractor:

_____   _____
(Authorized Representative)   (Authorized Representative)

Print Name _____   Print Name _____

Title _____   Title _____

Telephone _____   Telephone _____

Date _____   Date _____

Figure 18-5   A subcontract agreement *(continued)*

present a description of the change, including revised drawings and specifications to the contractor for review. This must be done before the change order is signed. If the contractor requires a contract change, he or she must present the change in writing to the owner and include revised prices and time schedules. This should be done before the change is initiated at the project site. Should the contractor carry out a change before all parties have signed the change order, the additional work may not be paid for. In Figure 18-6 a sample change order is illustrated.

CHANGE ORDER

At Your Service Landscape Company
Your Full Service Landscape Professionals
1515 West Roosevelt Parkway
Lincoln, Nebraska 12043

Client Name   Change Order Number

Address   Date of Change

                                               Telephone Number

Project Name and Address if Different   Job Number

                                               Date of Original Contract

CHANGES TO BE MADE
All changes become part of and in compliance with the original contract which these changes amend.

It is agreed that the above changes will be made at this revised price $

Previous Contract Price $

Revised Contract Total $

Client signature_____ Date _____

Company authorized signature _____ Date _____

Figure 18-6 A typical change order

Certificate for Payment

Large construction projects that require many months or even years to complete must provide for periodic payment to the contractor as the project proceeds. This permits the contractor to meet his payroll and pay his subcontractors and suppliers. The contract agreement will contain a payment clause allowing for partial payments as substantial portions of the project are completed. A request for payment will be submitted by the contractor to either the owner or his architect, engineer, or landscape architect. Following approval of the request, the owner's representative will issue the certificate for payment to the owner, who will then send the amount due to the contractor within an agreed-upon time period. Should the owner not pay by the time required, the contractor has the right to stop work on the project. Neither the issuance of a certificate for payment nor payment by owner represent acceptance of any part of the contractor's work by the owner. The owner is still free to claim defective work or default by the contractor at a later date.

Certificate of Final Completion

Two things are accomplished when a certificate of final completion is issued by the architect or other prime professional. It certifies that the project is ready for occupancy by the owner and ready to be used for its intended purpose. It also usually means that the contractor is entitled to most of the money remaining to be paid him. Usually 95 percent of the contracted amount is paid. The remainder is held by the owner until all guarantees, claims against the project, and other contingencies are settled. This latter withholding is called *retainage*. Final payment will return the retainage to the contractor.

The Complete Contract

The complete construction contract will always include the agreement form and may include some or all of the documents described previously. There is an unofficial correlation between the size of the project and the number of documents that comprise the contract. Large projects commonly involve more complex contracts than smaller projects. With the complexity comes the need for more documents to accomplish the project smoothly and on schedule. Some small landscape contractors never need more than the basic agreement form to establish a legal business relationship with their clients.

Where there are numerous documents involved in the contract, the agreement form will contain clauses that refer to each of the documents. The supplementary documents then serve to amplify the clauses.

Achievement Review

A. Explain why there may be numerous documents comprising the complete contract for a construction project.

B. Indicate which contract document is described by the characteristics below. Select your answers by number from these choices.

1. Standard form of agreement
2. Invitation to bid
3. Drawings
4. Specifications
5. Bid form
6. Noncollusion affidavit
7. Bid bond
8. Performance bond
9. Subcontract and materials bid invitations
10. Subcontract
11. Notice to proceed
12. Labor and material payment bond
13. Change order
14. Certificate for payment
15. Certificate of final completion

____ A standardized form used for the submission of bids, permitting all bidders to be treated equally

____ Provides monetary security to the owner of a project if a contractor defaults after the project has started

_____ A document that binds the principal parties of a contract. The form is acceptable to all parties because it has been used for past contracts and the possibilities of misunderstandings are reduced

_____ Documents that may include planting plans and construction schematics for landscape items

_____ Confirms the contractual relationship between the prime contractor and a specialty firm doing work on behalf of the contractor

_____ Surety-backed security for third-party suppliers

_____ An advertisement to all interested members of the construction industry that a project is seeking a builder

_____ Advertisements used by contractors to solicit bids from subcontractors and material vendors

_____ A written notice sent by the project owner to the contractor authorizing the commencement of work on a project

_____ Written explanations of requirements for the project

_____ A means of assuring the owner that bidders have not discussed the project prior to bidding and established a minimum bid among themselves

_____ Certifies that the project is ready for owner occupancy and that the contractor is entitled to most of the remaining money due

_____ A required security deposit to guarantee that the selected bidder will honor his/her bid

_____ A document that allows for amendments to the original contract

_____ Issued to the owner by the architect, engineer, or other principal, the document authorizes partial, periodic payments to the contractor

CHAPTER 19

Specifications

Objectives:

Upon completion of this chapter, you should be able to

- explain the concepts of project manual and Masterformat
- distinguish between standard and customized specifications
- explain the value and importance of uniformity
- identify the originators of specifications
- describe different methods of specifying
- list sources of specification data
- distinguish between closed and open specifications

As described earlier, specifications are the written instructions that describe how a project is to be built. Whereas the drawings for the project provide visualization and quantitative measurements, the specifications explain quality standards, describe materials, and establish the technical requirements of the project. Comparatively new to the construction industry, specifications have made an especially slow entry into landscape contracting. Until recently, landscape specifications were notes jotted down on the corners and edges of site plans and construction drawings. Even building construction relied upon this risky, but accepted, means of conveying the intent of the designers and engineers. However, recent years have seen growth in the use of synthetic building materials

and in uninhibited litigation. The entry of so many new construction materials and products into the marketplace requires carefully prepared descriptions of materials that are acceptable to the client and the means of using or installing them. Equally significant in explaining why specifications have taken on such importance in contracting is the unparalleled increase in recent years of lawsuits brought by owners against contractors, contractors against owners, and subcontractors against contractors. Fault, liability, and damages are at the center of most contract lawsuits, and the courts frequently refer to the specifications for guidance in attempting to determine who did what to whom.

Today, specifications rival drawings in their importance to construction contracting. As computer technology develops, drawings may lose their importance almost entirely, even as quantitative measures. With the computer able to calculate quantities from material descriptions and dimensions entered into it by architects and engineers, the drawing may serve little purpose except to aid the client's visualization of the project. Even then, the drawing may be a computer-generated monitor image.

Progressing from the scribbled notations of thirty years ago to bound volumes of several hundred pages today, specifications require careful preparation by writers who are experienced in construction, who understand the materials and methods they are describing, and who are accomplished technical writers. Specification writing is no job for amateurs. The specification writer is an important member of the architect-engineer-landscape architect team. While one of the design professionals may prepare the specifications for a project, it is more common for him or her to work with a professional specification writer who either works as

part of the office staff or is hired as an independent affiliate for the project.

Project Manuals

To aid in the organization of the paperwork and the dates, and comprehension of the chain of authority and responsibility, the documents pertaining to large projects are bound together in a **project manual**. The specifications are included in the manual, along with the bidding and contract forms. Frequently the manual is referred to as "the specs," which is a clue to the importance of the specifications to construction projects today, Figure 19-1.

A typical project manual is divided into four parts:
1. Bidding information and bidding documents
2. Contract forms (agreements, bonds, certificates)
3. General and supplementary conditions of the contract
4. The specifications

Figure 19-1 Project manuals include all of the contract documents, a set of plans, and frequently a computer disc of the details.

Masterformat

In an attempt to create order from potential chaos, professional specification writers have endorsed and initiated efforts to structure and standardize the mass of materials within each project manual. Although there is still no one system that is used by all branches of government and the private construction industry, one system enjoys widespread acceptance. Masterformat was developed by the Construction Specifications Institute (CSI) in counsel with Construction Specifications Canada (CSC), so the system is used throughout North America.

Project manuals that use the Masterformat system are first divided into two parts:

1. Bidding requirements, contract forms, and conditions
2. Specifications

All of the non-specification documents are included in the first part. The second part is the categorization of the specs into sixteen divisions. Each division is further subdivided into sections called "broadscope," "mediumscope," and "narrowscope." Using key words and a simple system of numbering, each division outlines the tasks and materials common to that area of construction. The numbering system is standardized, meaning that the same five-digit number is used for the same item whenever the Masterformat system is used for preparation of the specifications in broadscope and mediumscope formats. Further detailing of the specifications into narrowscope format is not as structured. Masterformat permits the individual specification writer to choose specific narrowscope sections, usually either by adding more digits or by adding a decimal with more numbers after the decimal. Other narrowscope identification techniques include use of a letter after the five-digit number or some combination of decimal, letters, and numbering.

Standard and Customized Specifications

While Masterformat is widely accepted as the industry standard for project manual organization, it does not provide text material from which actual specifications may be written. It is the skeletal form upon which the specification writer builds the body of qualitative descriptions that are the specifications.

A number of commercially prepared guide specifications are available for use by professional specification writers. These guide specifications offer extensive pre-written descriptions of the materials and techniques used in the construction industry. The guide specifications range in their quality and completeness from those that are fully detailed and totally comprehensive (thereby allowing their use with little or no editing), to those that are general fill-in-the-blank types. The two most widely used systems of prepared guide specifications are Masterspec and Spectext. Masterspec is a product of the AIA, and Spectext is produced and distributed by the Construction Sciences Research Foundation. Of the two, at present Masterspec enjoys a plurality of use and acceptance among the many commercially available guide specifications on the market. Both Masterspec and Spectext are available in both paper version and in computer usable form. Both systems generally follow the naming and numbering system of Masterformat.

While most landscape contractors do not write specifications, it is still worth noting that the federal and state governments, through their many agencies, are the sources of many additional guide specifications. Few are equal to Masterspec or Spectext in quality or completeness, being written primarily for the type of work done specifically by each agency. Landscape contractors who bid on government jobs may encounter these specifications.

The use of commercially prepared guide specifications permits something approaching standardization in the writing of specifications. Specifiers are able to copy and edit entire sections or selected paragraphs from the prepared guides.

Some specification writers prepare their own guide specifications or use those developed by and for the offices in which they work. For firms that specialize in certain types of construction projects, office guide specifications have usability. They permit faster and easier editing in preparation of new specifications since the specifier is able to reuse large portions of the specs prepared for similar projects in the past. The greatest problem in the use of office guide specifications is keeping them current.

Most office guides are not written from scratch by the specifiers. Instead, they are started from one of the commercially prepared guides, then edited to exclude those sections that are not relevant to the work of the office. As the commercial guides are updated, the office guide specifications must be updated too.

Uniformity in Preparation of Project Manuals

The construction industry at large is moving rapidly toward standardized methods and materials. Manufacturers of construction materials are uniting in associations that publish specification standards for the best use of their products. Even growers of plant materials present their products in conformance with standards established by the American Association of Nurseymen. See Figure 19-2.

Masterformat has been an important contributor to standardized naming and numbering of specifications sections. Using the Masterformat system in association with accepted industry methods and quality standards, specification writers are able to prepare project manuals that present no surprises to builders and suppliers. The predictability and uniformity are highly desirable. They have established the level of competency that courts expect construction professionals to display.

When specifications are written using standards and methods that are not uniformly understood and accepted, several undesirable results may occur. First, there may be fewer bidders on the project due to the added time required to research costs from the non-uniform, non-conforming specifications. Second, the probability of an error in the bidding process is increased as a result of the specification of unfamiliar methods. Finally, the specifier who does not use standardized methods of construction assumes a high risk of legal action against himself due to the increased probability of errors after construction is underway.

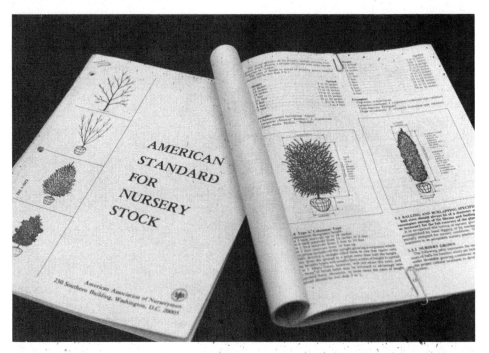

Figure 19-2 Industry trade organizations, such as the American Association of Nurserymen, establish their own standards for easy incorporation by specification writers.

The Writers

At one time specifications for projects were written by the architects, engineers, or landscape architects who designed the projects. This is still true for some projects. However, it is increasingly probable that the project manual's preparation and the actual writing of the specifications will be done by a full-time, professional specifier. While some specifiers work independently, most work as part of a larger multidisciplinary office. Working closely with the project designers, the specifiers must be experienced in as many aspects of the construction industry as possible. While competency as a technical writer is essential for a specifier, he or she must also know how things are built.

Methods of Specifying

There are different methods of specifying the requirements for a project. Most large construction projects use several methods of specification, although not within a single material or technique description. Specifications are usually technical, performance, or design specific.

Technical specifications are often used for portions of projects that are not easily verified for correctness after their construction has been completed. The specifications seek to provide written descriptions of the technical standards that the work must meet as construction proceeds.

Performance specifications allow the contractor much more freedom than do technical specs. Performance specifications establish the standard of performance or service that must be met by the feature being described. It is then left largely to the contractor to perform or provide the service to the standard established by whatever method he believes best. For example, the specifications might require a cantilevered deck to extend over an embankment and support a given weight and/or wind stress. While it would have to look like the plans described, the actual methods of construction, anchoring, and bracing would be left to the discretion of the contractor.

Performance specifications often result in lower construction costs, since each bidder is inclined to select materials and methods that he has used before and can fine-tune his bid more precisely.

Design specifications are nearly the opposite of performance specifications. All details regarding materials, methods, and performance standards are included in the specifications. Specific products may be cited by brand name and must be used by the contractor. The architect, engineer, or landscape architect is responsible for the finished product, allowing the contractor little or no autonomy on the project. The contractor is specified to carry out the directions of the design professional, down to the last detail.

Sources of Specification Data

Much specification data is available to the specifier from industry associations such as the American Institute for Timber Construction, the Brick Institute of America, and the American Concrete Institute. These organizations monitor the quality of their established products as well as the new products that are periodically introduced. The standards that the industry associations establish are reliable and are used with confidence by specification writers. Contractors must be familiar with these standards if they are to bid and compete successfully in the construction industry.

Additional sources of specification data include product manufacturers; professional associations; and federal, state, and local government agencies. It is not easy for specifiers or contractors to stay current with all of the new products and techniques of construction that enter the industry each year. However, staying up to date is critical.

Closed and Open Specifications

When the specifications require the use of specific products, identified by brand name, model number, manufacturer's name, and/or other unique characteristics, the specifications are termed **proprietary**. When the specifications do not allow for substitution with a similar product, they are termed **closed**.

Even when a brand name is not used in the specifications, they are regarded as closed if the product specified can only be obtained from one source. Since there is a suggestion of discrimination when only one supply source can be used, closed specifications are not common in public contracts. They are more commonly used in private contracts when the owner or designer wants to assure a certain quality standard. Closed specifications tend to reduce competitiveness among would-be suppliers.

Open specifications permit substitutions after approval by the design professional or owner. As expected, the proposed substitute must meet the same quality and performance standard of the spec-

ified proprietary product. The desire to substitute for the proprietary product may stem from the availability of a comparable product at a lower price, or from the unavailability of the specified material.

When Changes Are Needed

Sometimes it is necessary to make changes in the specifications, drawings, or bidding documents even while the project is still being bid. When that happens, an addendum is prepared in writing and sent immediately to all known bidders. The addendum then becomes part of the project manual and eventually part of the contract documents.

Achievement Review

A. Define the following terms:

1. project manual
2. CSI
3. Masterformat
4. specifications
5. closed specifications
6. open specifications

B. What is the value of uniform material and method specifications?

C. Who prepares specifications, and what qualifications must they possess?

D. List and describe three methods of stating project specifications.

E. What sources do specifications writers use for the uniform data they need?

CHAPTER

Installing
scape
s

ould be able to

tion of landscape plants

n of landscape plants

ntages of bare-rooted, balled and burlapped,

ng

of trees, shrubs, groundcovers, bedding

tages of organic and inorganic mulches

o the American Southwest

The
Prop

Hig
quality pl
installatio
and growt
contractor: ....e-dollar plants
set into se ....y-nve-dollar holes. In fact, a great
deal of labor and materials are often needed to
prepare a hostile planting site for a new plant.
Few sites offer a perfect combination of proper soil
texture, fertility, and pH with correct drainage and

mum water and humidity throughout the post-
splant period. All are necessary for successful
splanting.

e Necessary Tools

the plant material to be installed includes
, bedding plants, groundcovers, and trees
_hrubs of all sizes, a wide range of tools must
be available to accomplish the installation. The
hand tools most commonly used are shown and
described in Figure 20-1.

As plant materials increase in size, hand tools
must be supplemented with power tools. The tree
spade makes possible the successful transplant of

TOOL	FUNCTION
Spades	The spade performs obvious service in digging. It has a flat back and flat or pointed end. It has no sides and is not effective as a scoop. Spades are useful in the installation of all types of landscape plants.
Spading fork	Used for turning over the soil when it is not too hard or compacted, the spading fork is useful in preparing planting beds that need not be too deep, as for flowers, bulbs, and groundcovers.
Rounded **Shovel** Square	Used for cleaning loose soil from planting holes and for other scooping uses, a shovel has sides where a spade does not. The blade may be rounded or square. The handle may be long or short.
Spading shovel	A combination tool similar to both the spade and the shovel, it can be used for digging as well as scooping. It is useful in the installation of all types of landscape plants.
Scoop	Good for moving loose materials such as crushed stone, peat moss, and soil, a scoop has high sides. It is not used for digging. Aluminum scoops are available for lightweight materials. Steel scoops are used for heavy materials.
Manure fork	The best tool for moving coarse, lightweight materials, such as straw and wood chips.
Bulb planter	Used to install flowering bulbs, it is pressed into the soil, removing a plug of soil that is replaced after the bulb is inserted.

Figure 20–1   Tools for landscape installation

TOOL	FUNCTION
Hand trowel	Used to install bedding plants, groundcovers, and bulbs
Transplanting hoe	Its uses are similar to those of a hand trowel. It is less adaptable to other types of digging.
Garden hoe	Widely used for breaking up soil prior to planting, it is helpful in the installation of all types of landscape plants.
Grading hoe	Used to loosen hard or compacted soil during preparation for planting, it has a sharpened flat end.
Cutter mattock	Stronger than a grading hoe, its uses are similar. It has two flat ends.
Pick	Used for breaking up hard rocky soil, it has two pointed ends for gouging into the soil. A variation is the pick mattock, which has one pointed end and a flat mattock on the other end.
Wheelbarrow	The wheelbarrow has a multitude of functions. It can haul materials to and from the planting site. It can also be used at the site to hold soil conditioners for mixing.
Ball cart	Used to move balled and burlapped plants to the planting site.

Figure 20–1 *Continued*

large trees and shrubs, Figure 20-2. It operates on a hydraulic system, with each movement controlled from a set of levers that permits the machine to be operated by anyone, regardless of physical strength. Other power tools helpful in the installation of landscape plants include the power auger, power tiller, tractor, and front-end loader.

The Soil for Installation

It is best to use existing soil of the site to assure the most successful transplanting of trees and shrubs. Research has shown that plant roots adapt faster and grow deeper under those conditions. However, the soil removed from a planting hole may be unsatisfactory in its unaltered state for use as replacement (or backfill) soil after the new plant is set. It may be too heavy with clay and need the addition of sand and peat moss to provide better aeration and flocculation (aggregation of soil particles). There may be too much sand, requiring addition of humus to retain moisture around the new plant's roots. Landscapers may have to deal with the hard caliche layer if in the Southwest, salt-saturated soil next to roadways and walks, construction debris buried by builders, and the natural stoniness of rocky regions. Rich loam with ideal pH and good drainage is not common on most sites, where the landscaper is usually one of the last developers to be called in.

The soil to be filled into the planting hole must provide a medium in which the root system of the new plant can resume growth and develop fibrous root hairs to absorb water and nutrients for the new plant. If the soil does not drain adequately, the new plant may die from a lack of oxygen. If the backfill is too sandy, the new roots of the plant will stay within its soil ball and not grow out into the new soil.

Correct soil for plant installation has these qualities:

❦ Loamy texture (near equal mix of sand, peat, and soil)

❦ Good drainage

❦ Suitable pH

❦ Balanced nutrients

These qualities may be attained by blending conditioners such as sand, peat moss, compost, leaf mold, and manure with the soil before backfilling. In some situations, the original soil may be so unsatisfactory that it must be completely replaced, although recent research suggests that this is seldom necessary. In large landscape installations, the conditioning needs should be determined in advance so that necessary quantities of the additives can be ordered and available at the planting site. All members of the planting crew should be instructed in how to prepare the backfill mix. Periodic checks by the crew supervisor will assure that the new soil is mixed correctly and uniformly.

Figure 20-2 Large trees can be moved successfully with powered transplanters. There is limited disturbance to the root system. (Courtesy of Big John Tree Transplanter Mfg.)

Root Forms of Landscape Plants

Landscape plants are available in a variety of root forms. Bedding plants and groundcovers are usually grown in pressed peat pots or plastic packets that permit the root system to be transplanted intact. Trees and shrubs can be purchased as bare-rooted, balled and burlapped (B & B), or containerized plants. The advantages and disadvantages of each are compared in Figure 20-3.

Which root form is best to use depends upon the season of the year, the availability of stock, the size of the plants at the time of installation, and the budget of the project. Bare-root is a common root form for deciduous shrubs and a few trees that develop new roots quickly after transplanting. Evergreens are most often balled and burlapped or containerized. Deciduous trees and shrubs may also be obtained in B & B or containerized forms. Vines are usually containerized.

Root Form	Advantages	Disadvantages
Bare-rooted	• Comparatively inexpensive • Lightweight and easy to transport • Dormant at the time of planting	• Severely reduced root system • Transplant season limited to early spring • Usually small, requiring time to mature
Balled-and-burlapped	• Larger material can be transplanted • Less damage to the root system • Can be transplanted throughout spring and fall	• Usually the most expensive • Soil ball adds weight and bulk • For large plants, costly installation equipment is required
Containerized	• Less expensive than b & b material • Root system intact • Can be transplanted throughout spring, summer, and fall	• Seldom available in large sizes • Can become root-bound if kept in containers too long

Figure 20-3 Root forms of landscape plants

The Time to Transplant

The best season for transplanting depends upon the type of material being planted. Usually, the prime objective is to transplant at a time that will permit good root growth before shoots and leaves develop. For most plants in most parts of the country, that time is early autumn. Then the roots can grow as long as the soil remains unfrozen, while the cool air temperatures encourage the above-ground parts to go dormant. Early spring, when root growth exceeds shoot growth, is the second best season. Summer is not a good season unless containerized material is used, with its intact root system. Winter is not a good season for transplanting in northern regions because the roots cannot grow. In regions where the winter temperatures are milder and the ground doesn't freeze, winter can be a satisfactory alternate season for transplanting.

Flowering bulbs have definite transplant seasons. Hardy bulbs, which bloom in the spring, must be planted in the fall. Tender bulbs, which flower in the summer and will not survive the winter, are planted in the spring, dug up in the fall, and stored indoors over winter.

Annual flowers, purchased as bedding plants, are transplanted in the spring after all danger of frost is past.

Methods of Installation

The root form of the transplant will determine how the plant is installed. In certain cases, the size of the root mass may necessitate specialized equipment, but the principles behind the installation will remain the same.

Balled and Burlapped Plants

Plants with roots in a rounded soil ball wrapped in burlap are installed in a flat-bottomed, straight-sized hole that is wider and deeper than the ball. The extra volume of the hole leaves room for backfill soil to be added around the soil ball, Figure 20-4. Large plants need a smaller hole to help support the plant and prevent tipping.

Deciduous trees and shrubs may benefit from several handfuls of a slow-release, complete fertilizer mixed into the soil. This type of fertilizer provides nitrogen, phosphorus, and potassium gradually throughout the first year.

Application of fertilizer at the time of transplanting is seldom recommended for evergreens. Their roots will be severely harmed by the chemicals unless the plant has had at least a year to become established. If the plant appears to need fertilization, a slow-release fertilizer may be used; but it should be applied on the surface and beneath the

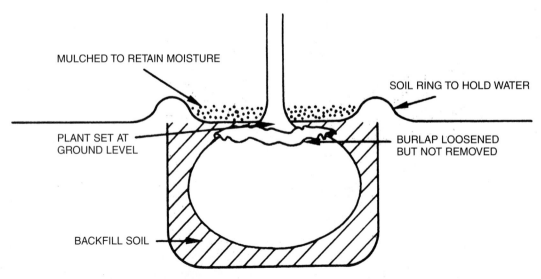

Figure 20-4 The finished planting

PLATE 1 Following construction of this home, a Maryland-area landscape firm began development of the public area setting.

PLATE 2 The entry drive was marked out and forms set where the pavers would be installed.

PLATE 3
Concrete was poured and leveled in sidewalk areas.

PLATE 4 ▶
The driveway pavers were set on top of a level and compacted subgrade of stone dust.

PLATE 5 Large trees had their soil balls wrapped in chains prior to being lifted into place.

PLATE 6 ▶
With a crane for lifting, and crew members for guidance, this large tree was placed in its prescribed location.

▲ **PLATE 7** Once the lawn was established, the public area was complete. The residence was ready to welcome guests.

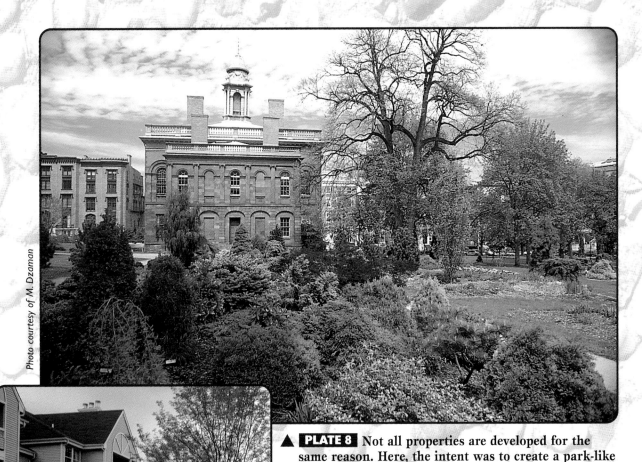

▲ **PLATE 8** Not all properties are developed for the same reason. Here, the intent was to create a park-like setting to accommodate the public near this commercial building.

▲ **PLATE 9** The purpose of this landscape is to visually link the separate units within this condominium complex.

PLATE 10 ▶
This landscape serves the traditional purpose of creating a welcoming approach to this private residence.

PLATE 11 Urban landscapes are frequently architectural in both style and function. This interior plantscape creates a setting for quiet conversation within a busy hotel lobby.

PLATE 12 ▶

In this multi-level outdoor room, a large deck overlooks the adjacent tennis court.

▼ **PLATE 13** Planted walks, a tree canopy, textured pavers, and comfortable furniture complete this outdoor room.

PLATE 14 Flowers are most dramatic when used in large quantities and in masses of few colors.

PLATE 15 ▶
Equally dramatic effects can be achieved by selecting and combining plants with sharp textural contrasts.

▲ **PLATE 16** Naturalized effects can be created with bulb flowers that are tossed randomly onto the ground and planted where they fall. Over time, they multiply and create drifts of color such as this planting of daffodils.

PLATE 17 ▶
When used as a small enrichment feature, water is similar to a mirror in the outdoor room. It reflects the light, color, and movement of other nearby components.

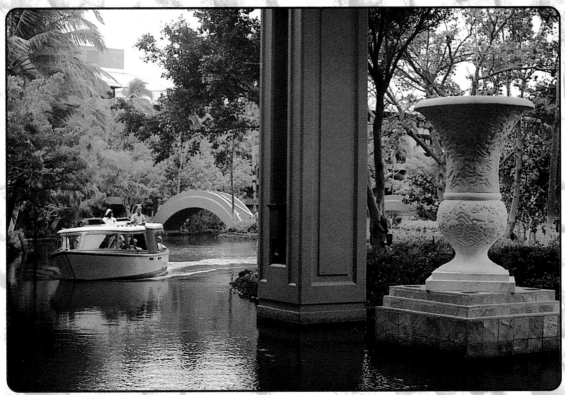

▲ **PLATE 18** Used over an extensive area, water becomes a visual adhesive, binding together the many diverse features of a large landscape.

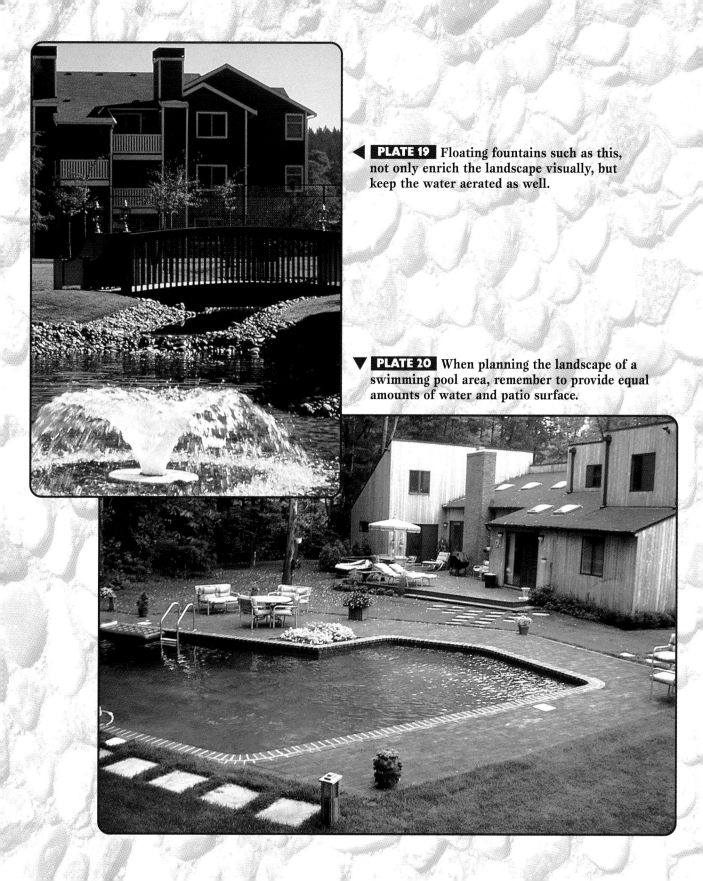

PLATE 19 Floating fountains such as this, not only enrich the landscape visually, but keep the water aerated as well.

PLATE 20 When planning the landscape of a swimming pool area, remember to provide equal amounts of water and patio surface.

▲ **PLATE 21** These twin waterfalls and boat landing are the designed centerpieces of a spectacular hotel lobby near arid Palm Springs, CA.

▼ **PLATE 23** Two repetitions develop the principle of unity here. Identical arching lines are evident in the water, turf, and pavement. Even spacing of the identical planters contributes as well.

▲ **PLATE 22** Unity is created by the repetition of lines, colors, fixtures, and other items. In this resort landscape, the repeating use of the same light fixture applies this principle.

mulch, not around the roots. Excessive amounts of any fertilizer applied to new plantings can cause serious damage to their root systems.

The backfill soil should fill the hole sufficiently to raise the plant to the level at which it grew in the nursery. That level is marked by the top of the soil ball. In regions of especially poor drainage, it is common to set the soil ball above the field level. Before the backfill soil is placed around the ball, the burlap should be loosened but not removed; then the soil can be added and tamped down with the feet to drive out the air pockets.

> *CAUTION:* If burlap has been replaced by plastic mesh or a wire basket, they should be removed before the backfill is added as they can stunt the development of the root system. Also, burlap that is left on should be thoroughly buried to hasten its decay and prevent it serving as a wick for moisture loss.

After the soil has been added, a mounded ring of compacted soil should be formed around the edge of the hole to catch and hold water. In large landscapes where frequent watering may not be possible (such as golf courses), the presence of the soil ring may determine whether the transplant survives during the first year, when moisture is especially critical.

Staking or guying may be necessary to hold the plant straight until the roots become established in the new location. A year or longer may be required. Depending upon the size of the plant, one of several techniques of support may be used. Smaller trees may be supported with stakes driven parallel to the trunk, Figure 20-5. Short wires wrapped in sections of hose brace the tree at or near the first set of branches.

Larger trees, since they have more sizeable foliage canopies, need stronger bracing. The technique utilizes three stakes driven into the ground beyond the soil ball. Wires extend upward to brace the tree. The wires should be wrapped in protective hose sections to prevent damage to the tree. Twisting the wires to keep them taut will assure that the tree stays straight and secure. To prevent someone from tripping over them, the wires should be flagged with reflective tape or cloth, Figure 20-6.

Figure 20-5 This two-stake technique is good for small trees. It is not a trip hazard, but it is not as sturdy as the three-stake method.

FLAGGING STRIPS ADDED FOR SAFETY

Figure 20-6 This three-stake method of securing a new transplant is strong and necessary for large trees. However, the wires are a potential trip hazard in populated areas.

Following the transplanting, the trunks of decid-uous trees can be wrapped with paper or cloth tree wrap to reduce desiccation from wind or direct sunlight, Figure 20-7. The tree wrap should be applied from the base of the trunk to the top of the trunk where the branching begins. The wrap may remain on the tree for a year or until it decom-poses, whichever comes first. Plants may also ben-efit from the application of an antitranspirant before and after transplanting. Antitranspirants are discussed later in this chapter.

Containerized Plants

The same type of large, flat-bottomed planting hole used for balled and burlapped plants is needed for containerized plants. The depth of planting, backfill with soil, tamping and watering to drive out air pockets, and formation of a ring of soil are the same for containerized plants as for B & B stock. Two additional steps are necessary in the installa-tion of containerized plants. First is the removal of the container, whether it be a clay or plastic pot, wooden basket, or other form of production holder. Second is preparation of the roots for normal orientation. Too much time in the container may have caused the roots to encircle themselves, a condition called "pot-bound." If the roots cannot be easily unwound, it is best to make vertical cuts through them at 2-inch intervals around the root mass, Figure 20-8. From these cuts, new roots will develop, growing naturally downward. Not making the cuts will allow the roots to continue their encir-cling growth even after the plant is removed from the container, eventually killing the plant.

Bare-Rooted Plants

If bare-rooted plants were placed in a flat-bottomed hole, the roots would be wadded and disoriented, and a stunted plant could result. The hole must have a mounded bottom to permit the plant's roots to spread over it and to encourage their natural orien-tation downward, Figure 20-9. Broken and damaged roots should be pruned away before planting.

As with other trees and shrubs, the planting hole for bare-rooted material should be wider and deeper than the root system it is to receive. As the

Figure 20-7 Tree wrap is applied from the base of the tree to the first set of branches. Each coil should slightly overlap the preceding one to seal out precipitation and insects.

backfill soil is added, there is a greater possibility of air pockets forming with bare-rooted material than with other root forms. Special care must be taken to prevent their formation.

Groundcovers and Bedding Plants

Both groundcovers and bedding plants are com-monly sold in strips of plastic or pressed peat moss pots or in flats. Certain plants such as geraniums or flowering perennials may also be marketed in small clay or plastic pots. Peat pots need not be removed, but other containers must be. Remember that the rim of the peat pot must be buried in the soil to prevent the wick-drying effect.

To install these plants, the entire bed is prepared rather than individual holes. This allows the ground-covers or flowers to be planted with a hand trowel

Figure 20-8 Containerized plants can become pot-bound. Before transplanting, they should be removed from the container and the root mass cut vertically at 2-inch intervals to promote new root growth that is oriented properly.

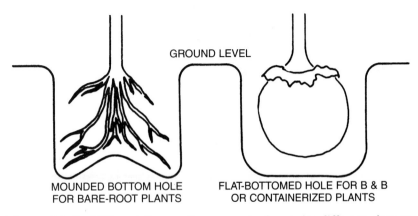

GROUND LEVEL

MOUNDED BOTTOM HOLE
FOR BARE-ROOT PLANTS

FLAT-BOTTOMED HOLE FOR B & B
OR CONTAINERIZED PLANTS

Figure 20-9 Different forms of nursery stock require different shapes of planting holes.

or hoe rather than a spade, Figure 20-10. The soil must provide good drainage and nutrients. It should also be as weed-free as possible at the time of planting. For large areas, a garden tiller may be used to loosen the soil and incorporate the necessary conditioners. The use of a preemergence herbicide prior to planting can reduce some of the maintenance requirements of the planting.

The spacing of groundcover plants depends upon the species and the speed of coverage desired. Naturally, closer spacing will result in more rapid coverage, but it will also increase the installation cost. To assure even and maximum coverage, groundcovers should be installed in a staggered planting pattern, Figure 20-11.

Due to the shallow root system of groundcovers and bedding plants, the plants can dry out easily. Prior to transplanting, each plant must be watered thoroughly. As each is set into the soil, it should be watered again. Thereafter, the new planting must be watered frequently and deeply to establish the plants successfully.

Groundcovers will benefit from mulching to reduce weeds and, most importantly, aid in preventing alternate freezing and thawing of the ground. Such ground activity can result in heaving the

Figure 20-10 With its plastic pot removed and the bed properly conditioned, this geranium is easily installed using a hand trowel.

groundcovers to the surface where their roots are exposed to the cold and drying air. To be successful, mulch should be applied to a new groundcover planting after the ground has frozen, to prevent premature thawing.

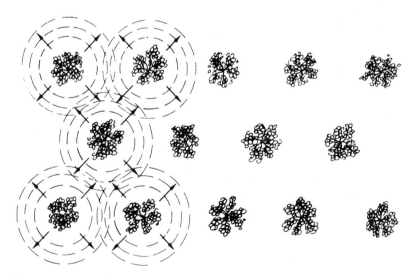

Figure 20-11 Alternating the placement of groundcovers fills space most efficiently.

Bulbs

Flowering bulbs require a rich, well-drained soil. They are planted either in flower beds and borders or as masses in the lawn. They may be gently tossed by the handful into open, turfed lawn areas to be planted wherever they land, in an irregularly spaced pattern. Bulbs are planted at differing depths and spacings, depending upon their species. Table 20-1 lists the depths and spacings of some common bulbs.

Bulbs are always set into the ground with the base oriented downward and the shoot pointed upward, Figure 20-12. Many can be installed with a bulb planter.

Other bulb-like structures, called tubers and rhizomes, are installed in mounded holes that permit the structure to be oriented horizontally and the roots directed downward, Figure 20-13. As always, the backfilling step should be done

carefully to assure that no air pockets form. Water collecting around the bulbs in air pockets can promote rotting.

Mulching

All plants benefit from mulching after installation. Mulching refers to the application of loose aggregate materials to the surface of a planting bed. The materials may be organic or inorganic. Examples of both types and their advantages and disadvantages are listed in Table 20-2.

The benefits of mulching a newly installed plant are that:

1. Water is retained in the soil around the root system and wilting is avoided.
2. Weed growth is discouraged.
3. The aesthetic appearance of the planting is enhanced.
4. Soil temperature fluctuation is minimized, preventing winter heaving of bulbs and groundcovers. Repeated freezing and thawing of the soil around a plant's base can also damage the bark and permit the entry of pathogens or insects.

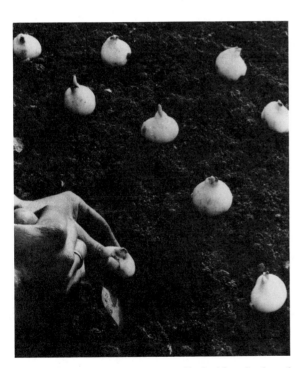

Figure 20-12 Bulbs are installed either by hand (as shown) or with a bulb planter. They are set into the ground "noses up" and are covered with soil pressed firmly to prevent air pockets. (Courtesy of USDA)

Figure 20-13 This iris rhizome is installed by spreading its roots evenly over a mound of soil, then backfilling to cover the roots, while allowing the leaves and top of the rhizome to remain exposed.

Table 20-1 A Guide to Bulb Installation

Plant	Depth to Top of Bulb	Spacing
Amaryllis	Leave upper 1/3 of bulb exposed	One bulb per pot
Anemones	2 inches	12 inches
Bulbous iris	2 inches	12 inches
Caladium	2–3 inches in North/1 inch in South	As desired for effect
Calla lily	Leave upper 1/3 of bulb exposed	One bulb per pot
Cannas	3–4 inches	1 1/2–2 feet
Crocus	3 inches	2–4 inches
Daffodil	4–5 inches	6–8 inches
Dahlia	5–6 inches	2–3 feet
Elephant ears	Just below soil surface	As desired for effect
Gladiolus	3–4 inches	6–8 inches
Grape hyacinths	3 inches	2–4 inches
Hyacinths	4–6 inches depending on size	6–8 inches
Lilies	Two to three times the thickness of bulb	1 foot
Paperwhite narcissus	Just below soil surface	One or two bulbs per pot
Ranunculus	2 inches	12 inches
Snowdrops	3 inches	2–4 inches
Summer hyacinths	4 inches	6–8 inches
Tuberous begonia	Just below soil surface	6–8 inches
Tulips	4–5 inches	6–8 inches

Table 20-2 Characteristics of Common Mulches

Organic Mulches (peat moss, wood chips, shredded bark, chipped corncobs, pine needles)	Inorganic Mulches (marble chips, crushed stone, brick chips, shredded tires)
• reduce soil moisture loss	• reduce soil moisture loss
• often contribute slightly to soil nutrition	• do not improve soil nutrition
• may alter soil pH	• seldom alter soil pH
• are not a mowing hazard if kicked into the lawn	• are a hazard if thrown by a mower blade
• may be flammable when too dry	• are nonflammable or fire-resistant
• may temporarily reduce nitrogen content of soil	• have no effect upon nitrogen content of soil
• require replacement due to biodegradation	• do not biodegrade
• may support weed growth as they decompose	• may retain excessive amounts of solar heat

For mulches to be effective, they must be applied 3 to 4 inches deep. A shallow layer of mulch does not reduce sunlight enough to discourage weed seed germination, retain moisture, or prevent changes in the surface temperature of the soil. If a more shallow layer of mulch is desired, weeds can still be controlled and water retained by spreading weed-barrier fabric around the plant base and adding 1 or 2 inches of mulch to weigh it down. The fabric prevents sunlight from penetrating and promoting weed growth. It also creates a physical barrier that weeds cannot grow through. This mulching technique works well on flat land but is less satisfactory for use on slopes, because rainwater tends to wash the mulch off the fabric surface. It must also be used cautiously on heavy clay soils because the fabric may cause the soil to hold too much water, drowning the new transplant.

Using Antitranspirants

Antitranspirants, also called **antidesiccants**, are chemicals that reduce the amount of water plants lose through transpiration. Antitranspirants are useful because excessive water loss can result in transplant shock. They normally act either to induce closing of the stomata or to cover the stomata with a water-impermeable coating. Several popular brand names are available. The antitranspirants are sprayed onto the plant before and after transplanting. Since most plant stomata are present in the greatest numbers on the lower surfaces of leaves, the underside of the canopy should receive thorough coverage.

Antitranspirants are of greatest benefit in the transplanting of deciduous trees and shrubs that are in leaf. They are also of benefit to evergreens, especially broad-leaved forms. Any evergreen will benefit from antitranspirants if it is transplanted in the fall, right before the dry winter period.

Problems of Arid Regions

The landscaper installing plants in the American Southwest encounters four distinct problems:

1. The soil quality is generally poor.
2. Irrigation is necessary throughout the year.
3. Higher altitudes can produce extremely hot daytime temperatures and very cool nights.
4. High winds dry out plants quickly and often damage them physically.

Arid soils generally fall into three categories: pure sand or gypsum, adobe, and caliche. Sand lacks both nutrient content and humus. **Adobe** is a heavy, clay-like soil that holds moisture better than sand but needs humus to lighten it and improve its aeration. **Caliche** soils are highly alkaline due to excessive lime content. They have a calcareous hardpan deposit near the surface that blocks drainage, making plant growth impossible. The hardpan layer may lie right at the surface or from several inches to several feet below ground level. The deposits may occur as a granular accumulation or as an impermeable concrete-like layer.

Generally, these are the characteristics of arid soils:

❦ Lack humus

❦ Require frequent irrigation

❦ Are nutritionally poor; nutrients are continually leached out by the irrigation water

❦ Are highly alkaline (pHs of 7.5 to 8.5 and higher)

❦ Are low in phosphate; phosphate may be rendered unavailable by the high pH

❦ Lack iron or contain it in a form unavailable to plants

❦ Have a high soluble salt content resulting from alkaline irrigation waters and from manures and fertilizers that do not leach thoroughly

When installing plants in the Southwest, the landscaper must add organic matter to the soils to improve their structure. Organic matter improves the water retention capability of light, sandy soils and breaks up heavy adobe soils. The only way to improve the drainage of caliche soil is to break through it and remove the impermeable layer. The excavated soil can be replaced with a conditioned mix that will support healthy plant growth.

To catch and retain the water so vital yet so limited in arid regions, the planting beds should be recessed several inches below ground level to create a catch basin, Figure 20-14. This method traps and holds applied water, preventing loss through runoff. In addition, organic mulches should be applied to a depth of 4 inches or more to slow moisture loss and create a cooler growth

Figure 20-14 A recessed planting bed creates a catch basin for moisture.

environment for the roots. Trunk wraps and white-wash paint are also applied to the trunks of trees to prevent water loss through their thin bark due to sun scald.

Cactus plants are sufficiently different from other plants to warrant special mention. They can be transplanted successfully by following these steps.

1. Before transplanting, mark the north side of the cactus. Orient this side of the plant to the north in its new location. The plant will have developed a thicker layer of protective tissue on its south side to withstand the more intense sunlight.

2. By trenching around the cactus, lift as much as possible of the root system.

3. Brush soil from the roots and dust them with powdered sulfur.

4. Place the cactus in a shaded area where air circulates freely, and allow the roots to heal for a week before replanting.

5. Plant the cactus in dry, well-drained soil. Stake the plant if necessary.

6. Water the plant in three or four weeks, after new growth starts. Thereafter, apply water at monthly intervals.

Whenever possible, native or naturalized plants should be selected for Southwestern landscapes. They have a better chance of surviving the transplant, and they keep maintenance costs down. In situations where the soil is especially unsuitable for planting, there may be little choice but to install the plants above ground in planters.

Achievement Review

A. Answer each of the following questions as briefly as possible.

    1. Identify the following tools, used in the installation of landscape plants.

a.

e.

b.

f.

c.

g.

d.

h.

i.

n.

j.

o.

k.

2. List the four qualities of good backfill soil.

3. Indicate whether the following are characteristic of bare-rooted (A), balled and burlapped (B), or containerized (C) plants.
   a. lightweight and easily transported
   b. retain the entire root system
   c. severely reduced root system
   d. usually the most expensive
   e. plants may become pot-bound
   f. permits large plants to be transplanted
   g. transplant season limited to early spring
   h. allows transplanting in any season

l.

4. What is the primary objective when deciding the timing of a plant transplant?

5. Why is early autumn the most desirable transplanting season for most plants?

6. Indicate whether the following are characteristic of inorganic (A) or organic (B) mulches, or both (C).

m.

a. reduce water loss from the soil

b. may alter soil pH

c. may temporarily reduce nitrogen content of soil

d. hazardous if thrown by a mower blade

e. do not biodegrade

f. sometimes are slightly nutritional

7. What benefit is gained by applying antitranspirants to new transplants?

B. Indicate if the following statements are true or false.

1. The root form of the transplant determines how the plant is installed.

2. Balled and burlapped plants are installed in a hole with a mounded bottom.

3. Burlap can be left around the ball of earth, but plastic mesh or wire baskets should be removed before planting.

4. Earthen rings to retain water and staking if necessary to assure alignment are used regardless of the root form.

5. Containerized plants are installed in a hole with a mounded bottom.

6. Plants are installed at the depth at which they were growing before relocation.

7. Bare-rooted plants are installed in a hole with a mounded bottom.

8. The planting hole for tree and shrub transplants should be wider and deeper than the root system it is to receive.

9. For rapid and maximum coverage, groundcover should be installed in staggered, alternating rows.

C. Installation of plants in the American Southwest is complicated by poor soil and the lack of adequate natural water supplies. What must a landscaper do to counter these threats to plant survival?

Selecting the Proper Grass

Objectives:

Upon completion of this chapter, you should be able to

- list six factors used in the comparison of different turfgrasses
- list the information required by law on grass seed labels
- explain the differences between single-species plantings, single-species blends, and mixtures of species

Turfgrasses are among the oldest plants used for landscaping. They are the most common choice for surfacing the outdoor room. A neatly trimmed lawn of good quality is not only comfortable to walk on, but ideal for many athletic and recreational activities. Because the growing point of the turfgrass is at the crown of the plant, near the soil, it is protected. This permits turfgrasses to be mown and walked upon repeatedly.

Comparison of Turfgrasses

Most turfgrasses used in landscapes are perennial, surviving from one year to the next. Nearly all species reproduce from seed, although several can be reproduced vegetatively, without pollination and seed production. A typical grass plant produces new leaves throughout the growing season. During the growing season, the turfgrasses will increase beyond the number of seeds sown. One of the objectives of good lawn development is to encourage turf growth as quickly and as evenly as possible.

Growth Habits

Grasses have differing growth habits, which result from the three different ways that new shoots are produced, Figure 21-1.

- Rhizome-producing (rhizomatous): A **rhizome** is an underground stem. New shoots are sent to the surface some distance out from the parent plant. Each new plant develops its own root system and is independent of the parent plant.

- Stolon-producing (stoloniferous): The shoots extending out from the parent plant are above ground. They are called **stolons**. New plants develop independently as described above.

RHIZOME-
PRODUCING

STOLON-
PRODUCING

BUNCH
TYPE

Figure 21-1 Growth habits of grasses

❦ Bunch-type: New shoots are produced from the sides of the plant, gradually increasing the plant's width.

Rhizome and stolon-producing grasses tend to reproduce more quickly and evenly than bunch-type grasses. Therefore, the bunch-type require more seed and closer spacing in order to cover an area quickly and without clumps.

Texture, Color, and Density

Grass **texture** is mostly a way of describing the width of the grass leaf (blade). The wider the blade is, the coarser will be the texture. Generally, fine-textured grasses are more attractive than coarse-textured grasses. They are also more expensive. The color of a grass and its density will also differ among species. Colors can vary from pastel greens to dark, bluish tones. **Density** refers to the number of leaf shoots that a single plant will produce. It can range from sparse to thick, depending upon the type of grass.

Size of Seed

The size of the seed is another reason for variation in the quality and quantity of grass seed mixes. Fine-textured grasses have very small seeds. Coarse-textured grass seeds usually are much larger. Thus, a pound of fine-textured grass seed contains considerably more seeds than a pound of coarse-textured grass seed.

Because of the greater number of seeds per pound, a pound of fine-textured grass seed plants a larger area of land. For example, a pound of fine-textured Kentucky bluegrass contains approximately 2,000,000 seeds. That number of seeds plants about 500 square feet of lawn. A pound of coarse-textured tall fescue contains 227,000 seeds; therefore, only 166 square feet can be planted with a pound of this particular seed.

Other comparisons can further point out the difference in seed sizes. For example, there are as many seeds in 1 pound of bluegrass as there are in 9 pounds of ryegrass; and as many seeds in a pound of bentgrass as there are in 30 pounds of ryegrass.

Soil and Climatic Tolerance

Most grasses, like almost all other plants, do best in good-quality, well-drained soil. However, every state contains landscape sites that fall short of the ideal conditions preferred for turfgrass success. Some grasses can adapt to a wide range of soil conditions, while others are very limited in their adaptability.

Similarly, some grasses tolerate high humidity and reduced sunlight; others do not. Some thrive in the subtropics and tropics; others are better-suited for temperate and subarctic regions.

Grasses are often grouped into two categories based upon the temperatures at which they grow best:

❦ Cool-season grasses are favored by daytime temperatures of 60 to 75 degrees F.

❦ Warm-season grasses are favored by daytime temperatures of 80 to 95 degrees F.

Figure 21-2 shows the peak growth rates of the two types of grasses. Knowledge of the optimum growing temperatures of grasses explains why northern lawns are often brown and dormant in midsummer when the days are very warm. Likewise, warm-season grasses do not really flourish in early spring and late fall when temperatures fall below the optimum temperature.

Common Warm-Season Grasses	Common Cool-Season Grasses
Bermuda grass	Kentucky bluegrass
zoysia grass	red fescue
centipede grass	colonial bentgrass
carpet grass	ryegrass
St. Augustine grass	
Bahia grass	

Figure 21-3 illustrates the climatic regions of the continental United States that are favorable for the growth of the grasses listed and others. Any seed purchased for planting in a certain climatic region should be composed of the appropriate grasses.

Use Tolerance

Under the same conditions of use, some grasses will survive and others will quickly wear away. Some will accept heavy use and recover quickly, while others will recover much more slowly. Some can accept the compaction of heavy foot traffic and still look good; yet others may discolor and slow their rates of growth.

Disease and Insect Resistance

Certain grasses possess greater resistance than others to insect and disease pests. The resistance may be natural or may have developed through the efforts of plant breeders. Many grasses are continually being improved by horticultural scientists searching for features such as better color, resistance to drought, shade tolerance, and pest resistance.

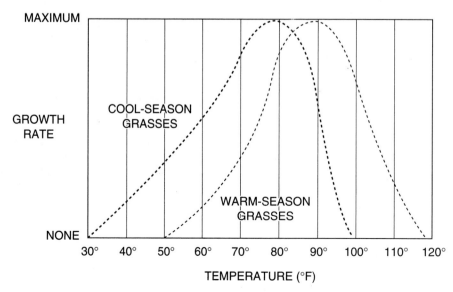

Figure 21-2 Relation of temperature to growth rate in cool-season and warm-season grasses

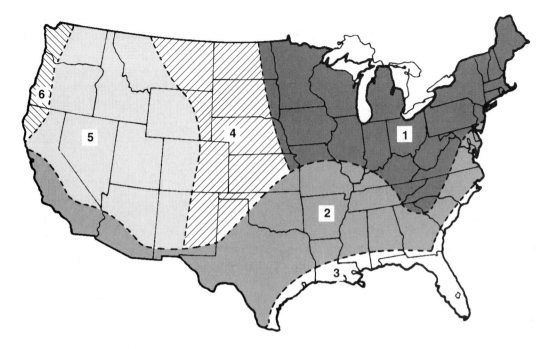

CLIMATIC REGIONS IN WHICH THE
FOLLOWING GRASSES ARE SUITABLE FOR LAWNS:

1. Kentucky bluegrass, red fescue, and colonial bentgrass. Tall fescue, Bermuda, and zoysia grasses in the southern part.
2. Bermuda and zoysia grasses. Centipede, carpet, and St. Augustine grasses in the southern part; tall fescue and Kentucky bluegrass in some northern areas.
3. St. Augustine, Bermuda, zoysia, carpet, and Bahia grasses.
4. Nonirrigated areas: crested wheat, buffalo, and blue grama grasses. Irrigated areas: Kentucky bluegrass and red fescue.
5. Nonirrigated areas: crested wheatgrass. Irrigated areas: Kentucky bluegrass and red fescue.
6. Colonial bentgrass, Kentucky bluegrass, and red fescue.

Figure 21-3 Regions of grass adaptations

In "A Comparison Chart for Turfgrasses" on pages 286–289 many of the most commonly used turfgrasses are compared.

Purchasing Grass Seed

Grass seed is sold in small quantities through retail outlets such as garden centers, supermarkets, hardware stores, and department stores. It is also sold in bulk amounts through wholesale suppliers. Professional landscapers usually purchase seed wholesale. However, most clients have purchased packaged seed from retailers in the past. They may not understand why the seed selected by the landscaper is priced higher than expected. Landscapers must be prepared to explain why all grass seeds are not alike and how the quality of seed is measured.

The key to determining the quality of grass seed is the seed analysis label. The **seed analysis label**, which by law must appear on every package of seeds to be sold, gives a breakdown of the contents of the seed package on which it appears. The analysis label may be on the package itself, or, if the seed is being sold in large quantities, on a label tied to the handle of the storage container.

A Comparison Chart for Turfgrasses

Grass Species	Cool Season or Warm Season	Growth Habit	Leaf Texture	Mowing Height/ Inches	Fertilization. Pounds of Nitrogen Per 1,000 Square Feet Per Year
Bahiagrass	Warm	Rhizomatous	Coarse	1 1/2 to 2	1 to 4
Bermudagrass	Warm	Stoloniferous and rhizomatous	Fine	1 to 2	4 to 9
Bentgrass, Colonial	Cool	Bunch-type (with short stolons and rhizomes)	Fine	1/2 to 1	2 to 4
Bentgrass, Creeping	Cool	Stoloniferous	Fine	1/2 or less	4 to 8
Bentgrass, Redtop (see Redtop)					
Bentgrass, Velvet	Cool	Stoloniferous	Fine	1/2 or less	2 to 4
Bluegrass, Annual	Cool	Bunch-type or stoloniferous	Fine	1	2 to 6
Bluegrass, Canada	Cool	Rhizomatous	Medium	Does not mow well	1 or less
Bluegrass, Kentucky	Cool	Rhizomatous	Fine	1 to 2 1/2	2 to 6
Bluegrass, Rough	Cool	Stoloniferous	Fine	1 or less	2 to 4
Bromegrass, Smooth	Cool	Rhizomatous	Coarse	Does not mow well	1 or less
Buffalograss	Cool	Stoloniferous	Fine	1/2 to 1 1/2	1/2 to 2
Carpetgrass, Common	Warm	Stoloniferous	Coarse	1 to 2	1 to 2
Carpetgrass, Tropical	Warm	Stoloniferous	Coarse	1 to 2	1 to 2

Soil Tolerances	Climate Tolerances	Uses	How Established. If Seeded, Pounds Per 1,000 Square Feet
Infertile, acidic, and sandy	Subtropical and tropical	Utility turf; good for use along roadways	Seeded at 6 to 8
Does well on a wide range of soils	Warm temperate and subtropical	Sunny lawn areas; good general purpose turf for athletic fields, parks, home lawns	Plugging or seeded at 1 to 1 1/2
Moderately fertile, acidic, and sandy	Temperate and seacoastal	Areas where intensive cultivation is practical	Seeded at 1/2 to 2
Fertile, acidic, and moist	Subarctic and temperate	Golf greens and other uses where intensive cultivation is practical	Sprigging or seeded at 1/2 to 1 1/2
Moderately fertile, acidic, and sandy	Temperate and seacoastal	Shaded, intensively cultivated areas	Seeded at 1/2 to 1 1/2
Fertile, neutral to slightly acidic	Temperate and cool subtropical	Not planted intentionally; but common in intensively cultivated turfs during spring and fall	Does not apply
Infertile, acidic, and droughty	Subarctic and cool temperate	A soil stabilizer	Seeded at 1 to 2
Fertile, neutral to slightly acidic	Subarctic, temperate, and cool subtropical	Sunny lawn areas; good general purpose turf for athletic fields, parks, and home lawns	Seeded at 1 to 2
Fertile and moist	Subarctic and cool, shaded temperate	Some use on shaded, poorly drained sites	Seeded at 1 to 2
Infertile and droughty	Dry and temperate	A soil stabilizer	Seeded at 1 to 2
Does well on a wide range of soils; tolerant of alkaline soils	Dry temperate and subtropical	Useful in semiarid sites as a general purpose lawn grass	Seeded at 3 to 6
Infertile, acidic, and moist	Subtropical and tropical	Utility turf; good for use along roadways and as a soil stabilizer	Seeded at 1 1/2 to 2 1/2
Infertile, acidic, and moist	Humid subtropical and tropical	Utility turf; good for use along roadways and as a soil stabilizer; can be used as a lawn grass in tropics	Seeded at 1 1/2 to 2 1/2

A Comparison Chart for Turfgrasses *(continued)*

Grass Species	Cool Season or Warm Season	Growth Habit	Leaf Texture	Mowing Height/ Inches	Fertilization. Pounds of Nitrogen Per 1,000 Square Feet Per Year
Centipedegrass	Warm	Stoloniferous	Medium	1 to 2	1 to 2
Fescue, Chewings	Cool	Bunch-type	Fine	1 1/2 to 2	2
Fescue, Creeping Red	Cool	Rhizomatous	Fine	1 1/2 to 2	2
Fescue, Hard	Cool	Bunch-type	Medium	Does not mow well	1 or less
Fescue, Meadow	Cool	Bunch-type	Coarse	1 1/2 to 3	1 or less
Fescue, Sheep	Cool	Bunch-type	Fine	Does not mow well	1 or less
Fescue, Tall	Cool	Bunch-type	Medium to Coarse	1 1/2 to 3	1 to 3
Gramagrass, Blue	Warm	Rhizomatous	Fine	Does not mow well	1 or less
Redtop (a bentgrass)	Cool	Rhizomatous	Coarse	1 1/2 to 3	1 to 2
Ryegrass, Annual	Cool	Bunch-type	Medium	1 1/2 to 2	2 to 4
Ryegrass, Perennial	Cool	Bunch-type	Fine	1 1/2 to 2	2 to 6
St. Augustinegrass	Warm	Stoloniferous	Coarse	1 to 2 1/2	2 to 6
Timothy, Common	Cool	Bunch-type	Coarse	1 to 2	3 to 6
Wheatgrass, Crested	Cool	Bunch-type	Coarse	1 1/2 to 3	1 to 3
Zoysiagrass (Japanese lawngrass)	Warm	Stoloniferous and rhizomatous	Medium	1/2 to 1	2 to 3
Zoysiagrass (Manilagrass)	Warm	Stoloniferous and rhizomatous	Fine	1	2 to 3
Zoysiagrass (Mascarenegrass)	Warm	Stoloniferous and rhizomatous	Fine	Does not mow well	2 to 3

Soil Tolerances	Climate Tolerances	Uses	How Established. If Seeded, Pounds Per 1,000 Square Feet
Infertile, acidic, and sandy	Subtropical and tropical	Utility turf; also usable as a low-use lawn grass	Seeded at 1/4 to 1/2
Infertile, acidic, and droughty	Subarctic and temperate	Shaded sites with poor soil	Seeded at 4 to 8
Infertile, acidic, and droughty	Subarctic and temperate	Shaded sites	Seeded at 3 to 5
Fertile and moist; not tolerant to droughty soil	Moist and temperate	A soil stabilizer	Seeded at 4 to 8
Widely tolerant of all but droughty soils	Moist and temperate	Utility turf; good for use along roadways	Seeded at 4 to 8
Infertile, acidic, well-drained, and droughty	Dry and temperate	A soil stabilizer	Seeded at 3 to 5
Does well on a wide range of soils	Warm temperate and subtropical	Utility turf; good for use along roadways; new cultivars (Brookston, Olympic, and Rebel) good for lawns	Seeded at 4 to 8
Does well on a wide range of soils	Dry and subtropical	Utility turf; good for use along roadways and in arid sites	Seeded at 1 to 2
Does well on a wide range of soils	Subarctic, temperate, and cool subtropical	Utility turf; good for use along roadways and in poorly drained areas	Seeded at 1/2 to 2
Fertile, neutral to slightly acidic and moist	Temperate and subtropical	Useful for quick and temporary lawns in the temperate zone and for winter color in the subtropic zones	Seeded at 4 to 6
Fertile, neutral to slightly acidic and moist	Mild and temperate	Used in mixed species lawns and as an athletic turf	Seeded at 4 to 8
Does well on a wide range of soils	Subtropical and tropical seacoastal	A good lawn grass with excellent shade tolerance	Sprigging
Fertile, slightly acidic, and moist	Subarctic and cool temperate	Utility turf; good for athletic fields in cold regions where preferable species won't survive	Seeded at 1 to 2
Does well on a wide range of soils	Subarctic and cool temperate	Useful as a general purpose turf on droughty sites	Seeded at 3 to 5
Does well on a wide range of soils	Temperate, subtropical, and tropical	Useful as a general purpose turf for home lawns, parks, and golf courses, especially in warmer regions	Plugging
Does well on a wide range of soils	Subtropical and tropical	A good lawn grass	Plugging
Does well on a wide range of soils	Warm subtropical and tropical	A soil stabilizer and groundcover	Plugging

While legal definitions vary somewhat from state to state, most analysis labels contain the following information:

Purity. The percentage, by weight, of pure grass seed. The label must show the percentage by weight of each seed type in the mixture.

Percent Germination. The percentage of the pure seed that was capable of germination (sprouting) on the date tested. The date of testing is very important and must be shown. If much time has passed since the germination test, the seed is older and less likely to germinate satisfactorily.

Crop Seed. The percentage, by weight, of cash crop seeds in the mixture. These are undesirable species for lawns.

Weeds. The percentage, by weight, of weed seeds in the mixture. A seed qualifies as a weed seed if it has not been counted as a pure seed or a crop seed.

Noxious Weeds. Weeds that are extremely undesirable and difficult to eradicate. The number given is usually the number of seeds per pound or per ounce of weed seeds.

Inert Material. The percentage, by weight, of material in the package that will not grow. In low-priced seed mixes, it includes materials such as sand, chaff, or ground corncobs. Inert material is sometimes added to make the seed package look bigger. At other times, the inert material is already present in the seed and is not removed because the cost involved would raise the price of the seed.

Three sample analyses follow. Study the contents of each mixture and determine which would probably cost the most and which the least.

It is likely that Mixture C would be the most expensive. It contains the highest percentage of fine-textured grasses, no coarse grasses, and the lowest percentage of weeds. Mixture A would probably cost the least, since it contains a high percentage of coarse-textured grasses, the lowest

percentage of fine grasses, and the greatest percentage of weeds. None of the mixtures is very poor in quality, since there are no crop or noxious weed seeds claimed by any.

Mixture A

Fine-Textured Grasses	
12.76 percent red fescue	85 percent germ.
6.00 percent Kentucky bluegrass	80 percent germ.
Coarse Grasses	
53.17 percent annual ryegrass	95 percent germ.
25.62 percent perennial ryegrass	90 percent germ.
Other Ingredients	
2.06 percent inert matter	
0.39 percent weeds—no noxious weeds	

Mixture B

Fine-Textured Grasses	
38.03 percent red fescue	80 percent germ.
34.82 percent Kentucky bluegrass	80 percent germ.
Coarse Grasses	
19.09 percent annual ryegrass	85 percent germ.
Other Ingredients	
7.72 percent inert matter	
0.34 percent weeds—no noxious weeds	

Mixture C

Fine-Textured Grasses	
44.30 percent creeping red fescue	85 percent germ.
36.00 percent Merion bluegrass	80 percent germ.
13.54 percent Kentucky bluegrass	85 percent germ.
Coarse Grasses	
None claimed	
Other Ingredients	
5.87 percent inert matter	
0.29 percent weeds—no noxious weeds	

Mixtures, Blends, and Single-Species Lawns

Grass seed is commonly purchased either as a mixture or a blend. It is also available as a single species (such as all Kentucky bluegrass or all Chewings fescue). A mixture combines two or more different species of grass. A blend combines two or more cultivated varieties of a single species. Both mixtures and blends have their places depending upon the site and circumstances. Mixtures are most common in temperate zone landscapes; single-species plantings are more common in subtropical and tropical landscapes.

Mixtures sometimes have the disadvantage of variegated color and texture. This is a result of the different species they contain. They have the advantage of being able to tolerate mixed environmental conditions and can recover from insect and disease pests that would wipe out a single species.

Single-species turf plantings offer a more uniform appearance than mixtures. However, a single-species planting is often unable to adjust to severe changes in environmental conditions. It can also be completely destroyed by a single insect or disease invasion.

Blends attempt to retain the advantages of both mixtures and single-species plantings. If the cultivated varieties of the blend are carefully selected, a blend offers these advantages: uniform color and texture, resistance to damage from environmental changes, resistance to wear, resistance to pest injury, and the varieties in the blend will have similar maintenance needs.

Achievement Review

A. Not all turfgrasses are alike. They can be compared using different factors. Insert the correct factor into each of the following sentences.

  1. Adapting to differences in pH, aeration, fertility levels, humidity, light, and temperatures measures a turf's _____.
  2. Rhizomatous, stoloniferous, and bunch-type are different _____ of grasses.
  3. Blade width, color variation, and the number of shoots per plant are measures of _____.
  4. The ability of turf to withstand the compaction of foot traffic indicates its level of _____.
  5. Certain grasses will suffer pest damage more than other grasses because of their _____.
  6. One pound of fine-textured grass differs from a pound of coarse-textured grass in many ways. One way is in the number and _____ of the seeds.

B. What could cause a very high-quality grass seed purchased in the south to be unsuitable for planting in the north?

C. Of the three seed mixtures A, B, and C shown in this chapter, which mixture is most likely to result in a sparse second-year lawn? Why?

D. Why is a grass seed mixture usually preferable to a pure, single-species seed?

E. List and define the important terms found on a grass seed analysis label.

Suggested Activities

1. Grow some grasses. Start flats or flower pots of pure grass species in the classroom. Compare fine-leaf and broad-leaf types. If possible, also grow samples of warm-season and cool-season grasses for comparison.

2. Obtain several grass seed mixtures from various sources and in as many price ranges as possible. Rank the mixtures on the basis of package appearance, advertised claims, and brand names. Rank the mixtures again, using the seed analysis labels as the measure. How closely do the package claims match the actual facts about the mixture as shown on the labels? How closely does the price ranking follow the quality ranking?

3. Make a seed count. Weigh 1/4-ounce quantities of a fine-textured grass and a coarse-textured grass. Be as accurate as possible. Count the number of seeds in each measure. Do the fine-textured seeds outnumber the coarse-textured seeds?

   NOTE: Do not use redtop for the coarse-textured grass in this exercise. Its seeds are atypically small for a coarse grass.

CHAPTER
22

Lawn Construction

Objectives:

Upon completion of this chapter, you should be able to

- describe four methods of lawn installation
- outline the steps required for proper lawn construction
- explain how to calibrate a spreader

Selecting the Method of Lawn Installation

There are four methods that may be used to install a turfgrass planting:

- Seeding
- Sodding
- Plugging
- Sprigging and stolonizing

The method selected depends upon the species of grass, the type of landscape site, and how quickly the turf must be established.

Seeding

Seeding is the most common and least expensive method of establishing a lawn. The seed can be applied by hand or with a spreader on small sites. On large sites, a cultipacker seeder (pulled by a tractor) or a hydroseeder (a spraying device that applies seed, water, fertilizer, and mulch at the same time) may be used. The hydroseeder is especially helpful for seeding sloped, uneven areas, Figure 22-1.

Figure 22-1 Use of the hydroseeder gives rapid stabilization to this steep embankment. (Courtesy of USDA)

Sodding

When a lawn is needed immediately, sodding may be selected as the method of installation. Sod is established turf that is moved from one location to another. A sod cutter is used to cut the sod into strips. These are then lifted, rolled up, and placed onto pallets for transport to the site of the new lawn, Figure 22-2A, B, and C. At the new site, the sod is unrolled onto the conditioned soil bed. The effect is that of instant lawn, Figure 22-3. Sod is produced in special nurseries where it can be grown and harvested efficiently and in large quantities. Sodding is much more

Figure 22-2A A sod cutter being used on a sod farm

Figure 22-2C Pallets of fresh cut sod are carried from the field for transport to landscape sites.

Figure 22-2B The sod cutter removes the growing turf along with a thin layer of soil.

Figure 22-3 The sod is unrolled at the new location and laid in place. The effect is one of instant lawn.

costly than seeding. However, the immediacy is important to some clients, and it is necessary on sites where seed might wash away.

Plugging

Plugging is a common method of installing lawns in the southern sections of the United States. Certain grasses, such as Bermuda, St. Augustine, and zoysia, are not usually reproduced from seed. Instead, they are usually placed into the new lawn as **plugs** of live, growing grass. Since the growing season in the southern regions is longer than elsewhere, the plugs have time to develop into a full lawn. Plugging is a time-consuming means of installing a lawn, which is its major limitation. However, plugging is necessary for many warm-season grasses that are poor seed producers. On large sites, some mechanization of the planting is possible.

Sprigging and Stolonizing

Like plugging, sprigging and stolonizing are more commonly used with warm-season grasses than cool-season grasses. A **sprig** is a piece of grass shoot. It may be a piece of stolon or rhizome or even a lateral shoot. Sprigs do not have soil attached and so are not like plugs or sod. They are planted at intervals into prepared, conditioned soil. Several bushels of sprigs are required to plant 1,000 square feet. If done by hand, the process is slow and tedious. Mechanization can lessen the time required.

Stolonizing is a form of sprigging. The sprigs are **broadcast** (distributed evenly) over the site and covered lightly with soil. Then they are rolled or disked. Since each sprig is not individually inserted into the soil, this method is faster.

Proper Lawn Construction

If the lawn is to be of the best quality, it must be given every possible chance for success. Proper construction of the lawn is vital. Six steps should be followed by the landscaper to assure a successful beginning for the lawn.

- ❦ Plant at the proper time of year
- ❦ Provide the proper drainage and grading.
- ❦ Condition the soil properly.
- ❦ Apply fresh, good-quality seed, sod, plugs, or sprigs.
- ❦ Provide adequate moisture to promote rapid establishment of the lawn.
- ❦ Mow the new lawn to its correct height.

Time of Planting

Lawns in southern sections of the country require warm-season grasses. Such grasses grow best in day temperatures of 80 to 95 degrees F. It is most effective to plant them in the spring, just prior to the summer season. In this way, they have the opportunity to become well established before becoming dormant in the winter.

Cooler northern regions require cool-season grasses to yield the most attractive lawns. Bluegrasses and fescues germinate best when temperatures are in the range of 60 to 75 degrees F. These lawns thrive in locations where days are cool and nights are warm. The best planting time for these grasses is early fall or very early spring, prior to the ideal cool season in which they flourish. If cool-season grasses are planted too close to the intensely hot or cold days of summer and winter, they will die or become dormant before becoming well established.

Grading and Draining the New Lawn

Each time the rain falls or a sprinkler is turned on, water moves into the soil and across its surface. **Grading** (leveling land so that it slopes) directs the movement of the surface water. **Drainage** allows the water to move slowly down into the soil to prevent erosion or puddling.

Even lawns that seem flat must slope enough to move water off the surface and away from nearby buildings. If a slight slope does not exist naturally, it may be necessary to construct one. A fall (grade) of between 6 inches and 1 foot over a distance of 100 feet is required for flat land to drain properly. Failure to grade lawns away from buildings can result in flooded cellars and basements.

Drainage of water into and through the soil is important. Without a supply of water to their roots, neither grasses nor any other plants can live. Without water drainage past their roots, turf grasses and other plants can be drowned. Depending upon the soil in the particular lawn area involved, good drainage may require nothing more than mixing sand with the existing soil to allow proper water penetration. In cases where the soil is heavy with clay, a system of drainage tile may be necessary.

If drainage tile is needed, it should be installed after the lawn's grade has been established, but before the surface soil has been conditioned.

Regular 4-inch agricultural tile is normally used, placed 18 to 24 inches beneath the surface. Tile lines are spaced approximately 15 feet apart, Figure 22-4. Each of the lateral tiles runs into a larger main drainage tile, usually 6 to 8 inches in diameter. This, in turn, empties into a nearby ditch or storm sewer, Figure 22-5.

Where the soil is naturally sandy, no special consideration for drainage may be necessary.

Conditioning the Soil

Proper soil preparation requires an understanding of soil texture and soil pH.

Soil texture is the result of differing amounts of sand, silt, and clay in the composition of the soil. A soil that contains nearly equal amounts of sand, silt, and clay is called a **loam** soil. Loam soils are

Figure 22-4 Drainage tile installation (Drawing not to scale)

Figure 22-5 The flow of water through tile drainage system (Drawing not to scale)

excellent for planting. Soil textures such as sandy loam, clay loam, and silty clay loam are named for the ingredient or ingredients that make up more than one-third of the composition of the soil. For example, the composition of sandy loam is more than one-third sand. Silt and clay each make up more than one-third of the composition of silty clay loam, and hence, it is less than one-third sand. In conditioning soil for lawn construction, clay, sand, silt, or humus (organic material) may be added to bring the existing soil closer to a medium-loam texture.

Soil pH is a measure of the acidity or alkalinity of soil. A pH measurement of less than 7.0 indicates increasing soil acidity. As the pH increases beyond 7.0, the soil becomes more alkaline or basic. Most turf grasses grow best in soil with a neutral pH (expressed as 7.0) to slightly acidic pH (6.5).

The measurement of soil pH is obtained from a soil test. Soil tests are usually available through county Cooperative Extension Services. Also, pH test kits can be purchased at a reasonable cost, allowing landscapers to make their own determination of pH much more quickly.

If the pH of soil is too acidic, it is usually possible to raise the pH by adding dolomitic limestone. The limestone should be applied in the spring or fall. As the following chart illustrates, the amount applied depends upon the texture of the soil and how far the natural pH is from 6.5 to 7.0.

Amount of Dolomitic Limestone Applied Per 1,000 Square Feet of Lawn

Natural Soil	Soil Texture		
	Sandy	Loam	Clay or Silt
pH 4.0	90 lb.	172 lb.	217 lb.
pH 4.5	82 lb.	157 lb.	202 lb.
pH 5.0	67 lb.	127 lb.	150 lb.
pH 5.5	52 lb.	97 lb.	120 lb.
pH 6.0	20 lb.	35 lb.	60 lb.
pH 6.5–7.0	None needed	None needed	None needed

Where it is necessary to lower the pH of the soil to attain the desired level of 6.5 to 7.0, landscapers commonly use sulfur, aluminum sulfate, or iron sulfate.

The attainment of a suitable texture and the proper pH are very important in the conditioning of lawn soil. Equally important is the removal of stones from the surface layer of the soil, the loosening of the soil to a depth of 5 or 6 inches, and the incorporation of organic matter into the soil.

Stones may be removed by hand, by rake, or by machine. If the lawn is to be smoothly surfaced, even the smallest surface stones must be discarded.

Decaying organic matter creates **humus**, a valuable ingredient of soil. Humus aids the soil in moisture retention. It also helps air reach the soil. Organic matter can be added to the soil during its conditioning with materials such as peat moss, well-rotted manure, compost, or digested sewage sludge. The landscaper may choose the material that is most easily available and relatively low in cost.

All necessary soil additives (pH adjusters, organic matter, sand, and fertilizers) can be worked into the soil at the same time. This is done most effectively with a garden tiller, which also loosens the soil surface and breaks the soil into small particles, Figure 22-6. Once the soil has been properly conditioned, it is ready to plant.

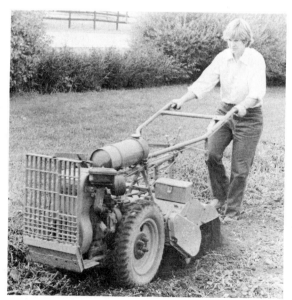

Figure 22-6 A large garden tiller turns the soil while working soil additives into it. For large lawn areas, the tiller is a necessity.

Planting the Lawn

Seed. Seed is applied to the prepared soil in a manner that will distribute it evenly. Otherwise, a patchy lawn develops. When applied with a spreader or cultipacker seeder, the seed may be mixed with a carrier material such as sand or topsoil to assure even spreading. The seed or seed/carrier mix is divided into two equal amounts. One part is sown across the lawn in one direction. The other half is then sown across the lawn at a 90-degree angle to the first half, Figure 22-7.

Placing a light mulch of weed-free straw over the seed helps retain moisture. It also helps to prevent the seed from washing away during watering or rainfall. On a slope that has not been hydroseeded, it is wise to apply erosion netting over the mulched seed to reduce the possibility of the seed washing away, Figure 22-8.

Sod. Sod must be installed as soon after it has been cut as possible. If not, the live grass will be damaged as a result of the excessive temperatures that build up within the rolled or folded strips of sod. Permitting the sod to dry out while awaiting installation can also damage the grass and result in a weak, unsatisfactory lawn.

The soil should be moist before beginning installation of the sod. The individual strips are then laid into place much as a jigsaw puzzle is assembled. The sod should not be stretched to fit, as it will only shrink back later, leaving gaps in the lawn surface. Instead, each strip should be fitted carefully and tightly against the other strips. Using a flat tamper or roller, the sod should be tamped gently to assure that all parts are touching the soil.

Plugs. Plugs are small squares, rectangles, or circles of sod, cut about 2 inches thick. Their instal-

Figure 22-7 Spreader application. Half the material is applied at a 90-degree angle over the other half.

Figure 22-8 Erosion netting is used here to prevent the grass seed from washing away until it can become established. Steep slopes such as this one are usually difficult to seed.

lation is similar to that of groundcovers. They are set into the conditioned soil at regular intervals (12 to 18 inches), in staggered rows to maximize coverage. The top of each plug should be level with the surface of the conditioned soil. The soil should be moist but not wet at the time of installation. This prevents some of the plugs from drying out while others are still being installed.

Sprigs. Sprigs are planted 2 to 3 inches deep in rows 8 to 12 inches apart. In hand installations, rows are not drawn. Instead, the sprigs are distributed as evenly as possible over the prepared soil surface and pushed down into the soil with a stick. As described earlier, stolonizing uses a top-dressing of soil over the sprigs and eliminates the need for individual insertion. The soil should be moist, but not overly wet, when planting begins. If the lawn area is large, planted areas should be mulched and lightly rolled as the installation progresses. Waiting until the entire installation has been completed could result in drying out of the sprigs.

The Importance of Watering. Water is essential to the growth of all plants. As long as grass is dormant in the seed, it needs no water. However, once planted and watered, the seed swells and germinates. At that point, an uninterrupted water supply is very important. The soil surface must not be allowed to dry until the grass is about 2 inches tall. Watering several times a day, every day for a month, may be necessary.

Caution should be taken to keep the new seedlings moist without saturating the soil. Too much moisture can encourage disease development. The use of a lawn sprinkler is a much better method than simply turning a garden hose onto the new grass. With a sprinkler, the water can be applied slowly and evenly.

The First Mowing. The first mowing of a new lawn is an important one. The objective is to encourage horizontal branching of the new grass plants as quickly as possible. This creates a thick (dense) lawn. The first mowing should occur when the new grass has reached a height of 2 1/2 to 3 inches. It should be cut back to a height of 1 1/4 to 1 1/2 inches. Thereafter, different species require differing mowing heights for proper maintenance. For the first mowing, it is a good practice to collect and remove the grass clippings. After that, clipping removal is usually unnecessary unless the grass has grown so tall between mowings that clumps of grass are visible on the lawn. Grasses used for soil stabilization do not require mowing.

Calibrating a Spreader

Two types of spreaders are used to apply seed, fertilizer, and other granular materials to lawns. These are the **rotary spreader** and the **drop spreader**. The rotary type dispenses the material from a closed hamper onto a rotating plate. It is then propelled outward in a semicircular pattern. The drop spreader dispenses the material through holes in the bottom of the hamper as it is pushed across the lawn, Figure 22-7. In both types, the amount of material applied is controlled by the size of the holes through which the material passes, and by the speed of application. Therefore, the spreader must be calibrated to dispense the material at the rate desired. For materials that are applied often and by the same person, the spreader need only be calibrated once, with the proper setting noted on the control for future reference. Different materials usually require different calibrations, even when the rate of application is the same.

The object of calibration is to measure the amount of material applied to an area of 100 square feet. A paved area such as a driveway or parking lot is an excellent calibration site. Afterwards, the seed or other material can be swept up easily for future use. Covering the area with plastic is also helpful in recollecting the material.

The spreader should be filled with exactly 5 pounds of the material being applied. Selection of a spreader setting near the center of the range is a good point at which to begin.

The material is applied by walking at a normal pace in a straight line. The spreader is shut off while it is being turned around. Each strip should slightly overlap the previous one. When an area of 100 square feet has been covered once, the spreader is shut off. The material remaining in the spreader is then emptied out and weighed. By subtracting the new weight from the original weight, the quantity of material applied per 100 square feet is determined. The spreader can then be adjusted to increase or reduce the rate of application.

Achievement Review

Briefly answer each of the following questions.

1. How does the cost of sodding compare to that of seeding?
2. Which method has a more immediate effect, seeding or sodding?
3. Which lawns are commonly started by plugging, sprigging, and stolonizing?
4. Which method of lawn installation and establishment requires the most time?
5. At what time of year should warm-season grasses be planted?
6. At what time of year should cool-season grasses be planted?
7. At what time of year should bluegrasses and fescues be planted? Why?
8. Why is it important that soil drain properly?
9. What size of agricultural drainage tile is recommended for lawn use, and how is it spaced?
10. Define soil texture.
11. What type of soil is considered ideal for planting?
12. Explain soil pH.
13. What is a neutral pH level?
14. If soil pH is raised, does the soil become more acidic or more alkaline?
15. If a sandy soil has a pH of 5.0, how many pounds of dolomitic limestone per 1,000 square feet are needed to raise the pH level to that required for a lawn?
16. If the soil mentioned in question 15 covers a lawn arexa of 3,000 square feet, how much limestone should the landscaper purchase?
17. What are the water requirements of a new lawn?
18. At what height should a lawn mover be set for the first mowing of a new lawn?
19. What is meant by the calibration of a spreader?
20. How is a spreader calibrated?

Suggested Activities

1. Invite a Cooperative Extension Service agent to visit the class for a discussion of soil testing. Ask the agent to demonstrate how a soil sample is collected and to explain how landscapers in the state can arrange to have soil tested.

2. Obtain several inexpensive pH testing kits. Bring in soil samples from gardens for testing.

3. Construct a lawn. If materials for proper lawn construction are available at the school, install a new lawn there. If budget restrictions prevent this, volunteer as a class to work for a nearby park, institution, or property owner in return for equipment and materials for the project.

4. Visit a sod farm if one is located in the area.

5. Borrow several spreaders for calibration. (Families of students might be one source.) If there is no budget for seed, substitute sand for demonstration purposes.

CHAPTER 23

Landscape Irrigation

Objectives:

Upon completion of this chapter, you should be able to

- describe historic and current uses of landscape irrigation
- distinguish between sprinkler and trickle irrigation use and understand key irrigation terms
- select and describe irrigation sprinkler heads
- explain precipitation rates and the determination of water needs by geographical area
- describe how irrigation pipe is sized and the available flow and pressure of water are determined
- explain drip tube and emitter placement
- explain simple irrigation designs

Irrigation is the supplying of water to land through artificial means. While the delivery systems have varied and changed over the years, the concept is not new. Once humans ceased to be hunters and gatherers and instead became farmers, the realization that water could be directed and relocated to help plants grow soon followed. Canals and channels that connect to nearby rivers have permitted water to flow across fields and plains yielding bounty from an earth otherwise incapable of bearing food. Runoff from the snow-capped peaks of distant mountains has been

collected and carried to farm fields and communities whose lifelines are literally the water pipelines that link their arid locales to the water sources.

As humans became garden builders, not just food producers, irrigation principles were found as applicable to the growth and maintenance of pleasure grounds as they were to agriculture. Princely potentates throughout Europe, Asia, and Africa commonly developed their gardens by redirecting the channels of streams and rivers to bring them into or near their properties. That nearby communities might be either flooded or deprived of their water supplies was often not a concern of the aristocratic garden builder.

Today, the irrigation of landscapes is common. Each year more properties receive designed irrigation systems for their lawns and plantings. These systems are gradually replacing the lawn sprinklers, garden hoses, and watering cans of years past. There are several reasons for the increased interest in landscape irrigation. Turf and ornamental plant research continually provides new information regarding the water requirements of specific plant species. Water needs are no longer guessing games. Automated irrigation systems also save water because they apply only the amount that can be absorbed, so there is no runoff. That savings is important for economic reasons, since water costs are rising. It is also important for environmental reasons. Controlled water application prevents water laden with fertilizers or pesticides from washing into waterways and sewers. Also, in regions of limited rainfall, landscape irrigation assures that none of the precious liquid is wasted.

Types of Landscape Irrigation Systems

Landscape irrigation water can be applied in two different ways. **Sprinkler irrigation** applies water under pressure through a delivery system that delivers the water over the tops of plants, Figure 23-1. **Trickle irrigation** supplies water directly to the root zone of the plants, Figure 23-2. The delivery system is low pressure and may be placed on top of or within the soil. Table 23-1 compares sprinkler and trickle irrigation systems. The table should be studied before proceeding further.

The Terminology of Landscape Irrigation

Since landscape irrigation is a somewhat specialized technical field, it has its own specialized vocabulary. An understanding of these terms can provide a basic comprehension of some of the technology of the field.

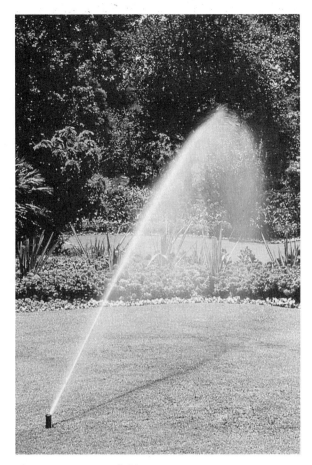

Figure 23-1 Sprinkler system

Sprinkler head: The device through which water leaves the pipe and is propelled onto the lawn or plantings.

Spray pattern: The specific distribution pattern of a specific sprinkler head.

Spray head: One of two types of sprinkler heads. Spray heads are stationary and made of two parts, the sprinkler body and the spray nozzle. The nozzles control the amount of water that leaves the spray head as well as the spray pattern.

Rotary heads: The other type of sprinkler head. Rotary heads move in a circle and propel water in tiny streams under pressure. Rotary heads can distribute water much further than spray heads are able to.

Figure 23-2 Trickle system (Photo courtesy of Edward J. Plaster)

Impact drive heads: One of two types of rotary sprinkler heads. As the pressurized water flows from the head it bumps a spring-loaded arm that slowly moves the head around in a circle. The repeating bumping of the arm makes this a somewhat noisy form of irrigation.

Gear drive heads: The other type of rotary sprinkler head. The head contains a number of gears that are turned by the water as it leaves the sprinkler. The movement of these internal gears is what causes the head to turn.

Precipitation rate: The amount of water placed over a landscape area. It is measured in inches of water per hour.

Geographical area: An area of the landscape with a specific water need, based upon the needs of the plants growing there and the rate of water absorption by the soil. A landscape may contain numerous and different geographical areas.

Emitter: The device that functions as a sprinkler head for trickle irrigation.

Discharge rate: The amount of water flowing from the irrigation system over a measured period of time. Sprinkler systems are measured and rated in gallons per minute (gpm). Trickle systems are slower and measured in gallons per hour (gph).

Pressure compensating emitters: Not all emitters can maintain the same discharge rate if the pressure of the water changes. Those that can are termed pressure compensating emitters. They are more expensive than noncompensating types.

Micro spray heads: Similar to the spray-head design described earlier, but with a discharge rate more like that of an emitter (in gph). The diameter of water throw is considerably reduced. It is used for small, specialized areas of the landscape such as flower beds and groundcover plantings.

Table 23-1 A Comparison of Sprinkler and Trickle Irrigation Systems		
Feature	**Sprinkler System**	**Trickle System**
Delivery System	Water is sent under pressure through polyethylene or polyvinylchloride pipe and delivered from either spray heads or rotary heads.	Water is sent under pressure through plastic or poly pipes or tubes and delivered from a small sprinkler head known as an emitter.
Delivery Site	Over the tops of the plant and onto the soil beneath	At the roots of the plants
Water Pressure at the Delivery Site	Varies with each system, but always greater than trickle systems. The pressure must propel the water through the air.	Low pressure resulting from either the small openings in the emitters or the long distance traveled through the pipes.
Speed of Delivery	Faster than trickle systems. Water delivery is measured in gallons per minute.	Slower than sprinkler systems. Water delivery is measured in gallons per hour.
Delivery of Pesticides and Fertilizers	As long as the materials are water-soluble, they can be delivered in the irrigation water. Some waste and/or chemical burning of foliage may result.	As long as the materials are water-soluble, they can be delivered in the irrigation water. Chemical burn is less likely. Plugging of the emitters is possible. Filtration of the water going into the system will help control the problem.
Water Runoff	Possible if the system is not set to deliver only what the soil can absorb	Little or none due to the slow speed of delivery
Weed Growth	Possible, due to the greater area of coverage	Less of a problem due to the more restricted area of coverage
Evaporated Water Loss	Water is lost to evaporation because it is thrown through the air, left atop foliage, and dispersed into soil outside the root area.	Little water loss because the water is never airborne or left on the tops of leaves to evaporate
Insects and Disease	Excessive moisture on foliage and soil can encourage pests to develop.	By eliminating the amount of wet foliage and soil, pest development is discouraged.
Installation Costs	More expensive due to the greater amount of hardware required	Less expensive due to the comparative simplicity of the system
Maintenance	Moderate maintenance. It is quickly apparent when all or parts of the system are malfunctioning.	High maintenance. Because the system cannot be seen working as easily as sprinklers are seen, it requires constant checking.

Irrigation pipe: Used for sprinkler irrigation systems, there are two types of pipe currently popular due to their superiority over metal pipe. Polyethylene pipe (PE) and polyvinyl chloride pipe (PVC) are widely used both separately and together.

Schedule rated pipe: A PVC pipe used as a covering or sleeve for thinner, weaker pipe when it passes beneath walks, roads, driveways, or walls. The higher the schedule rating is, the stronger the pipe will be.

Pressure rated pipe: A PVC pipe used to carry the water through an irrigation system. It is produced

in four forms. Each form is guaranteed to withstand a particular water pressure of either 125, 160, 200, or 315 pounds per square inch (psi).

Drip tubing: Used for trickle irrigation systems, the thin black tubing carries the water to the emitters. It is usually black in color and made of plastic.

Flow: The movement of water through the irrigation system.

Friction loss: The loss of pressure during water flow that results from the increasing speed of the water, the increasing length of the pipe, and the roughness of the lining of the pipe.

Velocity: The rate of flow, as expressed in feet of movement per second. For landscape irrigation, the ideal velocity of water is 5 feet per second.

Surge pressure: A shock wave type of pressure that results when water velocity is too high and then is suddenly stopped. Excessive surge pressure over time will weaken the pipe and connections within an irrigation system.

Static pressure: The pressure (in psi) that is present in a closed system, when the water is not flowing.

Dynamic pressure: The pressure (in psi) that is present at any one point in a system as a given quantity of water flows past that point. Dynamic water pressure varies within an irrigation system due to the rise and fall of water as the pipes move up and down across a landscape. Friction losses within the pipes also affect dynamic pressure.

Up to this point both sprinkler and trickle irrigation systems have been discussed simultaneously. Now the discussion will focus on each system separately.

Sprinkler Irrigation

Where large quantities of water must be applied in a controlled manner, sprinkler irrigation systems are the best delivery device. Golf courses and other large lawn areas as well as tree and shrub plantings are suitable for the rapid water delivery of sprinkler systems.

Sprinkler Heads

As noted earlier, there are two types of sprinkler heads, the spray head and the rotary. Spray heads are available as pop-up types that are recessed below mower level when not in use. The same water pressure that allows delivery also forces the spray head nozzle above ground level, giving the pop-up its name. Figure 23-3 illustrates several sizes of pop-up spray heads that are available.

Spray heads that do not recess are also available. They are needed for irrigating tall grasses and shrub beds. A plastic riser that may extend several

Figure 23-3 Several sizes of the pop-up spray heads that are available

feet above the ground permits a nozzle, specially adapted for shrubs, to deliver water to the plants.

Aside from popping up and down, spray heads have no moving parts. Nozzles are usually purchased separately from the spray head bodies. That permits the system designer to customize the irrigation system for each particular landscape. Nozzles may be selected to distribute water in a full circle, half circle, or quarter circle. They can also deliver water at standard volumes or at low gallonages if site conditions necessitate a slower rate of water application.

Spray heads have their limitations. They can only propel water outward to a distance of 14 to 16 feet from the nozzle before the wind disrupts the spray pattern. A disrupted spray pattern results in non-uniform water application. The overall effect can be inconsistent growth and troublesome wet or dry spots.

Rotary heads have moving parts. They may or may not pop up, depending upon the style, but all rotary heads move in full or partial circles and can propel water further than spray heads. Both the impact drive and gear drive heads are available in sizes that will throw water 110 feet or more from the nozzle. Figures 23-4 and 23-5 illustrate the impact drive and gear drive rotary heads.

In comparison with spray heads, rotary heads do not apply water as quickly. Like spray heads, rotary heads are also affected by the wind.

Figure 23-4 Impact drive rotary head

Figure 23-5 The Weather-matic Turbo J2 gear drive head

Precipitation Rates

Because spray heads apply water faster than rotary heads are able to do, it can be said that they have a higher precipitation rate (in inches of water per hour). Regardless of the type of landscape irrigation system selected, it is necessary to match the discharge rate of the system (in gpm) to the precipitation rate that will permit optimum absorption by the plants and meet their water needs, which are usually stated as so many inches of water per week. To convert from gallons per minute through inches per hour to arrive at inches per week and determine irrigation time is not impossible; but there are some necessary steps along the way.

STEP 1: Determine the precipitation rate (PR)

a. Calculate total gpm by totaling the gpm of all irrigation sprinkler heads in the area of the landscape that is of concern.

b. Determine the square footage (SF) of the area being covered.

c. Using a constant figure, 96.3, to aid the conversion, insert the data into the following formula:

$$PR = \frac{\text{Total GPM of an area} \times 96.3}{\text{SF of the area}}$$

STEP 2: Convert the precipitation rate from inches of water applied per hour to inches per minute using the formula:

$$\frac{PR}{60 \text{ minutes}} = \text{Inches of water per minute}$$

STEP 3: Determine how long the sprinklers must run to deliver the plants' weekly water needs.

    a. Learn the water needs of particular plants and plant types. See Table 23-2 for examples.

    b. Divide the plant's weekly water needs by inches of water per minute.

Example

Given: A 1,000 SF shrub installation needing 1.5 inches of water each week is irrigated by a system that delivers 5.2 gpm.

Question: How much irrigation running time is needed to deliver 1.5 inches of water?

Solution:

1. $PR = \dfrac{5.2 \text{ GPM} \times 96.3}{1,000 \text{ SF}}$

    $= .5 \text{ inch per hour}$

2. $= \dfrac{.5 \text{ inches per hour}}{60 \text{ minutes}}$

    $= .008 \text{ inches of water per minute}$

3. $= \dfrac{1.5 \text{ inches}}{.008 \text{ inches/minute}}$

    $= 188 \text{ minutes of operation}$

4. $= \dfrac{188 \text{ minutes}}{60 \text{ minutes}}$

    $= 3 \text{ hours and 8 minutes}$
    of operation

In the situation described in the example, the sprinkler irrigation would need to operate 3 hours and 8 minutes or 188 minutes each week in the shrub installation to deliver the amount of water needed by the plants.

Geographical Areas

Table 23-3 illustrates that not all plants require the same amount of water for optimum growth. To landscapers, it is no surprise that not all soils are alike either. They differ greatly in their ability to

Table 23-2	A Comparison of Plant Water Needs
Plant Type	**Water Needs**
Turfgrass	1.5 to 2.0 inches/week
Flowers	1.5 to 2.0 inches/week
Trees and shrubs	1.0 to 1.5 inches/week
Groundcovers	0.5 to 1.0 inches/week

absorb water. Generally, the more clay that a soil contains, the slower is its rate of absorption. The greater the sand content of a soil, the faster will be its rate of absorption. Table 23-3 illustrates some of the characteristics of different types of soil that affect their rate of water absorption.

A landscape that is large enough and of sufficient quality to warrant an irrigation system is likely to contain a mixture of plants. It may also have a mixture of soil conditions. Therefore one operational schedule that will serve all areas of the landscape is unlikely. Instead the landscape must be divided and grouped into geographical areas. Each geographical area will contain plants that have similar water needs and similar soil absorption rates. Additionally, allowance must be made for terrain factors such as slopes or depressions that will speed up or slow down water runoff. Other environmental conditions such as temperatures, drying winds, presence or lack of shade, and the number of competing plants can influence the availability of water to the plants' roots.

Once the landscape site has been analyzed and the number of geographical areas has been determined, such follow-up questions as these must be answered.

1. What sprinkler heads are needed for each geographical area?
2. How many are required and of what size?
3. What is the total gpm required by the sprinkler heads in each geographical area and for the total system?
4. How does the total system need compare to the total water flow available from the source of water?

Table 23-3 Soil Characteristics and Absorption Rates

Soil Type	Components	Water Intake Rate	Water Retention	Drainage	Water Absorption
Clay	Clay, with small amounts of silt and/or sand	Low to very low as the amount of clay increases	High to very high due to reduced air space for drainage	Poor to very poor	0.2 inches per hour (Low)
Loams	Equal or near-equal mixes of sand, silt, and clay	Moderately low to moderately high depending upon the amounts of clay or sand in the loam	Moderately low to moderately high depending upon the amounts of clay or sand in the loam	Good	1.0 to 1.7 inches per hour (Moderate)
Sandy	Sand, with lesser amounts of clay and silt	High to very high	Low to very low	Good to excessive	Greater than 1.7 inches per hour (High)

5. What sizes of pipe are needed, and what layout is best?

6. What static and dynamic water pressures are available?

Only after these questions and others are answered and the irrigation system is made operational will there be enough data to determine things such as the amount of running time for the system. Then, if it is found that a single weekly operation of the system for the calculated amount of time is more than the soil in a geographical area can absorb, it will be necessary to set up a schedule that will apply a lesser amount several times per week. The total amount of water applied will be the same.

Sizing Irrigation Pipe

Cost of the pipe is one of the most expensive factors in the price of an irrigation system. The larger the pipe diameter, the more costly it will be. Therefore, system designers must specify pipe that is large enough to carry the amount of water required by all sprinklers that lead off it, but it need be no larger. Pipe sizes will vary within an irrigation system, being larger in diameter near the water source and smaller as they are needed further from the source, Figure 23-6. Table 23-4 explains the relationship of irrigation pipe size to the amount of water that can be carried.

Matching Water Flow and Pressure with Pipe Size

To select and specify the most suitable pipe, the irrigation system designer must know the amount of water demanded by the sprinkler heads that are downstream from each section of pipe. *Downstream* is the term to describe the direction of water flow away from the source. Pipe that is upstream will be of wider diameter than pipe that is downstream because it must carry a greater volume of water. Figure 23-6A illustrates an irrigation pipe fitted with ten sprinkler heads. Each sprinkler needs 3 gpm to operate. In the illustration the total gpm that must be carried by each section of pipe is noted. In Figure 23-6B are seen the appropriate pipe sizes required to carry the volume of water at specific points.

Calculating Working Water Pressure

Complicating the match-up of water pressure and quantity with the appropriate pipe size is the problem that pressure varies, depending upon whether the water is static (still) or dynamic (moving). Pressure is also affected by the influence of gravity if water drops or rises from one level to another. Pressure will change as it is released downstream through valves along its route of flow and also in response to the roughness of the inside of the irrigation pipes (friction loss).

Figure 23-6A Pipe diameter diminishes as it moves downstream.

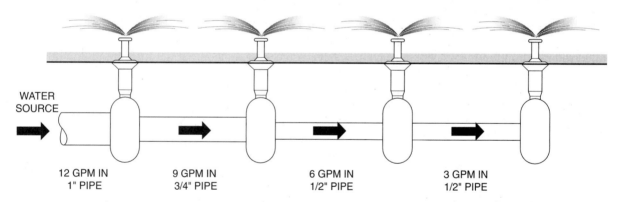

Figure 23-6B The pipe diameter must be able to carry the GPM required.

Table 23-4 Water Flow and Irrigation Pipe Size	
If the pipe must carry	**Select a pipe size of**
1 to 6 gallons per minute	0.50 inch
7 to 10 gpm	0.75 inch
11 to 16 gpm	1.00 inch
17 to 26 gpm	1.25 inches
27 to 35 gpm	1.50 inches
36 to 55 gpm	2.00 inches

Calculating Static Water Pressure

To calculate static water pressure at the system's main valve, the pressure at the mainline source must be adjusted by the pressure created by any elevation change that exists between the main line and the valve. This is done by multiplying the elevation change in feet by .433 pounds per square inch. This figure (.433 psi) is the constant weight of a 1-foot tall column of water. It is an accepted constant figure used in irrigation calculations. The following example will help clarify the calculation of static pressure.

Problem:

The water pressure supplied from the municipal source is 60 psi. From the main line, a pipe drops 15 feet to the valve that controls the flow of irrigation water. What will be the static pressure at the valve?

Solution:

Static pressure at valve

= Static pressure at main line
+ elevation change × .433 psi

Static pressure at valve

= 60 psi + 15 (.433 psi)

= 60 psi + 6.50

= 66.5 psi

Calculating Dynamic Water Pressure

To calculate moving water pressure it is necessary either to know or to calculate the friction loss as the water rubs against the inside of the pipe. It is also necessary to calculate the pressure lost as the water passes through valves, as it passes through the water meter, and as it rises and falls due to elevation changes.

As this chapter can only serve to introduce students to the subject of landscape irrigation, certain concepts must be explained with the understanding that further study and supplemental materials are needed if the student is to develop a true competency. Such is the case in describing the determination of friction loss and appropriate pipe sizes. Manufacturers of irrigation pipe provide irrigation system designers with pipe friction loss charts. There are different charts for different classes and types of pipes. Figure 23-7 illustrates a friction loss chart for a common class of plastic pipe. These charts record the friction loss or pressure loss for a particular size and material of pipe or piece of equipment at different flow rates and velocities. The charts also aid the systems designer in calculating the size of pipe needed, based upon the water flow through the pipe section.

Moving water pressure = Static pressure in the main line; minus the pressure loss in the water

meter, the friction loss in the pipe, the pressure loss in valves and fitting; plus or minus elevation pressure gains.

In the sample chart various pipe sizes are listed across the top. Beneath each size, two columns list the velocity in feet per second and the friction loss in pounds per square inch for various rates of water flow. The friction losses are for pipe lengths of 100 feet. For pipe lengths of less than 100 feet, the friction losses would be equivalently reduced.

Example

Given: Water flows through 100 feet of 2-inch pipe at the rate of 9 gallons per minute.

Question: What is the velocity of the water and the friction loss?

Answer: Velocity is 0.79 feet per second (fps). Friction loss is 0.06 pounds per square inch (psi).

Given: Water flows through 50 feet of 3/4-inch pipe at the rate of 10 gpm.

Question: What is the velocity of the water and the friction loss?

Answer: Velocity is 4.72 feet per second. Friction loss = 4.31 psi × .50 = 2.16 pounds per square inch.

It is important to know not only the friction loss but also the water velocity in order to minimize surge pressure in the pipes. Also known as the water hammer effect, surge pressure is likely to occur when the velocity goes above 5 fps. Since velocity and pressure loss increase within a pipe as it is required to carry increasing amounts of water, it is necessary to increase the pipe size when friction loss becomes great or when the velocity approaches the 5 fps that will trigger the water hammer effect.

Example

Question: If a 100-foot-long pipe has 20 gpm flowing through it, what is the smallest size of pipe that could be used?

Answer: From the chart in Figure 23-7, the first velocity below 5 fps is that for 1 1/4-inch pipe, with a velocity of 3.61 fps.

Question: What would be the friction loss in the above pipe?

(1120, 1220) SDR 21 C = 150   PSI Loss Per 100 Feet of Pipe (PSI/100 FT)   Sizes 3/4" thru 6"   Flow GPM 1 thru 600

	3/4"	1"	1 1/4"	1 1/2"	2"	2 1/2"	3"	4"	6"	
SIZE	3/4"	1"	1 1/4"	1 1/2"	2"	2 1/2"	3"	4"	6"	SIZE
OD	1.050	1.315	1.660	1.900	2.375	2.875	3.500	4.500	6.625	OD
ID	.930	1.189	1.502	1.720	2.149	2.601	3.166	4.072	5.993	ID
WALL THK	0.60	0.063	0.079	0.090	0.113	0.137	0.167	0.214	0.316	WALL THK

Flow G.P.M.	Velocity F.P.S.	P.S.I. Loss	Velocity F.P.S.	P.S.I. Loss	Velocity F.P.S.	P.S.I. Loss	Velocity F.P.S.	P.S.I. Loss	Velocity F.P.S.	P.S.I. Loss	Velocity F.P.S.	P.S.I. Loss	Velocity F.P.S.	P.S.I. Loss	Velocity F.P.S.	P.S.I. Loss	Velocity F.P.S.	P.S.I. Loss	Flow G.P.M.
1	0.47	0.06	0.28	0.02	0.18	0.01	0.13	0.00											1
2	0.94	0.22	0.57	0.07	0.36	0.02	0.27	0.01	0.17	0.00									2
3	1.42	0.46	0.86	0.14	0.54	0.04	0.41	0.02	0.26	0.01	0.18	0.00							3
4	1.89	0.79	1.15	0.24	0.72	0.08	0.55	0.04	0.35	0.01	0.24	0.01							4
5	2.36	1.20	1.44	0.36	0.90	0.12	0.68	0.06	0.44	0.02	0.30	0.01							5
6	2.83	1.68	1.73	0.51	1.08	0.16	0.82	0.08	0.53	0.03	0.36	0.01	0.24	0.00					6
7	3.30	2.23	2.02	0.67	1.26	0.22	0.96	0.11	0.61	0.04	0.42	0.01	0.28	0.01					7
8	3.77	2.85	2.30	0.86	1.44	0.28	1.10	0.14	0.70	0.05	0.48	0.02	0.32	0.01					8
9	4.25	3.55	2.59	1.07	1.62	0.34	1.24	0.18	0.79	0.06	0.54	0.02	0.36	0.01					9
10	4.72	4.31	2.88	1.30	1.80	0.42	1.37	0.22	0.88	0.07	0.60	0.03	0.40	0.01					10
11	5.19	5.15	3.17	1.56	1.98	0.50	1.51	0.26	0.97	0.09	0.66	0.03	0.44	0.01					11
12	5.66	6.05	3.46	1.83	2.17	0.59	1.65	0.30	1.06	0.10	0.72	0.04	0.48	0.02	0.29	0.00			12
14	6.60	8.05	4.04	2.43	2.53	0.78	1.93	0.40	1.23	0.14	0.84	0.05	0.56	0.02	0.34	0.01			14
16	7.55	10.30	4.61	3.11	2.89	1.00	2.20	0.52	1.41	0.17	0.96	0.07	0.65	0.03	0.39	0.01			16
18	8.49	12.81	5.19	3.87	3.25	1.24	2.48	0.64	1.59	0.22	1.08	0.09	0.73	0.03	0.44	0.01			18
20	9.43	15.58	5.77	4.71	3.61	1.51	2.75	0.78	1.76	0.26	1.20	0.10	0.81	0.04	0.49	0.01			20
22	10.38	18.58	6.34	5.62	3.97	1.80	3.03	0.93	1.94	0.32	1.32	0.12	0.89	0.05	0.54	0.01			22
24	11.32	21.83	6.92	6.60	4.34	2.12	3.30	1.09	2.12	0.37	1.44	0.15	0.97	0.06	0.59	0.02			24
26	12.27	25.32	7.50	7.65	4.70	2.46	3.58	1.27	2.29	0.43	1.56	0.17	1.05	0.07	0.63	0.02			26
28	13.21	29.04	8.08	8.78	5.06	2.82	3.86	1.46	2.47	0.49	1.68	0.19	1.13	0.07	0.68	0.02			28
30	14.15	33.00	8.65	9.98	5.42	3.20	4.13	1.66	2.65	0.56	1.80	0.22	1.22	0.09	0.73	0.02	0.34	0.00	30
35	16.51	43.91	10.10	13.27	6.32	4.26	4.82	2.20	3.09	0.75	2.11	0.29	1.42	0.11	0.86	0.03	0.39	0.01	35
40	18.87	56.23	11.54	17.00	7.23	5.45	5.51	2.82	3.53	0.95	2.41	0.38	1.62	0.14	0.98	0.04	0.45	0.01	40
45			12.98	21.14	8.13	6.78	6.20	3.51	3.97	1.19	2.71	0.47	1.83	0.18	1.10	0.05	0.51	0.01	45
50			14.42	25.70	9.04	8.24	6.89	4.26	4.41	1.44	3.01	0.57	2.03	0.22	1.23	0.06	0.56	0.01	50
55			15.87	30.66	9.94	9.83	7.58	5.09	4.85	1.72	3.31	0.68	2.23	0.26	1.35	0.08	0.62	0.01	55
60			17.31	36.02	10.85	11.55	8.27	5.97	5.30	2.02	3.61	0.80	2.44	0.31	1.47	0.09	0.68	0.01	60
65			18.75	41.77	11.75	13.40	8.96	6.93	5.74	2.35	3.92	0.93	2.64	0.36	1.59	0.10	0.73	0.02	65
70					12.65	15.37	9.65	7.95	6.18	2.69	4.22	1.06	2.84	0.41	1.72	0.12	0.79	0.02	70
75					13.56	17.47	10.34	9.03	6.62	3.06	4.52	1.21	3.05	0.46	1.84	0.14	0.85	0.02	75
80					14.46	19.68	11.03	10.18	7.06	3.44	4.82	1.36	3.25	0.52	1.96	0.15	0.90	0.02	80
85					15.37	22.02	11.72	11.39	7.50	3.85	5.12	1.52	3.45	0.59	2.09	0.17	0.96	0.03	85
90					16.27	24.48	12.41	12.66	7.95	4.28	5.42	1.69	3.66	0.65	2.21	0.19	1.02	0.03	90
95					17.18	27.06	13.10	13.99	8.39	4.74	5.72	1.87	3.86	0.72	2.33	0.21	1.07	0.03	95
100					18.08	29.76	13.79	15.39	8.83	5.21	6.03	2.06	4.07	0.79	2.46	0.23	1.13	0.04	100
110					19.89	35.50	15.17	18.36	9.71	6.21	6.63	2.45	4.47	0.94	2.70	0.28	1.24	0.04	110
120							16.54	21.57	10.60	7.30	7.23	2.88	4.88	1.11	2.95	0.33	1.36	0.05	120
130							17.92	25.02	11.48	8.47	7.84	3.34	5.29	1.29	3.19	0.38	1.47	0.06	130
140							19.30	28.70	12.36	9.71	8.44	3.84	5.69	1.47	3.44	0.43	1.59	0.07	140
150									13.25	11.04	9.04	4.36	6.10	1.68	3.69	0.49	1.70	0.08	150
160									14.13	12.44	9.64	4.91	6.51	1.89	3.93	0.55	1.81	0.08	160
170									15.01	13.91	10.25	5.50	6.91	2.11	4.18	0.62	1.93	0.09	170
180									15.90	15.47	10.85	6.11	7.32	2.35	4.42	0.69	2.04	0.11	180
190									16.78	17.10	11.45	6.75	7.73	2.60	4.67	0.76	2.15	0.12	190
200									17.66	18.80	12.06	7.43	8.14	2.85	4.92	0.83	2.27	0.13	200
225									19.87	23.38	13.56	9.24	9.15	3.55	5.53	1.04	2.55	0.16	225
250											15.07	11.23	10.17	4.31	6.15	1.27	2.83	0.19	250
275											16.58	13.39	11.19	5.15	6.76	1.51	3.12	0.23	275
300											18.09	15.74	12.21	6.05	7.38	1.78	3.40	0.27	300
325											19.60	18.25	13.22	7.01	7.99	2.06	3.69	0.31	325
350													14.24	8.05	8.61	2.36	3.97	0.36	350
375													15.26	9.14	9.22	2.69	4.25	0.41	375
400													16.28	10.30	9.84	3.03	4.54	0.46	400
425													17.29	11.53	10.45	3.39	4.82	0.52	425
450													18.31	12.81	11.07	3.77	5.11	0.57	450
475													19.33	14.16	11.68	4.16	5.39	0.63	475
500															12.30	4.58	5.67	0.70	500
550															13.53	5.46	6.24	0.83	550
600															14.76	6.42	6.81	0.98	600

Figure 23-7 A friction loss chart for a common class of plastic pipe

Answer: 1.51 psi

Given: Water passes through 100 feet of 1 1/4-inch pipe at the rate of 20 gpm. The static pressure in the main line equals 66.5 psi, the friction loss in the 1 1/4-inch pipe is 1.51 psi, and the friction loss as water passes through the valve is 6 psi.

Question: What is the working (moving) water pressure?

Answer: Working water pressure = Static pressure − pressure loss in the valve − friction loss in the pipe, = 66.5 psi − 6 psi − 1.51 psi = 58.99 psi.

Once through the valve, the water flows toward the first sprinkler. Again the pipe must be sized to assume enough pressure to pop the sprinkler out of the ground and deliver the desired amount of irrigation water, while avoiding surge pressure.

If the first sprinkler is 25 feet from the valve, reference to the chart finds the friction loss to be 0.38 psi (1.51 for 100 feet of pipe × .25) in 1 1/4-inch pipe with a flow of 20 gpm. Subtracting the friction loss from the working pressure at the valve reveals a working pressure at the first sprinkler of 58.61 psi.

Sprinkler heads operate within ranges of pressure. Different types of sprinkler heads have different ranges of operation. Continuing with the above example, assume that the sprinkler head being used has a pressure range of 35 to 80 psi. The 58.61 psi of pressure at the first sprinkler head is within the operating range, so the head would pop up and deliver its prescribed amount of irrigation water. As an example, assume that the sprinkler heads of the system deliver 10 gpm. After 10 of the 20 gallons of water per minute are used at the first sprinkler, the remaining 10 gallons per minute flow on toward the second sprinkler, spaced 25 feet further away. Again, the pipe must be sized. From the sample chart in Figure 23-7, the smallest pipe size that will carry the 10 gpm with a velocity of less than 5 fps is 3/4-inch pipe. In a real situation, with a higher rate of flow and more sprinkler heads, it is sometimes wise to use a size larger than the smallest pipe possible in order to reduce the friction loss that can reduce working pressure. Conversely, when work-

ing pressure needs to be reduced to fit within the pressure range of the sprinklers, the use of a smaller pipe will increase the friction pressure and reduce working pressure in the pipe. A pressure regulator can also be placed along the pipe to reduce water pressure.

Selecting and Locating Sprinklers

Upon leaving the sprinklers under pressure, the irrigation water falls across the landscape in a pattern determined by the type of sprinkler head. Either circular or rectangular patterns are possible. The size of the sprinkler head is partly responsible for the amount of area covered. The other factor controlling area of coverage is the trajectory of the spray. Sprinkler head size is matched to the amount of water being delivered. Trajectory is the path of the water as it is propelled through the air. Flat or low trajectories are used in groundcover and beneath shrubs in planting beds. Higher trajectories are used for lawns and for plantings that are watered from overhead.

Sprinkler Pattern

Rectangular patterns fill prescribed areas that are, obviously, rectangular in shape. Circular patterns may be full or partial circles, in radii ranging from several feet to over a hundred feet. Figure 23-8 illustrates common circular patterns. An irrigation system designer must have a thorough knowledge of the hardware available before he or she can design an efficient system.

Locating the Sprinklers

The objective in the positioning of sprinklers is simple: to obtain an even distribution of water. Uneven irrigation causes inconsistent growth and wet and dry areas within the landscape. Sprinkler heads can be set out in either a square pattern or a triangular pattern, Figure 23-9. Triangular patterns are preferred because they give more even water distribution and waste less water. Both patterns rely upon overlapping of the spray arcs because

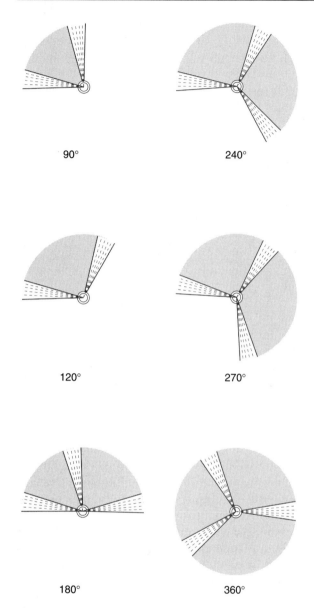

90°   240°

120°   270°

180°   360°

Figure 23-8 Common circular patterns

each sprinkler only delivers the full amount of water to 2/3 of its diameter. The outer third would be drier and the distribution of water uneven were it not for the overlapping.

In a triangular layout, the sprinkler heads are positioned as an equilateral triangle, with equal space between all sprinklers. Where wind is not a predictable factor, the sprinkler arcs should be

spaced at a distance that is 60 percent of the diameter of the arc. If wind is predictable and constant, spacing should be reduced. Manufacturers of sprinkler heads will provide recommendations for spacing of the sprinklers depending upon the various wind speeds. These recommendations should not be stretched. To do so will only result in uneven water distribution. In free-form or narrow lawn areas, the spacing may be crowded if necessary, but not stretched.

Problem:

To lay out the sprinkler heads in a square pattern for a rectangular lawn area that is 75 feet × 60 feet. Sprinklers will be used that require 15-foot spacing based upon wind conditions and the manufacturers' recommendations.

Solution:

1. Select the bordering edge that most needs to avoid overspray in order to prevent slippery walks or wet people. Divide the length of the border by the recommended spacing of the sprinkler heads. Round up to the next whole number if necessary.

    75 feet ÷ 15 feet = 5 spaces

2. Divide the other edge length by the recommended spacing of the sprinkler heads. Again round up to the next whole number if necessary.

    60 feet ÷ 15 = 4 spaces

3. Draw a grid across the scaled lawn, providing the calculated number of spaces. Locate sprinklers at all grid line intersections.

4. Select the appropriate sprinkler patterns to provide full coverage without overspray.

    NOTE: In the example the border length is divided evenly by the sprinkler spacing. When that is not possible, an additional calculation is needed to determine the actual spacing of the sprinkler heads.

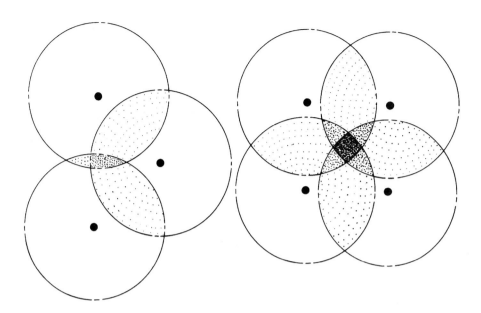

Figure 23-9 Triangular (left) and square (right) spray patterns

Example

If the border edge had been 79 feet, rather than 75 feet, the edge length would have been first divided by the ideal spacing:

.79 feet ÷ 15 feet = 5.27 = 6 (round up to a whole number)

then divided again by that result to determine the actual spacing of the sprinklers:

79 feet ÷ 6 = 13.17 feet

Problem:

To lay out the sprinkler heads for an 80 foot × 65 foot lawn area using a triangular pattern. Again, the sprinklers used will require 15-foot spacing to compensate for wind conditions.

Solution:

1. Select the border edge where overspray must be avoided. Divide the length by the recommended spacing of the sprinkler heads and round up to the next whole number if necessary.

   80 feet ÷ 15 = 5.33 = 6 spaces

2. Divide the border edge length by the number of spaces to determine the actual spacing of the sprinkler heads.

   80 feet ÷ 6 spaces = 13.33 feet

3. Divide the other edge length by the recommended spacing of the sprinkler heads, rounding up to the next whole number if necessary.

   65 feet ÷ 15 = 4.33 = 5 spaces

4. Divide the edge length by the number of spaces to determine the actual spacing of the sprinkler heads.

   65 feet ÷ 5 spaces = 13 feet

5. Draw a grid across the scaled lawn, providing the calculated number of spaces. Then locate the sprinkler heads in a pattern that approximates equilateral triangles as closely as possible.

6. Select the appropriate sprinkler patterns to provide full coverage without overspray.

Trickle Irrigation

For a quick review of the characteristics of trickle irrigation, consult Table 23-1. The system is useful in any situation where slow water application is preferable to sprinkler irrigation due to constraints of space, budget, or limited water supply. Instead of a sprinkler head, water is delivered from an emitter attached to a thin plastic or polyethylene tube. The delivery rate is measured in gallons per hour (gph), and is much slower than the gallons per minute delivery of sprinkler irrigation systems.

Emitters

There are two types of emitters: pressure compensating and non-compensating emitters. The difference is based upon their ability or inability to deliver a constant gph over a wide range of water pressures. In landscapes with rolling topographies, the rise and fall of the land causes drastic inconsistency in the water pressure in the thin tubes. Pressure compensating emitters are designed to allow for that variation and still deliver the intended amount of water and the desired rate. Non-compensating emitters cannot adjust to varied water pressures. If the pressure changes, their delivery rate changes too. In order to compensate for the effects of topographic variation on water pressure, it is common to use pressure regulators to reduce water pressure at points within the trickle system. Doing this permits the non-compensating emitters to deliver water at their intended rate of flow.

Tube and Emitter Placement

Drip tubing is not visually offensive and can be stretched on top of most planting beds or hidden inconspicuously beneath the mulch of the bed. Emitters may be attached directly to the tube or extended away from the tube using micro tubing. For large areas such as shrub plantings, multiple emitters are needed at a plant if sufficient water is to be provided.

In determining the number of emitters needed for a plant, the irrigation designer must consider the soil type and the subsurface wetting pattern that it promotes, the canopy area of the plant, and the percent of the plant's root zone that is to be wetted. It is only necessary to wet 50 percent of the root zone of most plants to provide the benefits of irrigation. Placement of the emitters should avoid direct contact with the trunk or stem of the plant. Failure to do so can result in soil pockets that are too wet or too long, and the health of the plant can be affected.

Subsurface wetting patterns are influenced by the type of soil being irrigated. In sandy soil, with its large particles, the water will move more vertically than horizontally. In fine soils like clay, with small soil particles, the movement of water is more horizontal than vertical.

Where the irrigation water contains high concentrations of sodium, or chloride, or both, trickle irrigation can encourage a dangerous salt build-up that can kill the plant. Improperly spaced emitters are the usual cause of the problem. Properly spaced, the emitters can both prevent the build-up and encourage the leaching of harmful salts away from the root zone.

Knowing the subsurface wetting patterns is important so that they can be overlapped and enough water applied to leach the salts away. Table 23-5 offers examples of the areas wetted by emitters in different soil types. A designer would need this information to plan for proper emitter placement.

The limitation of an introductory chapter to a complex subject is that it raises more questions in the reader's mind than it can answer. This chapter is not intended as a full coverage of landscape irrigation. It attempts only to introduce the subject to students and offer a few ideas and examples. Texts devoted solely to this subject should be consulted before attempting work with irrigation systems.

Table 23-5 Soil Types and Emitter Wetting		
Soil Type	Area Wetted Per Emitter	Diameter Wetted
Clay soil	65 to 160 square feet	9 to 14 feet
Loam soil	21 to 65 square feet	5 to 9 feet
Sandy soil	5 to 21 square feet	2½ to 5 feet

The Future

While landscape irrigation is not a technology still in its infancy, neither is it a technology standing still. In most parts of the country, water availability and quality are ongoing concerns. While the demand for landscape irrigation of home and commercial landscapes grows, the water resources needed to supply the systems do not grow. Many regions, especially in arid areas, are already using water faster than it can be cleansed and purified or replaced. In such regions the mere sight of an irrigation system in operation is, to some people, like waving a red flag in front of a bull. It is seen as a wasteful, aristocratic use of a precious and increasingly expensive resource.

For these and other reasons, the future of landscape irrigation will be shaped or altered by many factors. Among them:

❦ Turfgrasses, groundcovers, and other landscape plants will be bred, selected, and used on the basis of their low water requirements.

❦ Passive water conservation methods such as the proper placement of trees for shading buildings, lawns, and plantings will become more topical.

❦ Computer operation of irrigation systems as well as the use of computers to organize preventive maintenance programs for the system is now a reality. Computer-aided design of complex irrigation systems is also common today. The future will find even greater reliance upon the computer. It is not unlikely that a computer will enable the system to detect its own leaks, pinpoint the location, and send for repair assistance.

❦ Alternative sources of water for irrigation systems will be sought. Effluent water (treated sewage) and other types of reclaimed water may be possible sources.

❦ Irrigation systems will become more responsive to the actual need for water rather than merely using a timer that turns on the water whether it is needed or not.

❦ Systems may have sensors built in that will shut down the system when excessive winds cause inefficiency and waste.

❦ Systems may be powered by self-contained energy systems such as solar panels.

❦ The position of water manager will become a valued and professional field of career specialization.

Achievement Review

A. Briefly trace the development of irrigation from its agricultural beginnings to its present use in landscapes. Give examples of landscape irrigation systems in operation within your community.

B. Indicate if the following features are most typical of

(A) sprinkler systems or (B) trickle systems

1. Less expensive due to the comparative simplicity of the system
2. Maintenance is high since the system must be checked constantly to be certain it is working
3. Water delivery is over the top of the plants and onto the soil beneath
4. Water is delivered through an emitter
5. Water is delivered through spray heads or rotary heads
6. Water runoff is less due to the slow speed of delivery
7. Water pressure at the delivery site is the greatest

C. Insert the proper term to complete the following definitions.

1. _____ heads move in a circle and propel water in a circle.

2. The specific distribution pattern of a sprinkler head is its _____ .

3. _____ heads are driven by a spring-loaded arm that responds to water pressure and bumps the sprinkler around in a circle.

4. The amount of water placed over a landscape area by a sprinkler is the _____ .

5. The _____ is the delivery device for trickle irrigation.

6. The amount of water flowing from an irrigation system over a measured period of time is the _____ rate of the system.

7. The _____ of PVC pipe is a measure of its strength.

8. A PVC pipe used to carry water through an irrigation system is termed _____ pipe.

9. _____ is the loss of pressure during water flow resulting from the increasing speed of water, the increasing pipe length, and the roughness of the lining of the pipe.

10. Velocity is the rate of _____ , as expressed in feet per second.

11. When the water velocity is too high and is suddenly stopped, water hammer or _____ pressure can result.

12. The pressure present in a closed system, when the water is not flowing, is termed _____ pressure.

13. As water flows past a point and responds to rises or falls within the system, its pressure becomes _____ .

D. Indicate if the following characteristics of sprinkler heads apply to

    a. spray heads   c. both
    b. rotary heads   d. neither

1. The heads are available in pop-up styles.
2. They have no moving parts.
3. They are able to propel water across the greatest distance.
4. Water is distributed in full or partial circles.
5. Water is applied most rapidly.
6. Their spray pattern is affected by the wind.

E. State whether the following statements are (T) true or (F) false.

1. Spray heads have higher precipitation rate than rotary heads.

2. The discharge rate of the irrigation system must be matched with the rate at which plants can absorb the water.

3. Discharge rate is measured in gallons per minute.

4. Plant water needs are measured in gallons per week.

5. It is not possible to convert from gallons per minute to inches per week.

6. Not all plants require the same amount of water for optimum growth.

7. Sandy soil has a slower rate of absorption than clay soil.

8. Loams drain the best of all soil types.

9. Geographical areas contain a variety of plants and soils with a diversity of water needs and absorption rates.

F. As concisely as possible, answer the following questions in essay form.

1. Why does the size of irrigation pipe matter?

2. How do the terms *upstream* and *downstream* relate to water volume in an irrigation pipe?

3. What factors affect working water pressure?

4. What is a friction loss chart?

5. What relationship exists between water hammer, a water velocity of 5 fps, and the size of irrigation pipe?

G. Fill in the blanks to make the following paragraph read correctly.

In a trickle irrigation system, the water is delivered through _____. There are two types of these. They are pressure compensating and _____. Pressure compensating emitters deliver a constant rate of water over a wide range of _____. The use of _____ enables non-compensating emitters to do the same thing. In determining the number of emitters to specify, an irrigation designer must consider the _____ type and the _____ pattern, as well as the canopy of the plant and the percent of the _____ system that is to be wetted. It is only necessary to wet _____ percent of the root system of most plants to provide proper irrigation. Should the irrigation water contain high concentrations of sodium, chloride, or both, dangerous levels of _____ may build up in the soil.

CHAPTER 24

Interior Plantscaping

Objectives:

Upon completion of this chapter, you should be able to

- discuss the current status of the interior plantscape industry
- list problems unique to the interior use of plants
- describe the role of light quality, intensity, and duration
- list the characteristics of a good growing medium
- explain the steps in proper installation, watering, and drainage
- describe the interdisciplinary relationship required between architects, landscape architects, interior plantscapers, and maintenance professionals

Containerizing Plants, Past and Present

The human need and pleasure in having plants within our homes and places of business and play are rooted in antiquity. The ancient Greeks, Romans, and Egyptians commonly grew plants in containers to provide herbs for cooking, fragrance, and/or medicinal uses. To some extent most cultures, primitive or civilized, have used plants to add color, texture, softness, and a sense of life to their static, fabricated interior environments.

In America, the residential use of interior plantings has expanded from houseplants and holiday flowers to large trees and exotic desert and jungle species. The soaring popularity of interior plants began in the mid-1970s. America currently leads the world in the development and appreciation of interior plantscapes. Commercially, it is now expected that every new shopping mall, office building, hospital, or hotel will have plants featured prominently in the interior public areas. Most knowledgeable employers are aware of the increased productivity and reduced absenteeism of employees who have living, green plants as part of their working environment. Therefore, interior plantings are not limited to public areas; but are now evident within the nation's workplaces as well, Figures 24-1 and 24-2. Architects and interior decorators recognize that plants are as much a part of the indoor room as furniture, carpeting, and wall coverings. Properly selected and displayed, interior plantings duplicate the role and contribution of fine art to the mood and ambiance of a room.

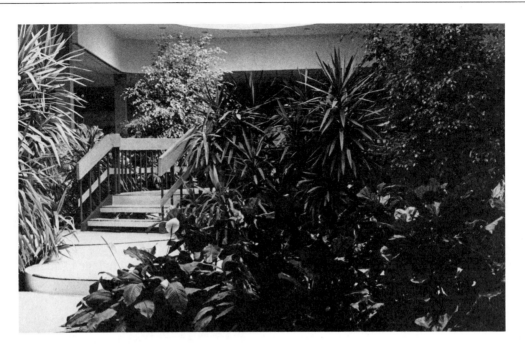

Figure 24-1 Interior plantscaping has expanded throughout the country. It enhances shopping malls and office complexes by providing a natural garden atmosphere.

Figure 24-2 Interior plantscaping creates attractive natural barriers to direct the flow of pedestrian traffic. It also provides an area where shoppers or office workers can relax in a peaceful environment.

So explosive was the demand for interior plantings in public buildings that the technology necessary to assure the plants' survival was not always able to keep pace. Some of the early efforts, heralded for their aesthetics, became corporate embarrassments a few months later. The lush green lobbies were replaced with dead stalks and leaf-littered carpets. Because the architects and decorators were unaccustomed to working with living materials, their failure to consider such foreign requirements as light, water, drainage, and fertilizer is understandable. Now they appreciate the differences between fabrics and philodendrons, and the mistakes of the past are less likely to be repeated.

The past decade has seen the technology needed to sustain plants indoors catch up to the public demand for lavish plantings. The new profession of interior plantscaping is still evolving and is still very labor-intensive. Certainly the next few years will find some advances being made in the way that interior plantings are prepared for installation, irrigated, and drained. New species will be tried.

Some will do well and gain public acceptance, while others will not adjust to the change from outdoors to indoors. Thus, while the profession is no longer in its infancy, it has not yet reached maturity. It is entering into a new stage of rapid and exciting growth; learning from past mistakes.

The Materials

As the profession has developed and changed, so has the array of plants used to create the interiorscapes. Yesterday's windowsill full of ivy and geraniums has been supplemented with containerized trees, in-ground installations, and lavish, changing displays of seasonal color.

Tropical foliage plants are the most popular and successful indoor plants for long-term installations. This is because they do not require the period of cool temperature dormancy that often makes temperate zone plants unsatisfactory. Originally collected from the jungles of South America, Africa, and other exotic locales, these plants are now grown in the greenhouses and nurseries of Florida, California, Texas, New York, Ohio, Pennsylvania, and Latin America, Figure 24-3.

Other plants finding their way into the residential and commercial interior plantscapes of America include subtropical plants such as cacti, traditional and short-lived flowering plants such as poinsettias and mums, and even turfgrasses, Figure 24-4.

Uniqueness of Interior Plantscapes

The transplanting of any plant from its production site to the plantscape site involves some risk even if it is an exterior plant. When the plant is intended for interior use, the relocation initiates problems that no exterior landscaper faces. Among them:

❦ A drastic reduction in the quality and intensity of light.

❦ Reduction and construction of the plant's root system.

❦ Replacement of natural rainfall by dependency upon humans for watering.

Figure 24-3 Tropical foliage plants are now grown in greenhouses and nurseries.

Figure 24-4 Poinsettia varieties grown for the winter holiday season

❦ A reduction in nutrient requirements and a potential for build-up of soluble salts (fertilizers).

❦ A lack of air movement and rainfall, allowing dust to accumulate on leaves, often plugging stomata and reducing photosynthesis.

❦ Potential damage by air conditioners, central heating systems, cleaning chemicals, water additives, and other irritants.

Under these conditions, plants may sustain themselves, but they seldom grow. Because of this, it is unnecessary to space plants for expansion; but it is necessary to install them in a manner that permits the replacement of dead or unsightly individuals as needed.

Light and Interior Plantings

Everyone knows that plants require light to survive and grow. Not everyone knows that there are differences in the kinds and sources of light. Also, few but horticulturists consider the amount of light that plants require. Yet despite how little some people know about the survival needs of interior plantings, expectations for survival of the plantings are always high and are frequently disappointed.

How long will plants live indoors? The answer depends upon the plant and the quality of installation and maintenance. However, in interior design as elsewhere, nothing lasts forever. Carpets wear thin, furniture nicks and sags; walls require fresh paint. Plants must be regarded as perishable furnishings as well. If installed correctly into a properly designed setting and maintained properly, the plantscape will serve satisfactorily for a time period that will unquestionably justify its cost. Then it will require at least partial replacement.

Light Intensity

Human activities do not require as much light as that required for the growth of plants. Even modern homes and buildings with extensive windows and skylights are unable to provide a light intensity equaling the outdoors. Glass filters the sun's light to the extent that an unshaded greenhouse still reduces the intensity of sunlight by at least 15 percent.

To understand light intensity requires knowing how light is measured. Light intensity is expressed in the units of **lux** or **footcandles**. A lux is the amount of illumination received on a surface that is 1 meter from a standard light source known as unity. A lux is an international measurement comparable in use to the metric system. In the United States, the footcandle unit is more commonly used and understood. One footcandle (fc) is equal to the amount of light produced by a standard candle at a distance of one foot. Direct-reading meters are manufactured that measure light intensity in footcandles up to 10,000. That is the intensity of sunlight on a typical clear, summer day in the temperate zone. In the south, intensity can approach 20,000 fc. A light meter is the only way to measure light intensity accurately and should be the first piece of equipment purchased by a beginning interior plantscaper, Figure 24-5.

The challenge of bringing plants accustomed to outdoor light intensities approaching 20,000 fc into a home or shopping mall is best appreciated through several examples. The average residential living room has a light intensity of 10 to 1,000 fc by day and as few as 5 fc at night. A good reading light provides 20 to 30 fc. A word processing operator may have 40 to 50 fc of illumination on the keyboard's surface. The average shopping mall provides 20 to 30 fc of light in pedestrian circulation areas and up to 100 fc in sales areas.

Acclimatization

Long before interior plantings are installed, the plants must be prepared to survive in their new, reduced light setting. The adjustment of an outdoor plant to interior conditions is known as **acclimatization**. Proper acclimatization changes the plant's

Figure 24-5 Light meter calibrated in foot candles

physiology (biological functioning) and morphology (physical structure). While there is still much to be learned about the adjustment of plants from outdoor to indoor growing sites, it is to the credit of forward-looking industry groups such as the Florida Foliage Association and the Interior Plantscape Division of the Associated Landscape Contractors of America that our present body of knowledge exists.

Acclimatization is directed at the adjustment of four vital factors that determine the survival of all plants: light, nutrients, moisture, and temperature. Under field or greenhouse conditions, the plants' metabolisms have been at or near optimum levels. As the times of harvest, transplanting, and interior installation nears, the metabolic activity must be slowed to minimal levels that permit survival and maintain an attractive appearance. Extensive additional growth is seldom an objective.

Light Intensity Acclimatization. The objective of light intensity acclimatization is to reduce the plant's light needs to a level where **photosynthesis** (the production of food) just slightly exceeds **respiration** (the use of food reserves for growth and maintenance). The point of exact balance is known as the **light compensation point** (LCP). Since a plant's ability to capture the energy of light declines as the plant ages, light levels in the interiorscape must be slightly above the LCP of the plant for long-term survival of the plant.

Light intensity acclimatization should begin at the greenhouse or nursery, prior to shipment to the interior site. Plant leaves that are produced in high light intensities are smaller and thicker than those produced under reduced light conditions. The smaller, thicker leaf reduces the potential of radiation damage from the sun. That danger does not exist in the interior setting. During the acclimatization period that may take as long as six months, the new leaves become thinner and larger to permit most efficient capture of the light energy that will sustain them in their interior location. The change in the leaf structure is both morphological and physiological.

Light intensity is reduced gradually over a period of several weeks or months. Each change reduces the light by 50 percent until the desired intensity (usually 100 to 200 fc) is reached. The acclimati-

zation process cannot be rushed without a severe reaction by the plant. That reaction may range from partial leaf drop to death.

Nutrient Acclimatization. Critical to the survival of the plant in its new interior setting is a healthy root system that is in balance with the leafy greenery of the plant. As the plant's ability to photosynthesize is scaled down by the reduced light intensity, so too is the amount of nutrients that need to be taken up by the roots in support of the plant's growth. Thorough soil **leaching** (flushing with water until it flows freely from the bottom of the container) at the beginning of the acclimatization and occasionally afterwards, will prevent a build-up of soluble salts (fertilizers). Soluble salt accumulations can damage the root system.

Following the leaching process, later fertilizations are less frequent than when the plant was growing under high light intensity. The reduction in fertilization must be coordinated with the reduction in light as closely as possible for successful acclimatization.

Moisture Acclimatization. Both the frequency of watering and the high humidity levels of the production area are gradually reduced as the plant is prepared for relocation. The interior site will be very stressful to the plant since office buildings and malls are kept at low humidity levels for greater human comfort.

Temperature Acclimatization. Production area temperatures are usually higher than human comfort levels in order to promote more rapid plant growth. During acclimatization, temperatures are gradually reduced to the 65 to 75 degree F range that is common to most interior areas.

Light Quality

Once acclimatized to the reduced light intensity of the interior site, the interior plantscape may still prove unsatisfactory if the quality of light is incorrect. **Light quality** is the color of light emitted by a particular source. The sun emits all colors of light, some of which the human eye can perceive and others that are unseen by humans but beneficial to plants. The green-yellow light most comfortable for humans is of little use in photosynthesis by plants. They depend on light from the blue and red bands of the visible light spectrum. Visible light is only a narrow region of the radiant light spectrum, Figure 24-6.

As long as both humans and plants can derive their light energy from the sun, the needs of each are satisfied. Indoors, however, where light energy is usually created or supplemented by artificial means, the quality of light can vary considerably. Light selected solely for its benefit to plants may cause the skin tones of human beings to appear ashen and deathly. In similar fashion, an interior decorator may specify a quality of lighting that gives human complexion a healthy glow while making nearby plantings appear brown and dead. Not all lamps provide the same quality of light, nor does any lamp provide a constant quality or intensity as the distance between the lamp and the plants changes. Categories and examples of lamps that have some use in interior plantscaping are shown in Table 24-1. Table 24–2 compares the lamps in all areas important to interior plant survival. You should study these tables thoroughly before proceeding further.

Table 24-1 Lamps for Interior Plant Illumination

Tungsten filament incandescent lamps
- Standard (the familiar household lightbulb)
- Reflector (spot or flood lights)
- Parabolic aluminized reflector (a weather-resistant type of floodlight with a more precise beam)
- Incandescent plant lamps (not proven to be any better than the standard incandescent)

Fluorescent lamps
- Cool white
- Warm white
- Plant lamps
- Wide spectrum plant lamps

High-intensity discharge lamps
- Mercury
- Metal halide
- High-pressure sodium

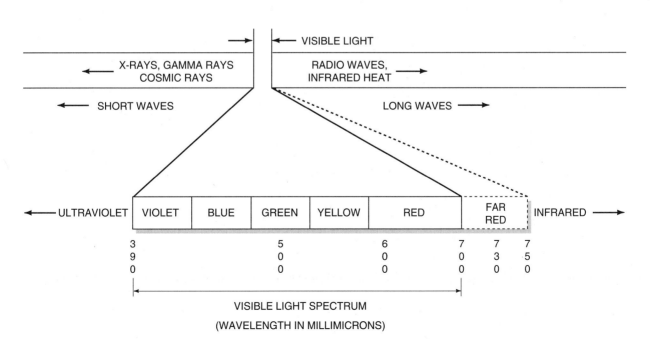

Figure 24-6 Electromagnetic spectrum and special distribution of visible light

Table 24-2 A Comparison of Artificial Lighting Sources for Interior Plantscapes

Lamp Type	How Light Is Produced	Quality of Light Produced	Percent of Visible Light Radiation	Color Rendition	Initial Cost	Operating Cost
Incandescent (all types)	Current flows through a tungsten filament heating it and making it glow.	High in red light; low in blue light	7–11	Good	Low	High
Cool White Fluorescent	Phosphor coating inside the glass tube is acted upon by radiation from a mercury arc.	High in blue and yellow-green light; low in red light	22	Good (blends with natural daylight)	Moderate	Moderate
Warm White Fluorescent	Phosphor coating inside the glass tube is acted upon by radiation from a mercury arc.	Low in blue and green light; more yellow and red light	22	Poor (blends with incandescent light)	Moderate	Moderate
Fluorescent Plant Growth Lamps	Same as other fluorescents. Special phosphors transmit most light energy in blue and red light regions of the spectrum.	High in red and blue light; low in yellow-green light	22	Average (enhances red and blue colors; darkens green colors)	Moderate	Moderate
Wide Spectrum Plant Growth	Same as other fluorescents. Special phosphors transmit most light energy in blue and red light regions of the spectrum.	Less blue and red than standard plant growth lamps; more far-red and yellow-green light	22	Average (favors red and blue colors; darkens green colors)	Moderate	Moderate
Mercury Deluxe white model, for interior plants)	An electric arc is passed through mercury vapor.	High in yellow-green light; less red and blue light, but still usable for plant growth	13	Poor (favors blue and green colors)	High	Moderate
Metal Halide	Similar to mercury lamps but with metal gas additives to produce a different spectrum	High in yellow-green light; less red and blue light, but still usable for plant growth	20–23	Good (similar to CW fluorescent)	High	Low

Life of the Lamp	Placement Height Above Plants	Plant Responses	Major Advantages	Major Disadvantages
750 to 2,000 hours	At least 3 feet to avoid foliage burn	Plants become long and spindly with pale foliage. Flowering is promoted, and senescence is accelerated.	• Good for special lighting effects • Compact source of light • Simple installation	• Energy inefficient; too much lost as heat • Light does not distribute evenly over a surface. • Glass blackens with time and light output is reduced. • Frequent replacement is needed.
Up to 20,000 hours	10 feet or less	Plants stay short and compact. Side shoots develop. Flowering extends over a longer period.	• Energy efficient • Heat is radiated over the length of the lamp, allowing closer proximity to plant foliage. • Light distributed more evenly over a flat surface	• Light does not focus well. • They are difficult to start when line voltage drops or humidity is high. • Installation is expensive. • Special fixtures are needed.
Up to 20,000 hours	10 feet or less	Same as CW fluorescent	• Same as CW fluorescent	• Same as CW fluorescent
Up to 20,000 hours	10 feet or less	Rich green foliage color. Large leaf size. Side shoots develop. Plants stay short. Flowering is delayed.	• Same as CW fluorescent • Light emission is from the region of the spectrum most important to photosynthesis.	• Same as CW fluorescent • Greater expense with little increase in benefit to the plants
Up to 20,000 hours	10 feet or less	Stems elongate. Side shoots are suppressed. Flowering is promoted. Plants age rapidly.	• Same as CW fluorescent	• Same as CW fluorescent • Growth may not be desired. • Poor color rendition on nonplant materials
Up to 24,000 hours	10–15 feet or more	Plants respond in a manner similar to CW fluorescent.	• Long life; useful for inaccessible fixtures • Medium energy efficiency	• Not interchangeable with other lamps • Warm-up time required
Up to 20,000 hours	10–15 feet or more	Plants respond in a manner similar to CW fluorescent.	• High energy efficiency, surpassing the mercury lamp • Good for both plant and general lighting	• Warm-up time required • Color and light quality change with operating hours.

Table 24-2 A Comparison of Artificial Lighting Sources for Interior Plantscapes

Lamp Type	How Light Is Produced	Quality of Light Produced	Percent of Visible Light Radiation	Color Rendition	Initial Cost	Operating Cost
High Pressure Sodium	Sodium is vaporized into an arc.	High in yellow-orange-red light	25–27	Poor (similar to WW fluorescent)	High	Low
Low Pressure Sodium	Sodium is vaporized into an arc.	High in yellow-orange-red light	31–35	Poor	High	Low

Interior plantings are seldom illuminated from a single source or type of light. For example, consider the plantscape of a typical office. The plants may be permanently set or in movable planters. Their functions may be to serve as room dividers, establish a mood, or relieve a cluttered desktop. The ceiling may be 8 or 10 feet high. Side windows or a skylight may admit some natural light. In such a setting, cool white fluorescent lighting would be ideal for both general lighting and the maintenance of the plants. People, plants, and furnishings look natural beneath cool white light due to its excellent color rendition. If additional task lighting is needed, small desk lights should be used. Special effects such as shadows or textured highlights can be created with incandescent lights installed beneath the plants and directed upward. (These are called uplights.) Some benefit will accrue to the plants from the addition of lighting at the base. However, if supplemental lighting is needed for photosynthesis, it is most efficient when applied from overhead.

A shopping mall presents a different situation. The corridors may have ceilings too high to permit the use of fluorescent lamps, which are not good for illumination when the ceiling is much beyond 10 feet. There may be skylights as well as decorative architectural lighting. Overall illumination by mercury or metal halide lamps would be best. As in an office, supplemental or decorative lighting might also be desirable for special effects or the health of the plants. When uplights are used, they should be installed directly into the planters and waterproofed. When supplemental lights are added for overhead illumination, they should be positioned to light the plants fully without shining in the eyes of viewers.

In both examples the total light radiation may be from several sources, some natural and some artificial. Add additional, perhaps uncontrollable, sources of light, such as store windows, electrical signs, and reflective surfaces, and the task of determining the total quality and quantity of illumination becomes complex.

Determining Total Radiation. Dr. Henry Cathey, director of the U.S. National Arboretum, and Lowell Campbell, an agricultural engineer at the U.S. Department of Agriculture, have developed conversion factors that enable light meter measurements of light from different sources to be converted to a common basis of measurement. The unit of measurement is watts per square meter, Table 24-3.

By taking independent light meter readings of the footcandles of illumination that fall upon the plants from each light source during the brightest periods of the day and during various times of the year, then multiplying by the conversion factor for that particular light source, the amount of radiation

(continued)

Life of the Lamp	Placement Height Above Plants	Plant Responses	Major Advantages	Major Advantages
Up to 24,000 hours	10–15 feet or more	Typical red-light plant responses; similar to fluorescent plant growth lamps when compared on equal energy	• High energy efficiency. When combined with blue light sources (such as metal halide), they provide good lighting for plants and people. • Long life	• Yellow color makes them unsatisfactory for general indoor lighting by themselves.
Up to 18,000 hours	10 feet or less	Plants respond in a manner similar to HP sodium	• The most energy efficient lamps available	• Yellow color makes them unsuitable where color rendition is important. They must be used at night after closing.

is determined. All readings should be taken at plant level. Each reading must be taken when the other light sources are off. Then the interior plantscaper can add the separate measurements together to determine if the proper total illumination is being provided. Cathey and Campbell report that most interior plantings can be maintained in a healthy state if given a minimum of 9 watts per square meter of illumination for 12 hours each day.

Table 24-3 Conversion Factors for Determining Watts Per Square Meter for Various Light Sources

Light Source	Conversion Factor
Daylight	0.055
High-pressure sodium lamps	0.034
Low-pressure sodium lamps	0.022
Metal halide lamps	0.034
Cool white fluorescent lamps	0.030
Warm white fluorescent lamps	0.030
Fluorescent grow lamps	0.044
Incandescent lamps	0.090
Mercury incandescent lamps	0.070

Natural Light

Most important of all light sources for interior plantscapes is natural sunlight when it can be planned for and depended upon. Each footcandle of illumination that nature provides is one less that has to be provided and paid for with artificial lighting. In our energy-conscious society, such savings are worth planning for. However, knowledge of how to maximize the benefits of natural light is vital; otherwise more heat energy is lost through inefficient windows than is gained in light energy.

Natural light is most helpful when it offers high levels of illumination throughout most of the year. Traditionally sunny areas like the Southwest can make better use of natural light than areas like the Northeast, where clouds block the sun and snow blankets the skylights through much of the winter season.

Sunlight entering from overhead is of greater use in the illumination of interior plantings than light entering from the side, although both are helpful. In neither situation will the natural light be as intense as outside light. It will be significantly reduced by the glass glazing through which it passes and the distance it travels between the point of entry and the leaf surface. Little usable light passes more than 15 feet beyond glass, so skylights in high lobby areas or shopping malls are of no benefit to plantings beneath them. Nevertheless, they can be of great benefit in single-story buildings with lower ceilings. In a similar fashion, a large interior plantscape may derive little benefit from side lighting, since usable light enters at a 45-degree angle, and plants must be placed within that narrow beam if they are to benefit, Figure 24-7. They cannot be too close to the glass or the foliage may burn, however. In a smaller room, as in a residence, natural side light can be of great value.

When skylights are used, they must be designed to permit the most light to enter while insulating against as much winter heat loss as possible. Within the limits allowed by heating and structural engineering, the ceiling well through which the light passes should be as wide and shallow as possible, with the sides painted a reflective white and beveled outward at 45-degree angles, Figure 24-8. As skylights become more narrow and the ceiling well deeper, there is less area through which the sunlight can enter and a narrower focus of illumination on the surface below. The difference between wide, shallow skylights and narrow, deep ones is similar to the difference in illumination between a floodlight and a spotlight. With natural light and interior plantings, the wide floodlight effect is most desirable.

Figure 24-7 Interior plantings receive the best quality of natural light within a 45-degree arc of the side windows.

Selecting the Correct Lighting

In summary, no single recipe for correct lighting can be given. There will be varied settings, needs, and objectives to accommodate. Plants will seldom be the only consideration in the selection of lamps

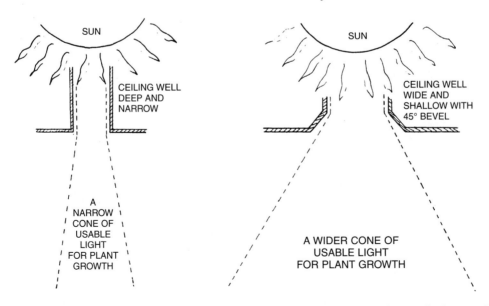

Figure 24-8 Cross-sectional view of skylights and the influence of their design on the amount of light available to plants

and the quality of illumination. When both plants and people are to be considered, a lamp should be selected that provides the yellow-green visible light needed to render human complexion, clothing, and furnishings attractive, while still providing sufficient blue and red light to allow photosynthesis to exceed respiration in the plants. The cool white fluorescent lamp is ideal for such a situation provided that the ceiling is not too high and that growth of the plants is not an objective. If growth is desired, additional incandescent lighting can be focused directly on the plants. The use of more expensive growth lamps is unnecessary since they have not been proven superior to the ordinary cool white fluorescent in maintaining plant health, and they do not render color well. Any natural light that can be used advantageously will reduce the cost of lighting the interior planting.

As for light intensity, not all plants available from growers have been carefully studied to determine the minimum at which they will survive attractively. The ALCA, the FFA, and the colleges of agriculture in the major foliage production states should be queried by anyone responsible for creating an interior plantscape. The many books that purport to give lighting specifications and that stock the shelves of libraries nationwide are satisfactory for homeowners but not for professional use. Their information is often dated, usually based upon greenhouse lighting, and may be questionable in its accuracy. Recommendations should be sought that give lighting minimums in footcandles, not in general terms such as high, medium, and low. When plants are installed without documented knowledge of their lighting requirements, the situation is risky at best and the plants must be watched carefully to determine if additional lighting is needed.

To assure that lighting is of the right intensity and duration, a simple timing device may be necessary. The lights must shine on the leaves for enough time each day (twelve hours minimum) to allow adequate photosynthesis to occur. Should the hours of lighting have to be reduced for some reason, the intensity of the lighting must be increased to compensate, Figure 24-9. It does not seem to matter whether foliage plants receive their needed lighting over a short or long period as long as the cumulative photosynthetic activity balances and slightly

exceeds respiration. With flowering plants, day length often plays a critical role in determining if and when the blossoms will appear. Since interior plantings are currently valued most for their foliage, day length is of limited importance.

The Growing Medium

No less important than light to the successful acclimatization of interior plants is the provision of a proper growing medium. The roots must be placed in an environment that provides structural support, allows the roots to absorb water, and provides essential minerals. Further, the growing medium must allow rapid drainage of water past the root zone and provide the correct pH for growth.

The medium that serves when plants are growing in a nursery field or production container is likely to be inappropriate for an interior installation. Natural field soil may be:

❦ too heavy to permit rapid drainage

❦ too heavy for the floor to support if the container is large

❦ inconsistent in composition, making standardized maintenance of separate planters difficult

❦ infested with insects, pathogens, or weeds

Although pasteurized natural soil may be a component of the growing medium for interior plantings, additives will probably be needed. It is even possible that the growing medium selected for the interior plantscape will have no natural soil

Figure 24-9 A timing system can be used to provide supplemental light for an interior plantscape.

in it for reasons of improved drainage, hygiene, pH balance, or nutritional consistency. The use of synthetic soils, whose composition is as controlled as a cake recipe, is becoming the rule rather than the exception.

Cornell University and the University of California have been leaders in the formulation of synthetic soils for interior plant production. The Cornell mixes are made from vermiculite or perlite and sphagnum moss. The University of California (U.C.) mixes are made from fine dune sand and sphagnum moss. The ratio of components varies with the species of plants and the maintenance program to be followed.

In addition to the U.C. and Cornell mixes (called the *peat-lite* mixes) there are bark mixes, composed of pine bark, sand, and sphagnum moss. These are especially acidic growing media and may require buffering (with dolomitic and hydrated lime) to sustain healthy plants.

All of the synthetic soils are mixed and sold commercially. In large interior installations, it may be more economical to mix the medium at the site rather than purchase the premixed commercial products. All decisions about the growing medium should be made before the planters are filled and the plants installed. Once planted, errors in the medium's composition are difficult to correct without removing the plant.

Because the artificial media have low nutritional value for the plant, complete fertilizers are required as well as periodic application of the minor elements. Regular soil testing is necessary to assure that the fertilization program is correct.

Installing the Plants

The acclimatization of a plant's root system involves establishment of the correct relationship between roots and foliage. Outdoors, a sizable root system is needed to supply adequate amounts of water and minerals to the leaves for near-maximal photosynthesis. Indoors, photosynthesis is reduced to survival-maintenance levels, so the root system need not be as large. Thus, one of the first steps at the time of installation is to remove the production

medium from around the roots and prune away excess roots, Figure 24-10.

Several methods can be used for setting plants into an interior plantscape, depending upon whether the plants are to be placed in ground beds or raised planters. The anticipated frequency of plant replacement will partially determine the method of installation. For example, large trees are often permanently planted in the floor of an enclosed shopping mall because they are too large and expensive to replace, Figure 24-11. The growing medium around them must permit good drainage and retain nutrients, while drainage tiles beneath the plant carry away excess water. Plants in such a setting must not be subjected to detergents and waxes used for floor maintenance, so special design provisions, such as raised edging, may be necessary.

Plants that are in raised planters or that require frequent replacement are usually not installed permanently. Instead, they are often planted in a growing medium within a nursery container with good drainage, Figure 24-12. The containerized plant is then placed in a support medium of peat, perlite, sand, or similar well-drained material. The support medium serves to retain the plants as well as permitting excess water to drain away from the

Figure 24-10 Excess roots are pruned off the foliage plant before transplanting to establish a more balanced relationship between the foliage and the root mass.

Figure 24-11 The grate around this permanently installed fig tree allows watering and fertilizing while protecting the root system from being trampled or compacted.

root zone and insulating the root system against abrupt temperature or moisture changes. A separating sheet (usually of fiberglass or a rot-resistant fine-meshed material) is placed between the support medium and the coarse gravel lining the bottom of the planter. The gravel is needed to facilitate drainage, and without the separator the growing medium and support medium would gradually wash into the gravel and plug it. A plastic tube inserted into the planter permits dip-stick testing to determine if the planter is being overwatered. It also provides a means of pumping out excess water if the planting is endangered by overwatering. Mulch on the surface of the planter serves to discourage moisture loss, provides a decorative appearance, and conceals the rims of the plant containers.

The same techniques used for planting containerized plants can be applied to in-ground plantings. If the planter drains directly into the building's drainage system, then the vertical plastic tube is unnecessary. All else remains the same except that there is no outer container.

PVC PIPE FOR MEASUREMENT OF WATER NEEDS OR EXCESSES

GROWING CONTAINER

DECORATIVE OUTER CONTAINTER

MULCH TO RETAIN MOISTURE

FILLER MATERIAL SUCH AS PEAT OR SAND OR FIRED CLAY TO SUPPORT AND CONCEAL GROWING POT WHILE ALLOWING AIR AND MOISTURE EXCHANGE

SEPARATOR TO PREVENT MIXING OF FILLER AND GRAVEL

GRAVEL TO PERMIT DRAINAGE OF EXCESS WATER

Figure 24-12 Installation of a containerized plant within an indoor planter

Watering and Drainage

As noted above, the ability of the growing medium to drain off excess water is critical. More interior plant deaths result from overwatering than for any other reason. All planters must permit the removal of standing water. The layer of coarse gravel already referred to is one method. Other methods include setting containerized plants on top of inverted pots within the larger planter and incorporating drains and spigots into planter bottoms. Where drainage from the planter base is planned, additional planning must assure that carpets, tiles, and other floor surfacings are not damaged by the runoff water. Sitting the plants on gravel beds into which the water can drain and then evaporate is one method.

Hand-in-hand with planning for drainage goes provision for proper watering. Some interior plantings require continuous moisture; others do better if permitted to dry out between regular waterings. Obviously the two types of plants would not coexist compatibly in the same planter.

An interior plantscape must be watered according to a schedule, and not according to the judgment of a custodian or other unqualified individual. The right watering frequency is most easily determined in controlled environments such as enclosed shopping centers. It is most difficult in locations where environmental variables are not stable, such as by open windows or in drafts near doors.

The need to water a planting can be determined by feeling the soil and observing its color. A gray surface color and failure of soil particles to adhere to the fingers indicate dryness. Moisture meters are also available for a more carefully controlled reading of the growing medium's water content.

Although some automated watering systems exist, there is surprisingly limited use of them in large installations. Nationwide, interior plantscapes utilizing hundreds of plants are all watered by hand. Most definitely, the technology of watering is still in the developmental stages.

When water is applied, it must be in a quantity adequate to wet the soil deeply, not shallowly. Shallow watering encourages shallow rooting and increases the vulnerability of the plants to damage from drying. Deep watering promotes deep, healthy rooting while providing the soil leaching necessary to prevent soluble salt buildup.

The quality of water used on the interior planting may vary with the location. The most likely source will be municipal water lines. Most public drinking water contains chlorine and often fluoride, as germicides and tooth-decay deterrents, respectively. Neither additive will harm plants under normal conditions. Although chlorine is potentially harmful, the amounts used in drinking water are dissipated by aeration as the water bubbles from the faucet or hose nozzle. More heavily chlorinated swimming pool or fountain water can damage plants and should never be used as a watering source.

NOTE: Plants grown around enclosed pools require good air exchange in the room or the chlorine gas from the pool may damage them.

Water is a source of soluble salts. Water that has been softened by means of cation-exchange softeners may be dangerously high in sodium, which can be toxic to plants. In buildings with such water softeners, alternate sources of water should be sought for the plantings. In regions of the country where the need for water conservation causes recycled water to be used on interior plantings, a chemical analysis of the water should be made to determine if any toxic chemicals are present that could damage the planting.

Other Concerns

As with any planting, certain routine procedures are needed to keep the plants healthy and attractive. The problems that are common to plants grown outdoors are similar to those that trouble plants grown indoors. In addition, indoor plantings often encounter stresses not common in outdoor landscapes.

Fertilization

Fertilization is needed to provide the mineral elements required for photosynthesis. Interior plantings need a complete fertilizer, but not as often as outdoor plantings, because the rate of plant growth is greatly reduced indoors. The ratio of the complete fertilizer should be fairly uniform; for example, 1-1-1 or 2-1-2. Too much nitrogen may lead to spindly, succulent, and unattractive vegetative growth. Trace elements will also need to be applied, especially if the growing medium contains no real soil. Frequent soil testing is necessary, regardless of whether the growing medium is true soil or synthetic. Excessive fertilizing follows only slightly behind overwatering as a major reason why interior plants fail to survive satisfactorily.

Either organic or inorganic fertilizers will work well, but the organics are generally slower in releasing their minerals for use by the plants. This means that soluble salt buildup is less frequently a problem. However, the odor of decomposing organic fertilizers may not be welcomed in such settings as shopping malls or library lobbies. Therefore, the use of organic fertilizers may not be practical in all situations.

Inorganic fertilizers are applied to large interior plantscapes in liquid form. This is faster and easier than applying dry granulars to each pot. When there are hundreds of pots and planters to maintain, speed is vital.

Humidity

Humidity is seldom a problem for plants growing outdoors, but it can cause problems for interior plantscapes. Because of the drying effects of central heating and air conditioning, interior plantings must adapt to an air environment that may contain half or less of the relative humidity outside. Interior humidities of 40 percent or less are common.

Preparation for the dry air must begin during the acclimatization process. Gradual drying of the plant's atmosphere will usually allow the plant to survive after transplanting indoors. In some cases, attempts to increase the humidity around plantings by misting the foliage during the day have proven to be of little or no value. Misting may also cause damage to carpeting or furnishings. Proper acclimatization is the best solution at present.

Air Pollution

Air pollution cannot be escaped by bringing the landscape indoors. Pollutants in the exhaust of cars and trucks, the smoke from cigarettes, the chlorine gas escaping from swimming pools, and the chemical soup that passes for air in our major metropolitan areas are all harmful to plants when sufficiently concentrated. Some plant species are more susceptible than others.

Good ventilation is an important element in the health of interior plantings. Proper ventilation will carry away chlorine vapors, fumes from smokers and chemical cleaning agents, or the ethylene that may be present if a building is heated by some form of hydrocarbon combustion (fossil fuel).

Ethylene can also damage plants as they are shipped to the site. This toxin is present in vehicle exhaust, and if permitted to seep into the cabin where plants are stored may result in injury. If packed tightly or in restrictive packaging, the plants also may injure themselves, since their own tissue produces ethylene. Because of the potential harm from ethylene, plants should be unpacked and ventilated immediately upon arrival. At the time the plants are purchased, the grower or shipper should guarantee that the transport vehicles will be ventilated and sealed against exhaust fumes.

Dust

Dust is an air pollutant different from the others in that it is a particulate, not a vapor. When the leaves of a plant are coated with dust, they are not only unattractive, but gas exchange may be reduced due to plugged stomata. Air filtration reduces the amount of dust. Regular cleaning of the plants can also prevent dust buildup. Most plants in a residential interior can be rinsed off under the shower or set outside during a rainfall to wash away the dust. Commercial plantings can be kept clean with regular feather dusting and periodic washing. Cleaning should be a regular task within a total maintenance program for the interior plantscape.

Pruning

Pruning will not be extensive in an interior planting because of the plants' reduced rate of growth. Most pruning will be done to keep the plants shaped for an attractive appearance. Broken or damaged branches will also require removal. If the plants are not intended to grow, in order to avoid crowding, the roots as well as the foliage must be pruned back. Excessive root growth in containers can result in strangulation of the root system.

For pruning the plantings, hand pruners will be suitable for most of the herbaceous material and much of the woody material. Lopping shears will be helpful with larger materials, and pruning saws may be needed for indoor trees.

Interior plants should be pruned to a shape that suggests the appearance of a full canopy. Due to the restricted lighting in an interior planting, a full canopy seldom forms, however, and the plant's full branching structure may be visible. The pruning must enhance and take advantage of this sparse foliage covering.

Repotting

Repotting of plants is necessary in plantings where growth is allowed. Containers need to be removed from the planter and from the plant, the next larger size selected, and the excess space filled with growing medium. If roots are matted, they should be loosened before repotting. If they have started to grow around themselves, the large and excess roots should be pruned away. The pot should not be filled to the rim with the growing medium. An inch of unfilled space below the pot rim will give water room to flow into the container.

Insects and Diseases

Insects and diseases are not as common to interior plantings as to exterior ones, but they do occur. Insect problems are more common than diseases.

The initial pest presence may be introduced by the plants themselves as they arrive from the grower. Insects or pathogenic inoculum may be present in the foliage, roots, soil, or containers. All should be checked carefully upon arrival, at an area away from the installation site. The same careful check should be made each time replacement plants arrive. Obviously a reputable grower is a first defense against pests.

The most common pest problems of interior foliage plants are:

- ❦ Aphids
- ❦ Mealybugs
- ❦ Spider mites
- ❦ White flies
- ❦ Scale
- ❦ Thrips
- ❦ Nematodes
- ❦ Root mealybugs
- ❦ Root rots
- ❦ Leaf spots
- ❦ Anthracnose
- ❦ Mildews
- ❦ Blights

These pests and others are not uncontrollable when modern pesticides and methods of application can be used against them. However, the interior location and the presence of people make the use of sprays, dusts, and fumigants difficult. Whenever there is a possibility of people making contact with a pesticide, it is dangerous to use it. Even furnishings and carpeting can be damaged by many of the corrosive or oil-based chemicals. Where practical, plants can be wrapped in loose plastic bags and sprayed within the bag. This helps reduce the drift and the danger to people and furnishings.

Control measures are restricted to:

- ❦ Chemical pesticides approved for application indoors
- ❦ Removal of infected plant parts
- ❦ Washing away of insects and inoculum from plant foliage
- ❦ Replacement of plants with healthy new ones

Vandalism and Abuse

Vandalism and abuse to interior plantings may necessitate replacement earlier than expected. Certain locations will bring certain predictable types of abuse. Plantings in cocktail lounges or nightclubs may have alcoholic drinks poured into their soil. Those in college snack bars and dorms may have cigarette holes burned through their leaves. Plantings in shopping malls may be mutilated by home propagators who take cuttings faster than the plants can replace them. Sometimes entire plants are stolen from the planters. The planters themselves are used as litter bins for paper cups, cigarettes, chewing gum, and assorted other debris.

The only real defense against such damage and abuse is public education and cooperation. By keeping the plantings attractive and well maintained, they are less likely to be deliberately damaged. Replacing or repairing damaged plants as soon as they are noticed displays the concern of the owners for an attractive planting that all can enjoy.

Grouping Compatible Species

Grouping compatible species simplifies the maintenance of the interior plantscape. Within any one planter the species selected should have the same requirements for light, moisture, fertilization, and soil mix. While a large plantscape may effectively combine tropical species, desert species, and sometimes even temperate species, they should be in separate planters. When plant replacement is necessary, the new plants should be compatible with existing ones.

The Interdisciplinary Team

The popularity of interior plantings has grown faster than the ability of any one profession to stay abreast of it all. A successful interior planting in a commercial building requires the expertise of interior plantscapers, plant growers, interior decorators, landscape architects, architects, maintenance professionals, and building management. Each professional brings his or her point of view to the project. The architect sees the plants as architectural features, not as living organisms. Building management personnel want the plants to attract customers and please employees, but they too do not have a horticultural sense of plant needs. Maintenance professionals regard the plants as something to be dusted, watered, and fertilized. They are concerned about the proximity of water, the difficulty of changing lamps, and the ease of reaching the plants. Each has concerns and contributions to the planning process that the others need to know.

The success of the plantscape is measured by its appearance and health. The interior plantscaper should be consulted while the building is still in the planning stage, as errors in lighting quality and intensity can be difficult to correct later. Drainage of planters directly into the building's drainage pipes requires that plantings be sited permanently near the pipes. Watering of the plantings necessitates a nearby supply of water to which a hose can be attached. A shopping mall or office lobby should have appropriate water outlets every 50 feet along the wall and preferably within each planter. Otherwise hoses will be stretched across walkways, endangering pedestrians and creating puddles. Maintenance of the plantings should be the responsibility of a contracted professional plant maintenance firm, which should work closely with the architect to assure sufficient water outlets, storage space, accessibility for equipment, and so on. Failure to involve the maintenance firm in the early stages of planning can result in an unattractive plantscape shortly after installation. The managers of the building must be made to understand why specific lamps are needed for plant survival even if less expensive ones are available.

In short, fewer problems will develop for the interior planting if all professionals whose work affects it work together from the beginning. At present, such interdisciplinary cooperation is the exception rather than the rule. The failure to use the team approach is often a case of architects not realizing their own limitations with horticultural materials.

The Future

No career field in ornamental horticulture holds greater promise than interior plantscaping. The current technology and knowledge of growing plants indoors is comparable to automobile technology when Henry Ford and the Wright Brothers began working in their professions. Fresh approaches are needed to move us beyond Ficus trees and hanging baskets. More research is needed to introduce new species and varieties suitable for interior use. Also needed is better data on how to acclimatize plants with less foliage drop and transplant shock. Further study of soil mixes, fertilizer needs, and lighting requirements is needed to replace the guesswork of today. Finally, more professional plant maintenance firms are needed. They need dependable studies of such matters as the time required to dust and clean various leaf sizes and textures. Buildings need to be designed as carefully for their planted occupants as for their human ones. Automated watering and fertilization systems are needed to ease current labor-intensive methods. There is ample work for young people who wish to train themselves for it.

Achievement Review

A. Indicate if the following statements are true or false.

1. Interior use of plants is a new concept.
2. Architects and interior decorators are trained to write correct specifications for interior plantscapes.
3. Tropical plants have proven to be better suited for interior use than temperate plants.
4. Interior plantings require less fertilizer than outdoor plantings.
5. Soluble salts originate from fertilizers and water.
6. Clear glass transmits 100 percent of the sunlight that shines upon it.
7. Light intensity is measured in nanometers.
8. Light wavelength is measured in foot-candles.
9. Most plants will survive after being acclimatized to a light intensity of 100 to 200 fc.

B. Select the answer from the choices offered for each question.

1. The most efficient natural light for interior plantscapes enters from _____.
   a. the sides of buildings
   b. skylights 20 feet overhead
   c. skylights 15 feet or less overhead
   d. incandescent bulbs

2. The most efficient skylight has _____.
   a. a shallow, wide well with vertical sides
   b. a shallow, wide well with 45-degree beveled sides
   c. a deep, narrow well with vertical sides
   d. a deep, narrow well with 45-degree beveled sides

3. In most single-story settings the best lamp for lighting an indoor planting is _____.
   a. an incandescent
   b. a cool white fluorescent
   c. a warm fluorescent
   d. a fluorescent plant growth lamp

4. Interior plantings illuminated with 100 to 200 fc of light require _____ hours of lighting each day.
   a. eight
   b. ten
   c. twelve
   d. fourteen

5. The most important reason for using a synthetic soil in an interior planting rather than natural soil is _____.
   a. drainage
   b. nutrient content
   c. structural support
   d. cost

6. An interior planting that reflects seasonal changes through use of such plants as daffodils, mums, and poinsettias, would be best if designed _____.
   a. as an in-ground planting
   b. as a containerized planting

7. Deep watering of interior plantings will _____.
   a. encourage deep rooting
   b. leach away soluble salts
   c. do both of these
   d. do none of these

8. Watering of interior plantscapes is best done _____.

a. irregularly
b. weekly
c. when convenient
d. according to an established schedule

9. The two major reasons why interior plants die are overwatering and _____.
   a. too much light
   b. too much fertilizer
   c. too little fertilizer
   d. incorrect pH

10. Chlorine gas, dust, and ethylene are all possible _____ affecting interior plantings.
   a. pesticides
   b. air pollutants
   c. pathogens
   d. toxins

C. Answer each of the following questions as briefly as possible.

1. Place *X*s where appropriate to compare the different lamp types.

Characteristic	Incandescent	Cool White Fluorescent	Fluorescent Plant Lamps	Mercury	Metal Halite	High Pressure Sodium
High in red light						
Low in red light						
High in blue light						
Low in blue light						
Good color rendition						
High initial cost						
Low initial cost						
Moderate initial cost						
Low operating cost						
Moderate operating cost						

2. Label the parts of this containerized plant.

3. Indicate whether the following are characteristic of interior or exterior plantings.
   a. Rapid growth is encouraged.
   b. Pruning is minimal.
   c. Dusting of foliage is needed.
   d. High-analysis fertilizers are used regularly.
   e. Ventilation of the planting site is needed.
   f. Plants must be acclimatized to low humidity.
   g. Plants are most affected by insects and diseases.
   h. Plants suffer most from vandalism and abuse.

Landscape Maintenance

SECTION

3

Maintaining Landscape Plants

Objectives:

Upon completion of this chapter, you should be able to

- water, fertilize, edge, and mulch tree and shrub plantings
- prune trees and shrubs correctly
- maintain annual and perennial flower plantings

Sustained Care of Plantings

As important as the design and proper installation of landscape plantings is their ongoing maintenance. Many gardens do not attain the appearance envisioned by the landscape designer until the plants have time to mature. Aiding those plants in their healthy maturation requires attentive and knowledgeable maintenance. Also, some designs require specific plant effects such as clipped formal hedges or espalier training to create the garden as envisioned. In both instances, the value of skillful landscape maintenance is apparent. A maintenance program generally will include the following tasks: watering, fertilizing, mulching, edging, pruning, pest control, and winterization.

Watering

Depending upon the region of the country, supplemental watering may be an infrequent task or so regular that it requires an automatic irrigation system. The need for a good water supply to the plantings has been discussed in earlier chapters. Here the subject is the objectives of proper watering, its frequency, and its quantity.

Initially, watering must promote deep root development by the plant to establish it securely in its location. Later, watering must keep the plant healthy and growing actively even during dry summer weather. Much winter damage to evergreens can be avoided if the plants are kept well watered throughout the summer and autumn.

Not all plants require the same amounts of water. Neither will all plant root systems grow to the same depth in the soil. While nearly all landscape trees and shrubs will die if kept in either arid or waterlogged soil for long, certain species are especially sensitive to sites that are too dry or too wet.

Infrequent and deep watering is preferable to frequent, shallow watering. Enough water should be applied to wet the soil to a depth of 12 to 16 inches. In the Southeast and Southwest, supplemental water may be required nearly every day. In other regions, supplemental watering may be weekly or even less frequent.

Fertilization

Trees will grow without fertilization in most soils once they are established. However, they will grow with greater health and vigor if they are fertilized annually. Shrubs will respond to proper fertilization with lush growth, greater resistance to pests, and less winter damage.

For shrubs growing in cultivated beds, fertilizer may be applied in later March or early April. Depending upon the plants involved, each 100 square feet of bed area should receive between 1 and 3 pounds of a low-analysis, complete fertilizer. The fertilizer should be distributed uniformly over the soil beneath the shrubs, with most of the fertilizer under the outer edge of the shrub where the fibrous roots that absorb the nutrients are located. The fertilizer should not be allowed to touch the foliage, or **foliar burn** (a reaction to the chemicals) may result. If the soil is dry, the fertilizer should be worked into the soil with a hoe. If the weather has not been abnormally dry, the fertilizer can be left untilled and the next rainfall or irrigation will wash it into the soil.

Traditionally, trees have been fertilized in a somewhat similar fashion by drilling holes (2 inches in diameter and 12 inches deep) at intervals of 18 to 24 inches around the tree beneath the drip line of the canopy. Into these holes is poured a dry mixture of 50 percent high-analysis, complete fertilizer and 50 percent sand as a carrier. The holes can be drilled by hand or with a powered auger. In recent years, tree spikes have been promoted, especially to homeowners. These are complete fertilizers manufactured in torpedo shapes and sold at a price that is prohibitive if used in large quantities. They are pounded into the ground, eliminating the need to drill holes. For use with a few small tress, they are convenient.

Landscape maintenance specialists most commonly fertilize trees using an automated process that drills the holes and injects the fertilizer quickly and neatly. Recently, a method for injecting fertilizer (and pesticides) directly into trees' vascular systems has been introduced to the industry. It is rapidly gaining favor as more people become familiar with the technique and skilled in its use, Figure 25-1.

The objectives of fertilization change as plants, especially trees, mature. Initially, plants are fertilized to promote their health and growth. Later, as the plants approach or attain maturity, the purpose of fertilization is not to enlarge them but to keep them healthy, attractive, and strong.

The one most common mistake in the fertilization of landscape plants is the application of fertilizer too late in the growing season. The result is often a flush of vegetative growth in response to the nitrogen that leaves the plant ill-prepared for winter. Great damage can result to plants from well-meaning but ill-timed fertilization.

Mulching

The objectives, advantages, and disadvantages of mulching, and examples of commonly available products, were outlined earlier. For extended maintenance of landscape plantings, mulches require replacement. Organic mulches decompose, forming humus and an ideal medium for the germination of newly deposited weed seeds unless mulch is replaced annually. Inorganic mulches do not decompose, but decline in appearance if not freshened periodically.

Figure 25-1 The patented Mauget system releases liquid pesticides and/or fertilizers into trees through tubes injected into the vascular system. (Courtesy J. J. Mauget Company)

CAUTION: Old mulch should be removed before new is added. Otherwise, the soil level over the roots is gradually deepened and the plants may die.

One important use of mulch is as a protective divider to prevent lawn mower damage to the base of plants, Figure 25-2. A gouge from a mower creates a site for pathogen or insect invasion of the plant. If hit repeatedly, a tree may become partially or completely girdled. In large landscapes with numerous trees and an understaffed maintenance crew who rely on riding mowers for grass cutting, frequent injury to trees results from attempts to mow as closely as possible, thereby eliminating hand trimming. A square of mulch around the base of each tree can protect the tree, speed the overall mowing time, and create a neat appearance, Figure 25-3.

Edging

The term *edging* has a double use. As a verb, *edging* refers to cutting a sharp line of separation, usually between a planting and the adjacent lawn. Reference may be made to edging a bed or edging the lawn along a walk. The term can also be used to describe a product, usually a steel or plastic strip that can be installed as a physical separator between planting beds and lawns or between lawns and paved areas.

To edge a bed requires an edging tool or a flat-back spade. The edger or spade is dug into the ground to a depth of 6 to 8 inches and a wedge of sod removed, Figure 25-4. The process continues along the edge of the bed or the paved area. Cutting through the sod sharply and vertically discourages the roots of the turf from growing into the bed. If the landscape is large, edging can be accelerated by using a power edger.

The installation of edging material against the cut edge further discourages the horizontal spread of grass roots into the bed. It also retains a sharp turf line along walks and drives. The best edging materials are firm, not easily bent and crushed. The corrugated foil edging promoted to the home

Figure 25-2 Typical damage resulting from a lawn mower when grass is allowed to grow next to a tree

Figure 25-3 The mulched area around the tree protects the plant from mower injury. The square shape is preferable to a round shape because it makes mowing easier.

Figure 25-4 A sharply beveled edge around a planting bed will slow rhizome grasses and stop bunch grasses.

Figure 25-5 Plastic edging should be anchored at the time of installation to prevent winter heaving. Here a metal rod is driven through the edging into the soil and clipped over the lip of the edging.

garden market is not satisfactory for professional use and should be avoided. Heavy-gauge steel and polyvinyl-chloride edging is available from numerous manufacturers. Satisfactory edges can also be created with wood, bricks, and other modular manufactured materials.

In regions where winter heaving of the soil is common, it is best to select an edging material that can be anchored, Figure 25-5. Most anchoring techniques are only moderately effective, however, and resetting heaved edging is a common spring activity in northern landscapes. It is worth the effort, considering the advantages that the material offers in retention of the bed line and separation of the mulch from the lawn mower.

Pest Control

There are both biological and chemical ways to control plant pests. Here the subject is how to choose a chemical product, and in what formulation, what strength, and when to apply it. Remember, chemical control of plant pests is not the only or even the best approach to control; nevertheless, it may be determined to be appropriate.

Regardless of the state in which the landscaper practices, certification is required of all who would purchase or supervise the application of any chemical registered with the United States Environmental Protection Agency (EPA). A few pesticides in dilute formulations are available for general purchase, but most of those needed by professional landscapers are restricted in use.

Every state has a college of agriculture that is responsible for approving the use of every pesticide within the state. Scientists at the colleges of agriculture are also responsible for determining the appropriate product and formulation for use against a particular disease, weed, or insect on a particular host at a particular time of year or stage of development. Their recommendations are updated and published each year for use by the states' professionals. Copies of the recommendations are available by direct mail from the university and often from the local Cooperative Extension office. Landscapers should realize that although pest problems may be much the same in different states and in different regions of a single state, the environmental conditions can vary in subtle ways that are not always apparent. Therefore, if a pesticide is not approved for use within a state or region, it is dangerous, unethical, and illegal to purchase the product elsewhere and use it in the restricted zone. It is equally dangerous and illegal to increase the dosage or frequency of application beyond those recommended. These are the abuses of pesticide application that the certification program of the EPA seeks to eliminate.

Pruning Trees and Shrubs

Pruning is the removal of a portion of a plant to improve its appearance and health and to control its growth and shape. It is easily done, but not so easily done correctly. Each time a bud or branch is removed from a plant, it has both a short-term and long-term effect. The short-term effect is the way the plant looks immediately after pruning, and perhaps through the remainder of the growing season. The long-term effect is the way the plant appears after several seasons of growth without the part that has been pruned.

Parts of a Tree

The following parts of a tree are important to an understanding of proper tree pruning, Figure 25-6. The **lead branch** of a tree is its most important branch. It is dominant over the other branches, called the **scaffold branches**. The lead branch usually cannot be removed without destroying the distinctive shape of the tree. This is especially true in young trees.

The scaffold branches create the **canopy**, or foliage, of the trees. The amount of shade cast by the canopy is directly related to the number of scaffold branches and the size of the leaves. When it becomes necessary to remove a branch from a tree, removal usually occurs at a **crotch**, the point at which a branch meets the trunk of the tree or another, larger branch.

It is always desirable to leave the strongest branches and remove the weakest. Where the

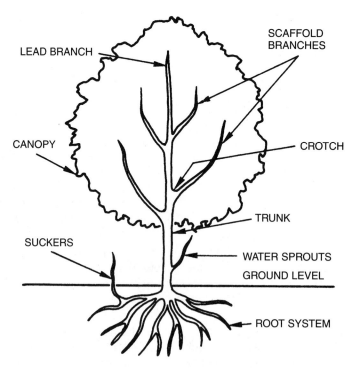

Figure 25-6 The parts of a tree

crotch union is wide (approaching a right angle), the branch is strong because there has been no crowding and pinching of the new wood produced each year by the cambiums of the trunk and branches. Where the crotch union is narrow, the branch is weak due to a pinch point forming where the expanding trunk meets the expanding branch, Figure 25-7. Growth in that area becomes compressed and dwarfed, and the branch may snap off at that point during a heavy wind or in response to the weight of a person climbing on it.

Two other types of branches often found on trees are suckers and water sprouts. **Suckers** originate from the underground root system. **Water sprouts** develop along the trunk of a tree. Neither is desirable for an attractive tree, and both should be removed.

Parts of a Shrub

A shrub is a multistemmed plant, Figure 25-8. Its branches and twigs differ in age, with the best flower and fruit production usually on the younger branches. The younger branches are usually distinguished by a lighter color, less bark, and smaller diameter. The point at which the branches and the root system of a shrub meet is the **crown**. New branches originate at the crown, causing the shrub to grow wider. New shoots, called **stolons**, may spread underground from existing roots to create new shrubs from the parent plant. In some grafted plants, the **graft** union may be seen at or near the crown. Shoots originating from the **stock** (or root portion) of a grafted plant are cut away since the quality of their flowers, fruit, and foliage is inferior. Only shoots originating from the **scion** (or shoot portion) are allowed to develop.

The Proper Time to Prune

Landscapers who design and install as well as maintain landscapes usually prefer to prune when they have little other work. This distributes their work and income more evenly throughout the year. Some plants can accept this off-season attention and remain unaffected by it. Other species accept pruning only during certain periods of the year.

There are advantages and disadvantages to pruning in every season. Since seasons vary greatly from

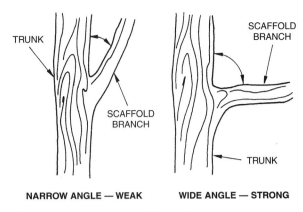

Figure 25-7 Tree crotch structure

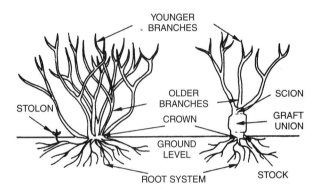

Figure 25-8 The parts of a shrub

region to region, the following can be used only as a general guide to the timing of pruning.

Winter Pruning. Winter pruning gives the landscaper off-season work. It also allows a view of the plant unblocked by foliage. Broken branches are easily seen, as are older and crossed branches. The major disadvantage of winter pruning is that without foliage it is difficult to detect dead branches. Because of this, plants can become seriously misshapen if the wrong branches are removed. An additional disadvantage is the damage that can be done through cracking frozen plant parts.

Summer Pruning. Summer pruning also provides work during a slower season for the landscaper. An advantage of summer pruning is that it allows time for all but very large wounds to heal before the arrival of winter. The major limitation of summer pruning is that problems of plants may be concealed by their full foliage. Branches that should be removed are often difficult to see. Especially with trees, it is difficult to shape the branching pattern unless all the limbs are visible.

Autumn Pruning. Pruning during the autumn season may conflict with more profitable tasks for the landscaper. In terms of the health of the plant, autumn pruning is acceptable as long as it is done early enough to allow cuts to heal before winter. Autumn pruning should not be attempted on plants that bloom very early in the spring, however. These early bloomers produce their flower buds the preceding fall. Thus, fall pruning cuts away the flower buds and destroys the spring show. Autumn pruning should be reserved for plants that bloom in late spring or summer, producing their buds in the spring of the year.

Spring Pruning. Since spring is the major planting season, most landscapers do not welcome pruning requests unless maintenance is their principal business. However, most plants are most successfully pruned during the spring. As buds begin to swell, giving evidence of life, it is clear which are the live and dead branches. Furthermore, there is little foliage to block the view of the complete plant. Spring pruning provides the plant with maximum time for wounds to heal. In addition, the unfolding leaves conceal the fresh cuts from the viewer's eye.

If the plant is an early spring bloomer, it is best to prune it immediately after flowering. Plants that have a high sap pressure in the early spring, such as maples, birches, walnuts, and poinsettias, should not be pruned until summer or fall, when the sap pressure is lower. Otherwise, the excessive exudation becomes unsightly.

Parts of the Plant to Prune

The reason for pruning will determine which limbs and branches should be removed from a tree or shrub. If the objective is to remove diseased portions, the cut should be made through healthy wood between the trunk or crown and the infected part, Figure 25-9. The cut should never be made through the diseased wood or very close to it. This contaminates the pruning tool, which may transmit the pathogen to healthy parts pruned later.

If the objective is to improve the overall health and appearance of the plant, branches growing into the center of the plant should be removed. Limbs and twigs that grow across other branches can crowd the plant and cause sites of infection to form by rubbing abrasions through the bark. If more than one limb originates at a tree crotch, the strongest should be left and the others removed, Figure 25-10. Major structural limbs and twigs must be left so that no holes appear in the plant. Often overlooked is the fact that many secondary branches can stem from one older branch. The removal of one key branch from a young tree can result in an entire side missing as the tree ages.

If the purpose of the pruning is to create denser foliage, as with evergreens, the center shoot is shortened or removed. This encourages the lateral buds to grow and create two shoots, where there had been only one, Figure 25-11.

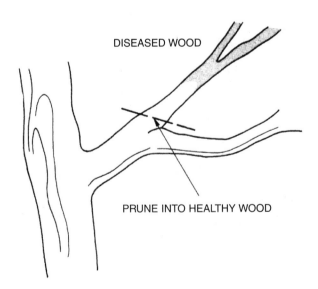

Figure 25-9 The correct way to remove diseased limbs or twigs

REMOVE BRANCHES GROWING TOWARD THE
CENTER OF THE PLANT.

REMOVE BRANCHES GROWING
ACROSS OTHER BRANCHES.

REMOVE EXTRA BRANCHES
AT TREE CROTCH.

Figure 25-10 Selecting branches to be pruned

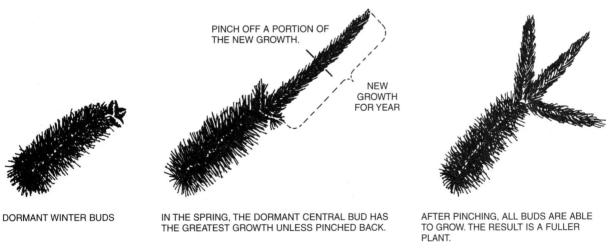

PINCH OFF A PORTION OF
THE NEW GROWTH.

NEW
GROWTH
FOR YEAR

DORMANT WINTER BUDS

IN THE SPRING, THE DORMANT CENTRAL BUD HAS
THE GREATEST GROWTH UNLESS PINCHED BACK.

AFTER PINCHING, ALL BUDS ARE ABLE
TO GROW. THE RESULT IS A FULLER
PLANT.

Figure 25-11 Evergreens are pruned in the spring if denser foliage is desired.

Pruning Methods

The method of pruning a tree or shrub depends upon the size and number of branches to be removed. Limbs are pruned from trees with a technique called **jump-cutting**. This method allows a scaffold limb to be removed without taking a long slice of bark with it when it falls. A jump-cut re-

quires three cuts for the safe removal of a limb, Figure 25-12. The final cut should remove the stub of the limb as close to the trunk as possible. The wound may then be covered with a wound paint to make it less conspicuous until the plant has time to heal. Wound paints are available in aerosol cans for small-scale use and in quart and gallon cans for

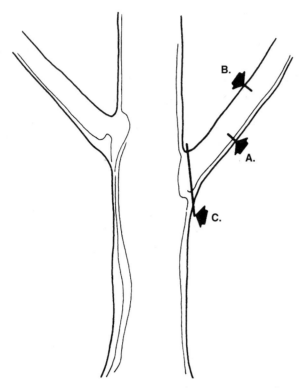

Figure 25-12 The removal of large limbs using the technique of jump-cutting. The cut at A allows the limb to snap off after a cut at B without stripping bark from the trunk as it falls. The final cut at C removes the stub.

THINNING OUT

As its name implies, this method involves selection of an appropriate number of strong, well-located stems and removal at the ground level of all others. This is the preferred method for keeping shrubs open and in their desired shrub size and form. With most shrubs, it is an annual task; with others, it is required twice a year.

HEADING BACK

This method involves trimming back terminal growth to maintain desired shrub size and form. It encourages more compact foliage development by allowing development of lateral growth. This is the preferred method for controlling the size and shape of shrubs and for maintaining hedges.

Figure 25-13 The techniques of thinning out and heading back

professional use. There is little evidence to support the claims of manufacturers that some formulations offer greater protection than others. Some recent research has suggested that wound paints may actually delay healing of plant tissue. Where this is suspected, the landscaper may choose not to use it.

When shrubs are pruned, one of two techniques is used. **Thinning out** is the removal of a shrub branch at or near the crown. It is the major means of removing old wood from a shrub while retaining the desired shape and size. **Heading back** is the shortening, rather than total removal, of a twig. It is a means of reducing the size of a shrub. In cases where shrubs have become tall and sparse, a combination of thinning out and heading back can rejuvenate an old planting, Figures 25-13 and 25-14.

CUT BACK

SELECT SIX OR MORE WELL-PLACED
VIGOROUS SHOOTS.

HEAD BACK

COMPLETE RENEWAL

This method involves complete removal of all
stems at the crown or ground level. Two to three
months later the suckers or new growth that
emerges is thinned to the desired number of stems.
These, in turn, are headed back to encourage lateral branching. Unpruned, seriously overgrown, or
severely damaged shrubs are prime prospects for
this treatment.

GRADUAL RENEWAL

This pruning method involves removal of all mature
wood over a 3- to 5-year period. Approximately
one-third of the mature wood is removed each season. This is the preferred method for shrubs that
have not been recently pruned and are somewhat
overgrown.

Figure 25-14 Two techniques used to rejuvenate old shrubs

In heading back, the location of the cut is
important, Figure 25-15. If too much wood is
left above the bud, the twig will die from the
point of the cut back to the bud, but the cut
may not heal quickly enough to prevent insect
and pathogen entry. Also, the woody stub itself
may decay later. A cut below the bud will cause
the bud to dry out and possibly die. The cut
should be made just above the bud and parallel
to the direction in which the bud is pointing.
The cut should be close enough to the living
tissue to heal over quickly but not so close to
the bud that it promotes drying. The direction
in which the branch of a plant grows can be
guided by good pruning techniques. Branches

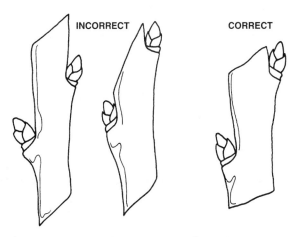

INCORRECT   CORRECT

Figure 25-15 Where to prune the twig

growing into the plant can be discouraged by the selection of an outward-pointing bud when heading back, Figure 25-16. If the twig has an opposite bud arrangement, the unnecessary one is removed.

How to Prune Hedges

The creation of a hedge requires close spacing of the shrubs at the time of planting and a special type of pruning. The landscaper must shear the plant so that it becomes as dense as possible. This is usually done with hedge shears. The hedge shears easily cut through the soft new growth of spring, the season when most hedges are pruned. For especially large hedges, electric or gasoline powered shears are available. However, practice and skill are required for the satisfactory use of power shears.

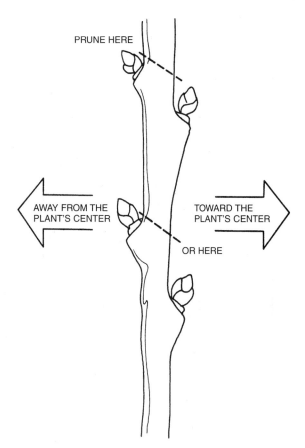

Figure 25-16 Twigs should be pruned to leave an outward-pointing bud.

Damage can occur quickly if the landscaper does not keep the shears under control.

A properly pruned hedge is level on top and tapered on the sides, Figure 25-17. It is important that sunlight be able to reach the lower portion of the hedge if it is to stay full. Otherwise it becomes leggy and top-heavy in appearance.

Flower Plantings

No components of the landscape require more maintenance time than flower plantings. This is why landscapes with a low budget for maintenance must minimize the use of flowers. Reasons for the high maintenance costs are that:

❧ Weeds must be pulled by hand or controlled with costly selective herbicides.

❧ Flowers are more susceptible to insects and diseases that mar their aesthetic appearance, therefore requiring additional expenditure for pesticides.

❧ Flowers tend to go to seed or get leggy with age if not pinched back frequently during the growing season. (This subject is discussed later in the chapter.)

❧ Many perennials bloom only once a year but must be cared for throughout the growing season to assure a good flower display the following year.

❧ Some perennials, notably the tender bulbs, must be dug up, put into storage for the winter, and set out again each spring.

Weed Control

Mulches help control weeds in flower plantings and retain surface moisture for the planting as well. Due to the short height of some flowers, the thickness of the mulch may be reduced to 3 inches rather than 4. Black plastic beneath the mulch can be beneficial in annual beds. The plants can be set into the soil through holes in the plastic and mulch put on top of the plastic. In perennial beds, where the flowers go dormant each fall and revive in the spring, the plastic could do more harm than good by trapping new flower growth beneath it.

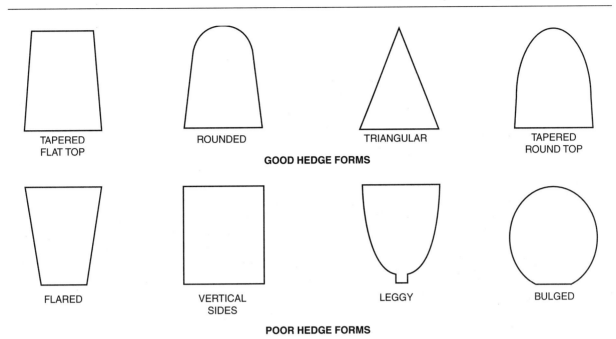

Figure 25-17 Good and poor hedge forms as seen from the side

Several good herbicides are approved for use in flower plantings. If applied for preemergence and post-emergence control, a weed-free flower planting can be attained. One danger of herbicides in flower beds is that the chemicals are always selective; that is, they kill either grasses or broadleaved weeds, but seldom both. Since flower plantings can have characteristics of both, damage can be done to some flowers even when others are uninjured.

Most herbicides used in flower plantings lose their effectiveness if the soil is disturbed after application. In such cases, landscapers must avoid cultivating the soil surface if they wish the full benefit of the herbicide.

Watering

All flowers should be watered frequently and deeply during dry periods. Their shallow roots quickly react to drought conditions, and they reach a critical wilting point much sooner than the woody plants of the landscape. Flowers planted beneath trees and shrubs must compete with the woody plants for surface water, and they will dry out faster than flowers not in such competitive locations.

Fertilization

Annuals can be fertilized in midsummer with a low-analysis fertilizer to keep them lush and healthy. Bulbs should be fertilized immediately after flowering with a high-phosphorus fertilizer such as bone meal. Nonbulbous perennials grow best if fertilized in the early spring. Summer or fall fertilization of perennials can harm the plants by keeping them too succulent as winter approaches.

Pinching

Annuals are most likely to benefit from pinching back, but certain perennials such as hardy mums will also do better if pinched. Pinching removes the terminal shoot on each branch of the flower and allows the lateral shoots to develop, thereby creating a fuller plant. Soft pinching is done with the thumb and forefinger and removes the terminal bud or, at the most, the terminal bud and first set of laterals. Hard pinching may shorten each stem by one-third or more. Most flowers benefit by a hard pinch soon after being set out, followed by one or two soft pinches during the summer. With perennials such as mums whose flower bud initiation is tied to a photoperiod response, the last pinch should not be after mid-July if a good flower display is to be seen in the autumn.

Flowers such as petunias or rose moss, which are valued for their profuse blossoms, can be kept from getting leggy and going to seed by severely cutting them back about midsummer. A tool as indelicate as a pair of grass clippers can be used to provide a hard pinch if the planting is extensive. A period of several weeks with few flowers will follow until new reproductive growth begins. However, the fresh look and new flowers that result will carry the annuals right into the fall season.

Achievement Review

A. From the choices given, select the answer that best completes each of the following statements.

1. The main objective when watering trees and shrubs is to _____.
   a. keep the humidity high
   b. promote deep root development
   c. keep the foliage clean and moist
   d. prevent wilting

2. The proper time to fertilize shrubs is in _____.
   a. early spring
   b. late summer
   c. early fall
   d. winter

3. Trees should be fertilized _____.
   a. near their anchor roots
   b. at the nursery before digging
   c. near their base
   d. beneath the drip line of the canopy

4. The purpose of edging a bed is to _____.
   a. divert surface water
   b. discourage turf roots from entering the bed
   c. hold the mulch within the bed
   d. improve drainage

B. Answer each of the following questions as briefly as possible.

1. You discover a leaf spot disease on the English ivy plantings that are the major groundcover of a large landscape. How can you determine what control to use, what products to select, and when and how much to apply?

2. Complete the following statements.
   a. Removing a portion of a plant for better appearance, improved health, controlled growth, or attainment of a desired shape is termed _____.
   b. The most important branch on a tree is the _____.
   c. Scaffold branches create _____ of the tree.
   d. The point at which a branch meets the trunk or a larger branch is termed the _____.
   e. Undesirable branches originating from the root system of a tree are termed _____.
   f. Undesirable branches originating from the trunk of a tree are termed _____.
   g. The point of a shrub at which the branches and root system meet is termed the _____.
   h. Generally, the younger branches in a shrub will have a _____ color.
   i. Shoots originating from the stock of a grafted plant are _____ compared to those originating from the scion.
   j. Flower and fruit production occur most profusely on _____ branches.

3. Place Xs in the following chart to compare the advantages and disadvantages of pruning during different seasons.

Characteristic	Spring	Summer	Autumn	Winter
A full view of the plant's branching structure is possible.				
View of the plant's branching structure is blocked.				
Dead branches cannot be detected.				
The timing may conflict with more profitable activities.				
The most time is allowed for cuts to heal.				
This is the best season to prune for the good of most plants.				
This is the best season to prune shrubs that flower in early spring.				
Plants may be damaged by unintentional breakage.				

4. Which three branches shown in this drawing should be removed, and why?

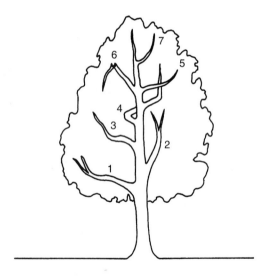

5. What is the difference between thinning out and heading back in the pruning of shrubs?

C. Write a short essay on the maintenance of flower plantings. Outline the care needed by both annuals and perennials to keep them healthy and attractive throughout the growing season. Assume the environmental conditions to be those in your own area.

Care of the Lawn

Objectives:

Upon completion of this chapter, you should be able to

- describe the operations needed to repair a lawn in the spring season
- explain the meaning of fertilizer analysis statements
- determine the amount of fertilizer needed to cover a specific area of lawn
- compare revolving, oscillating, and automatic sprinklers
- compare reel, rotary, and flail mowers
- explain the ways lawns can be damaged

Of all aspects of the landscape that require maintenance during the year, lawns consume the most time. A lawn necessitates both seasonal care and weekly care. Like any other feature of the landscape, it is easier to maintain if it is installed properly. Thus, the landscaper who is hired to install and maintain a lawn may have an easier job than the landscaper hired to maintain a lawn that was poorly installed by someone else.

Spring Lawn Care

Cleanup

Spring operations begin the season of maintenance. In areas of the country where winters are long and hard, lawns may be covered with compacted leaves, litter, or semi-decomposed thatch. The receding winter may leave behind grass damaged by salt injury, disease, or freezing and thawing.

Small areas can be cleaned of debris with a strong rake. Larger areas of several acres or more require the use of such equipment as power sweepers and thatch removers to accomplish the same type of cleanup.

Rolling of the Lawn

In many central and northern states, the ground freezes and thaws many times during the winter. Such action can cause **heaving** of the turfgrass. Heaving pulls the grass roots away from the soil, leaving them exposed to the drying wind. Heaving also creates a lumpy lawn.

Where heaving occurs, it is advisable to give the lawn a light rolling with a lawn roller in the spring.

Rolling presses the heaved turf back in contact with the soil. The roller is applied in a single direction across the lawn, followed by a second rolling at an angle perpendicular to the first.

Two precautions should be observed before rolling a lawn. One is that clay soil should never be rolled, since air can be easily driven from a clay soil and the surface quickly compacted. The other precaution is that no soil should be rolled while wet. The roller can be safely used only after the soil has dried and regained its firmness.

The First Cutting

The first cutting of the lawn each spring removes more grass than the cuttings that follow later in the summer. The initial cutting at 1 1/4 to 1 1/2 inches is done to promote horizontal spreading of the grass. This, in turn, hastens the thickening of the lawn. An additional benefit of the short first cutting is that fertilizer, grass seed, and weed killer that are applied to the lawn reach the soil's surface more easily. Cuttings done later in the year are usually not as short.

Figure 26-1 Patch seeding of thin areas in established lawns is done by breaking the soil surface and applying a small handful of seed.

Patching the Lawn

If patches of turf have been killed by diseases, insects, dogs, or other causes, it may be necessary to reseed them or add new sod, plugs, or sprigs. Widespread thinness of the grass does not indicate a need for patching. It indicates a lack of fertilization, improper mowing, disease, or insects.

Patching is warranted when bare spots are at least 1 foot in diameter. Seed, sod, plugs, or sprigs should be selected to match the grasses of the established lawn. Plugs can be set directly into the soil using a bulb planter or golf green cup cutter to cut the plug, and then to remove the soil where it is to be planted. With seed, sod, and sprigs, it is best to break the soil surface first with a toothed rake. A mixture of a pound of seed in a bushel of topsoil is handy for patching where seed is to be used, Figure 26-1. Mulch and moisture must then be applied, as stated earlier.

The timetable for patching is the same as for planting and is related to the type of grass involved.

Aeration

Aeration of a lawn is the addition of air to the soil. The presence of air in the soil is essential to good plant growth. If the lawn is installed properly, the incorporation of sand and organic material into the soil promotes proper aeration. However, where traffic is heavy or the clay content is high, the soil may become compacted. The groundskeeper can relieve the compaction by use of a power aerator, Figure 26-2. There are several types of aerators. All cut into the soil to a depth of about 3 inches and remove plugs of soil or slice it into thin strips.

A topdressing of organic material is then applied to the lawn and a rotary power mower run over it. This forces the organic material into the holes or slits left by the aerator. The plugs of soil left on top of the lawn may be removed by raking. If the soil plugs are not too compacted, they can be broken apart and left as topdressing. Equipment is made that can aerate and convert the soil plugs to topdressing in a single operation.

Figure 26-2 The aerator is used to remove plugs of soil from compacted lawns, allowing air to enter the soil.

Vertical mowing is a technique that can break up the soil plugs left by an aerator or even remove excessive thatch if necessary. It requires a power rake or a mower whose blades strike the turf vertically. It is done when the lawn is growing most rapidly and conditions for continued growth are favorable. For cool-season grasses, late summer or early autumn is the best time. For warm-season grasses, late spring to early summer is best.

The blades of the vertical mower are adjusted to different heights depending upon the objectives of the operator. A high setting is used to break up soil plugs. A lower setting gives deeper penetration into the thatch layer. This makes it easier to remove and relieves compaction of the soil. Deep vertical mowing is only practiced on deep-rooted turfs. Shallow-rooted turfs often grow mainly in the thatch layer. They can be harmed more than helped by vertical mowing.

Lawn Fertilization

Much like grass seed, lawn fertilizer is sold in an assortment of sizes and formulations and priced accordingly. Stores selling fertilizers range from garden centers to supermarkets. The professional groundskeeper needs to have a basic knowledge of fertilizer products prior to their purchase. Otherwise, it is difficult to choose among the many brands available.

Nutrient Analysis and Ratio

The fertilizer bag identifies its contents. It displays three numbers that indicate its **analysis**, that is, the proportion in which each of three standard ingredients is present. These numbers, such as 10-6-4 or 5-10-10, indicate the percentage of total nitrogen, available phosphoric acid, and water-soluble potash present in the fertilizer, Figure 26-3. The numbers are always given in the same order and always represent the same nutrients.

With simple arithmetic, fertilizers can be compared on the basis of their **nutrient ratios**. For example, a 5-10-10 analysis has a ratio of 1-2-2. (Each of the numbers has been reduced by dividing by a common factor, in this case 5.) A fertilizer analysis of 10-20-20 also has a ratio of 1-2-2. As the example in the following column illustrates, a 5-10-10 fertilizer supplies the three major nutrients in the same proportion as a 10-20-20 fertilizer, but twice as much of the actual product must be applied to obtain the same amount of nutrients as is contained in the 10-20-20 fertilizer.

50 pounds of 5-10-10 fertilizer contain:	50 pounds of 10-20-20 fertilizer contain:
2½ pounds of N (nitrogen)	5 pounds of N
5 pounds of P_2O_5 (phosphoric acid)	10 pounds of P_2O_5
5 pounds of K_2O (potash)	10 pounds of K_2O

The ratio of the fertilizers is the same, but the amount of nutrients available in a bag of each

Figure 26-3 How to interpret fertilizer analysis figures. The nutrients are always shown in the same order.

differs. The 5-10-10 mixtures should be less expensive than the higher analysis material.

Thus, one measure of the quality of a fertilizer is its analysis. The higher the analysis, the greater is the cost. Whether or not a high analysis fertilizer is needed depends upon the individual plant. Generally, residential lawns do not need a high analysis fertilizer.

Forms of Nitrogen Content

Another factor influencing the quality and cost of fertilizers is the form of nitrogen they contain. Some fertilizers contain nitrogen in an organic form. Examples include peat moss, peanut hulls, dried blood, tobacco stems, sewage sludge, and cottonseed meal. The nitrogen content of these materials ranges from 1 1/2 to 12 percent, depending upon the particular material. While sewage sludge is used to some extent on golf course turf, organic fertilizers are not widely used for fertilization of grasses because they are too low in nitrogen. Often, the

nitrogen present is not in a form that can be used by plants. The best use of organic fertilizers is as soil conditioners that greatly improve the water retention and aeration of the soil.

Chemical forms are the most commonly used fertilizers. They contain a higher percentage of nitrogen. The nitrogen may be quickly available or slowly available; this determines the timing of the nitrogen's release into the soil and uptake by the grass or other plants. It also influences the cost of the fertilizer.

Quickly available fertilizers usually contain water-soluble forms of nitrogen. This means that the nitrogen can be leached (washed) through the soil before the plants take it in through their root systems. Slowly available fertilizers (also called **slow-release**) make their nitrogen available to the plant more gradually and over a longer period of time. The slow-release effect is possible because the nitrogen used is in a form that is insoluble in water. This gives the plants more time to absorb the nitrogen and prevents fertilizer burn of the plant. Slow-release fertilizers are therefore more expensive than the quickly available forms. Slow-release fertilizers are usually labeled as such. This helps the landscaper to know what is being purchased and what to expect as a response from the plants.

Fillers

A final factor affecting the price and quality of a fertilizer is the amount of filler material it contains. This is directly related to the analysis of the product. In addition to the three major nutrients, fertilizers may contain additional **trace elements** (nutrients that are essential, but needed in smaller amounts) and filler material. **Filler material** is used to dilute and mix the fertilizer. Certain fillers also improve the physical condition of mixtures. However, filler material adds weight and bulk to the fertilizer, thereby requiring more storage space.

The following listing compares high analysis fertilizers (those with a high percentage of major nutrients) and low analysis fertilizers (those with a low percentage of major nutrients) on various plants.

High Analysis Fertilizer	Low Analysis Fertilizer
Contains more nutrients and less filler	Contains fewer nutrients and more filler
Cost per pound of actual nutrients is less	Cost per pound of actual nutrients is greater
Weighs less; less labor is required in handling	Is bulky and heavy; more labor is required in handling
Requires less storage space	Requires more storage space
Requires less material to provide a given amount of nutrients per square foot	Requires more material to provide a given amount of nutrients per square foot
Requires less time to apply a given amount of nutrients	Requires more time to apply a given amount of nutrients

In summary, fertilizer cost is determined by three major factors: analysis, form of nitrogen, and amount of bulk filler material. The higher the analysis and the greater the percentage of slow-release nitrogen, the more expensive is the fertilizer.

When to Fertilize Lawns

Lawns should be fertilized before they need the nutrients for their best growth. Cool-season grasses derive little benefit from fertilizer applied at the beginning of the hot summer months; only the weeds benefit from nutrients applied during the late spring. Cool-season grasses should be fertilized in the early spring and early fall. This supplies proper nutrition prior to the seasons of greatest growth. Landscapers should never practice late fall fertilization; it encourages soft, lush growth, which is damaged severely during the winter.

Warm-season grasses should receive their heaviest fertilization in late spring. Their season of greatest growth is the summer.

Amount of Fertilizer

The amount of fertilizer to use is usually stated in terms of the number of pounds of nitrogen to apply per 1,000 square feet. The number of pounds of nitrogen in a fertilizer is determined by multiplying the weight of the fertilizer by the percentage of nitrogen it contains.

Examples

Problem:

How many pounds of actual nitrogen are contained in a 100-pound bag of 20-10-5 fertilizer?

Solution:

100 pounds × 20% N = pounds of N

100 × 0.20 = 20 pounds of N

Problem:

How many pounds of 20-10-5 fertilizer should be purchased to apply 4 pounds of actual nitrogen to 1,000 square feet of lawn?

Solution:

Divide the percentage of N into the pounds of N desired. The result is the number of pounds of fertilizer required.

4 pounds of N desired

÷ 20% = pounds of fertilizer required

4 ÷ 0.20 = 20 pounds of 20-10-5 fertilizer required

"A Comparison Chart for Turfgrasses" in Chapter 21 lists general fertilizer recommendations for various grasses.

When applying fertilizer to lawns, the recommended poundage should be divided into two or three applications. For example, the 4 pounds of nitrogen per 1,000 square feet for bluegrasses and fescues might be applied at the rate of 2 pounds in the early spring and 2 pounds in the early fall. Another possibility is to apply 1 pound in early spring, 1 pound in midsummer, and 2 pounds in early fall. A spreader must be used to assure even distribution of the fertilizer. It is applied in two directions with the rows slightly overlapped, Figure 26-4.

Watering the Lawn

Turfgrasses are among the first plants to show the effects of lack of water, since they are naturally shallow rooted, compared to trees or shrubs. The groundskeeper should encourage deep root growth

Figure 26-4 Distributing fertilizer with a spreader

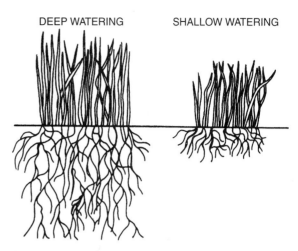

Figure 26-5 Deep watering promotes deep, healthy root growth. Shallow watering promotes shallow rooting and leaves the grass susceptible to injury by drought.

by watering so that moisture penetrates to a depth of 8 to 12 inches into the soil. Failure to apply enough water so that it filters deeply into the soil promotes shallow root growth, Figure 26-5. Such shallow root systems can be severely injured during hot, dry summer weather.

Infrequent, deep watering is much preferable to daily, shallow watering. The quantity of water applied during an irrigation will depend upon the time of day and the type of soil. Clay soils allow slower water infiltration than coarser textured sandy soils, but clay soils retain water longer. Therefore, less water may need to be applied to clay soils, or the rate of application may need to be slower, or both.

The amount of water given off by a portable sprinkler can be calibrated once and a notation made for future reference. To calibrate a portable sprinkler, set several wide-topped, flat-bottomed cans with straight sides (such as coffee cans) in a straight line out from the sprinkler. When most of them contain 1 to 1 1/2 inches of water, shut off the sprinkler. The amount of time required should be noted for future use. Figures 26-6 and 26-7 illustrate two

Figure 26-6 One type of lawn sprinkler. An arch of water is cast from side to side. This sprinkler requires periodic relocation.

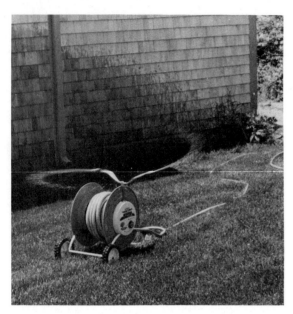

Figure 26-7 This revolving sprinkler covers a limited area and must be manually moved to each new location. Wind gusts can affect the evenness of the coverage.

Figure 26-8 An automatic lawn irrigation system in operation (Courtesy of Weather-Matic Irrigation)

types of portable sprinklers. In addition, permanently installed irrigation systems are available (at considerable cost) for large turf plantings, Figure 26-8.

The best time of day to water lawns is between early morning and late afternoon. Watering in the early evening or later should be avoided because of the danger of disease; turf diseases thrive in lawns that remain wet into the evening. Watering prior to evening gives the lawn time to dry before the sun sets.

If watering is done at the proper time and to the proper depth, it is necessary only once or twice each week.

Mowing the Lawn

Three types of mowers are available for the maintenance of lawns: the reel mower, the rotary mower, and the flail mower. The **flail mower** is used for turfgrasses that are only cut a few times each year. Reel and rotary mowers are used to maintain home, recreational, and commercial lawns. On a **reel mower**, the blades rotate in the same direction as the wheels and cut the grass by pushing it against a nonrotating bedknife at the rear base of the mower, Figures 26-9 and 26-10. The blades of a **rotary mower** move like a ceiling fan, parallel to the surface of the lawn, cutting the grass off as they revolve, Figure 26-11. Reel mowers are most often used for grasses that do best with a shorter cut, such as bentgrass. Rotary mowers do not cut as evenly or sharply as reel mowers. However, they are satisfactory for lawn grasses that accept a higher cut, such as ryegrass, bluegrass, and fescue. The riding mower, so popular with homeowners, is a rotary mower. Many campuses, parks, and golf course fairways are mown with a large bank of reel mowers, called a gang mower. It is pulled behind a tractor that has been fitted with tires that will not rut the lawn.

In every situation, the blades must be sharp to give a satisfactory cut. Dull or chipped mower blades can result in torn, ragged grass blades that will die and give the lawn an unhealthy color of gray or brown. The sharp blades of any lawn mower should be respected. When powered, they can cut through nearly any shoe. Workers should never mow unless wearing steel-toed work shoes. Hands should never be near the blades while the mower is running. No inconvenience caused by

Figure 26-9 A powered reel mower

Figure 26-11 A rotary mower

shutting off a mower can equal the instant injury that a power mower can cause to a worker's hand or foot.

"A Comparison Chart for Turfgrasses" in Chapter 21 illustrates the wide range of tolerable mowing heights that exists between and within species. Within the range, the mowing height selected often depends upon how much care the lawn can be given and what surface quality is expected. Shorter heights require more frequent mowing and watering, and, often, greater pest control efforts. In turn,

Figure 26-10 Reel mowers provide the best quality cut.

shorter heights give greater density and finer texture to a lawn. A taller lawn surface will have a slightly coarser texture and take longer to thicken. However, it will not require cutting as often. Longer cutting heights will also withstand hot and dry periods better, since the extra blade length will cast cooling shadows over the soil's surface. Often, fewer weeds are an added benefit of a taller lawn surface. Since not all species are mowed to the same height, mixed-species lawns should be made up of grasses that have similar cutting requirements.

The frequency of mowing is variable. Because the rate of growth of a lawn can vary with the temperature and with the moisture provided, the frequency of the mowing cannot always be precisely specified. Ideally, a lawn should be mown when it is needed, not because a contract specifies cutting on a certain day.

A long-standing rule of thumb is that mowing should remove about one-third of the length of the grass blade. Thus, if the turf is being kept at 1 1/2 inches, it should be mowed when it reaches a height of 2 1/4 inches. If the grass gets too long before cutting, the dead clippings can mar the appearance of the lawn. Then the only alternative is to collect the clippings either in a grass-catcher or with a lawn sweeper or rake. If the grass is cut properly, the clippings will not be excessive. They will decompose rapidly and will not require collection.

The pattern of mowing should be varied regularly to prevent wheel lines from developing in the lawn. Varying the pattern also encourages horizontal growth of the shoots. An easy variation is to mow at a 90-degree angle to the last mowing. If done on the same day with a reel mower, an attractive checkerboard pattern develops. The pattern is not so apparent if done with a rotary mower, but the practice is just as healthy for the lawn.

Damage to Lawns

Like all plants, turfgrasses are susceptible to assorted injuries. Damage can be visual or physical and frequently is both. Causes of lawn injury include:

❦ Weeds

❦ Pests

❦ Drought

❦ Vandalism

Weeds

There are many definitions of the word *weed*; student landscapers undoubtedly have their own definitions. For discussion purposes, a **weed** may be defined as a plant that (1) is growing where it is not wanted, and (2) has no apparent economic value. Beyond these two qualities, weeds have broad leaves (such as dandelion, plantain, and thistle); others are grasses. Some are annuals, others are perennials.

In lawns, weeds are professionally controlled almost entirely by chemical means. A chemical that kills weeds is called an **herbicide**. Some herbicides are selective, meaning that they kill only certain weeds and do not harm other plants. Other herbicides are nonselective; these kill any plant with which they come into contact. Certain herbicides are classified as preemergence types. They are applied before the weed seeds germinate (sprout) and kill them as they sprout. Postemergence herbicides are applied to weeds after they have germinated.

Because the regulations governing the purchase and use of herbicides and other pesticides vary among states, this text contains no recommendations of specific weed controls. Professional groundskeepers in all states can obtain current recommendations for pesticides approved for use in their states by consulting a college of agriculture or the Cooperative Extension Service. The current trend is toward registration of pesticide applicators and restriction of the more poisonous pesticides to use by professionals only. Many herbicides are poisonous, and the landscaper should handle them cautiously and with a serious attitude.

Good landscaping practices are the best defenses against weeds. However, the need for some herbicides is almost certain to exist in most lawns. They are available in liquid or powdered form. Sometimes herbicides are mixed with fertilizer and sold as combination products, which save time and labor.

If an herbicide is applied in liquid form, the sprayer should be cleaned afterward and set aside to be used only for that purpose. Small amounts of herbicide that remain in the sprayer can kill valuable ornamental plants if the landscaper uses the same sprayer to apply another liquid material directly to plants.

Pests

The following table, "Common Turf Problems," contains a partial list of the pest problems that can injure a typical lawn.

The best defense against most pest injury is the selection of resistant varieties and the creation of a growth environment that favors the grass more than the pest. For example, watering at night promotes the growth of many fungus plants that cause turf diseases. Watering should be done earlier in the day, allowing the grass to dry before nightfall. Another example: allowing a thick thatch layer to develop provides a good habitat for certain insects. Minimizing the thatch layer reduces these insect populations. When soil insects are reduced or eliminated, the turf is not as attractive to rodents, such as moles, that burrow through the soil to feed on them. Thus, solving one problem may indirectly solve another.

As was noted for weeds, the local Cooperative Extension Service can provide information about the appropriate pesticides needed for insects, disease, and rodent control.

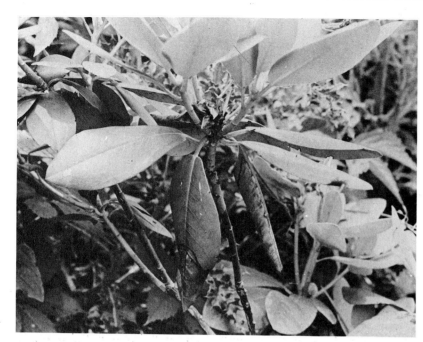

Figure 27-1 Rolled and discolored leaves show the effects of windburn on this rhododendron.

Needled evergreens can also burn. If burn has occurred, brown-tipped branches are apparent in the spring when new growth is beginning. As with broad-leaved forms, windburn on conifers is likely to be confined to the outermost branches on the most exposed side of the plant.

Temperature extremes can also cause injury to plants. Plants that are at the limit of their hardiness (termed *marginally hardy*) may be killed by an extended period of severely cold weather. Others may be stunted when all of the previous season's young growth freezes.

After some cold winters, certain plants may show no sign of injury except that their spring flower displays are absent. This results if the plant produces its flowers and leaves in separate buds. The weather may not be cold enough to affect the leaf buds, but freezes the more tender flower buds. This is especially common with forsythia and certain spireas in the northern states.

Unusually warm weather during late winter can also cause plant damage. Fruit trees may be encouraged to bloom prematurely, only to have the flowers killed by a late frost. As a result, the harvest of fruit can be greatly reduced or even eliminated. Spring flowering bulbs can also be disfigured if forced into bloom by warm weather that is followed by freezing winds and snow.

Sunscald is a special type of temperature-related injury. It occurs when extended periods of warm winter sunshine thaw the above-ground portion of a plant. The period of warmth is too brief to thaw the root system, however, so it remains frozen in the ground, unable to take up water. Above-ground, the thawed plant parts require water, which the roots are unable to provide. Consequently, the tissue dries out and a scald condition results.

Sunscald is especially troublesome on evergreens planted on the south side of a building. It also occurs on newly transplanted young trees in

---

Common Turf Problems		
Turf Insects	**Turf Diseases**	**Other Problems**
• Ants	• Anthracnose	• Dogs
• Army worms	• Brown patch	• Gophers
• Bill bugs	• Copper spot	• Ground squirrels
• Chinch bugs	• Dollar spot	• Mice
• Cut worms	• Fairy rin•	• Moles
• Grubs	• Fusarium blight	• Human vandalism
• Leaf hoppers	• Leaf spots	• Vehicles and
• Mites	• Net blotch	equipment
• Mole crickets	• Nematodes	
• Periodical cicadas	• Powdery mildew	
• Scale	• Pythium blight	
• Sod webworm	• Red thread	
• Weevils	• Rots	
• Wireworms	• Rusts	
	• Slime molds	
	• Smuts	
	• Snow molds	

Drought

Periods of severe water shortage can harm a lawn. The grass will turn brown and enter a stage of **dormancy** (nongrowth). If the drought continues, the entire lawn can be killed. Irrigation is the major defense against drought. In areas of predictable drought (arid regions), tolerant varieties of grasses should be selected at the time of installation. A greater lawn height is also helpful in withstanding periods of drought.

Vandalism

Vandalism is impossible to control if the vandals are determined. Lawns rutted by vehicles are unattractive and are common abuses of residential and recreational landscapes. Locked gates and the strategic placement of trees can sometimes help by making vehicular access difficult for the would-be vandal. Education to increase public awareness of the value of the landscape and the responsibilities of good citizens is the only real solution of vandalism.

Achievement Review

A. Define the following terms.

   1. heaving
   2. aeration
   3. fertilizer analysis
   4. low analysis fertilizer
   5. slow-release fertilizer
   6. herbicide

B. What does 10-6-4 on a bag of fertilizer mean?

C. Would 10-6-4 fertilizer be considered a high analysis or low analysis product? Why?

D. Of the three fertilizers listed below, which two have the same ratio of nutrients?

   10-20-10
   5-10-5
   5-10-15

E. Why might the prices of two 50-pound bags of fertilizer differ greatly?

F. At what time of the year are warm-season grasses fertilized? Cool-season grasses?

G. How many pounds of actual nitrogen are contained in a 50-pound bag of 12-4-8 fertilizer?

H. How many pounds of 10-10-10 fertilizer should be purchased to fertilize a 2,000 square-foot lawn of bluegrass and fescue if it is applied at the recommended rate?

I. How much water is needed to deeply soak an average lawn?

J. What is the best time of day to water lawns?

K. Describe how to repair a lawn by seeding, sodding, sprigging, and plugging.

L. Match the type of sprinkler with its characteristic:

   a. revolving   1. casts water in an arched pattern

   b. oscillating   2. permanently installed system for large landscapes

   c. automatic   3. casts water in a circular pattern

M. Match the type of mower with its characteristic:

   a. flail   1. blades move like a ceiling fan, parallel to the lawn's surface

   b. reel   2. used for grasses that are only cut a few times each year

   c. rotary   3. the blades rotate in the same direction as the wheels

N. List four different ways that lawns can be damaged.

Suggested Activities

1. Study "A Comparison Chart for Turfgrasses" in Chapter 21 and match grasses growing in your local area that could be blended and grown together successfully. Compare grasses on the basis of texture, frequency of fertilization, mowing height, and preferred soil type.

2. Calibrate one or more types of portable sprinklers following the directions given in this chapter. Determine the length of time each requires to apply 1 inch of water and measure the area of coverage.

3. Visit a lawn equipment dealership. Ask the owner to show the various models of mowers, spreaders, sprayers, rollers, and rakes that are stocked for lawn maintenance.

4. Have an equipment field day, perhaps in association with other nearby schools. Invite equipment dealers to bring selected pieces of power equipment to the school for demonstration and/or student use.

CHAPTER 27

Winterization of the Landscape

Objectives:

Upon completion of this chapter, you should be able to

❦ list those elements of the landscape that require winter protection

❦ describe eight possible types of winter injury

❦ explain nine ways to protect against winter injury

Winter injury is any damage done to elements of the landscape during the cold weather season of the year. The injury may be due to natural causes or to human error. It may be predictable or totally unexpected. At times winter injury can be avoided, while at other times it can only be accepted and dealt with.

Winter injury attacks most components of the outdoor room. Plants, paving, steps, furnishings, and plumbing are all susceptible to damage from one or more causes.

many different examples within these two categories.

Natural Injuries

The severity of winter weather can cause damage to plant materials in the landscape.

Windburn results when evergreens are to strong prevailing winds throughout the months. The wind dries out the leaf tissue the dehydrated material dies. Windburn brown to black discoloration of the leaves windward side of the plant. Very often further into the plant or on the side opposite wind show no damage. Broad-leaved evergreens are the most susceptible to windburn because have the greatest leaf surface area exposed drying winds. To protect themselves, many leaved evergreens roll their leaves in the reduce the amount of exposed surface. Figure 27-1.

Types of Winter Injury

While the types of winter damage are almost unlimited, there are several that commonly occur. The landscaper should be especially aware of these. Injuries are caused by one of two agents: nature or human beings. There are a similar location. The young, thin bark scalds easily and the natural moisture content of the tissue is low because of the reduced root system.

Heaving affects turfgrass, hardy bulbs, and other perennials when the ground freezes and thaws repeatedly because of winter temperature fluctuations. The heaving exposes the plants' roots to the drying winter wind, which can kill the plants.

Ice and snow damage can occur repeatedly during the winter. The sheer weight of snow and/or ice on plant limbs and twigs can cause breakage and result in permanent destruction of the plant's natural shape, Figures 27-2 and 27-3. Evergreens are most easily damaged because they hold heavy snow more readily than deciduous plants. Snow or ice falling off a pitched roof can split foundation plants in seconds.

Plants that freeze before the snow settles on them are even more likely to be injured. Freezing reduces plant flexibility and causes weighted twigs to snap rather than bend under added weight.

Unfortunately, the older and larger a plant is, the greater is the damage resulting from heavy snowfalls and ice storms. There are numerous recorded accounts of the street trees of entire cities being destroyed by a severe winter storm.

Animal damage to plants results from small animals feeding on the tender twigs and bark of plants, especially shrubs and growing trees. Bulbs are also susceptible. Entire floral displays can be destroyed by the winter feeding of small rodents. Shrubs can be distorted and stunted by removal of all young growth. In cases where the plant becomes **girdled** (with the bark around the main stem completely removed), the plant is unable to take up nutrients and eventually dies.

Figure 27-2 Evergreen trees can be broken and even suffer permanent damage from a heavy snow.

Human-Induced Injuries

Certain types of injury are created by people during wintertime landscape maintenance. Some types of injury are due to carelessness on the part of groundskeepers. Other types are the predictable result of poor landscape design. A large number are injuries created because the landscape elements are hidden beneath piles of snow.

Salt injury harms trees, shrubs, bulbs, lawns, and paving. Often the damage does not appear until long after the winter season passes. Thus, the cause of the injury may go undiagnosed.

The salt used to rid walks, streets, and steps of slippery ice becomes dissolved in the water it creates. The saline solution flows off walks and into nearby lawns or planting beds. Paving sometimes crumbles under heavy salting. Poured concrete is especially sensitive to this treatment, Figure 27-4.

Figure 27-3 The weight of ice on the branches of deciduous plants can break and misshape them.

Figure 27-4 Excess salt had a damaging effect on these concrete steps. This type of damage is unnecessary.

Salt is toxic to nearly all plant life. The resultant injury to lawns appears as strips of sterile, barren ground paralleling walks, Figure 27-5. Injury can also be seen on the lower branches of evergreens.

Snowplow damage can occur to plant and con- struction materials for several reasons. A careless plow operator may push snow onto a planting or a bench. This often results when those unfamiliar with the landscape are hired to do the snowplow- ing. Other injuries from plowing are the result of

Figure 27-5 Dead grass edging the walk in the summer is a symptom of excessive winter salting.

design errors. Plants, outdoor furniture, and light fixtures should not be placed near walks, parking areas, or streets where they will interfere with winter snow removal.

Damage to lawns can result when the plow misses the walk and actually plows the grass, scraping and gouging the lawn, Figure 27-6. The grass may not survive if this occurs repeatedly.

Rutting of lawns is the result of heavy vehicles parking on softened ground. When the surface layer of the soil thaws but the subsoil remains frozen, surface water is unable to soak in. Users of the landscape accustomed to finding the ground firm may be unaware of damage caused by vehicles temporarily parked on soft lawns. The soil becomes badly compacted, resulting in unsightly ruts.

Figure 27-6 A scraped lawn is the result of a snowplow, which has missed the unmarked sidewalk.

Reducing Winter Injury

Some types of winter damage can be eliminated by properly winterizing the landscape in the preceding autumn. Other types can be reduced by better initial designing of the grounds. Still other winter injuries can only be minimized, never totally eliminated.

Windburn can be eliminated in the design stage of the landscape if the planner selects deciduous plant materials rather than evergreen materials. This problem can be avoided in the garden by the use of deciduous shrubs on especially windy corners. If evergreens are important to the design or already exist in the garden, windburn can be reduced by erecting burlap shields around shrubs, Figure 27-7. The use of an antitranspirant may also reduce water loss from plants and thereby reduce the effects of windburn. The antitranspirant must be applied in the autumn and again in late winter. While antitranspirants are fairly expensive, they are more practical for the protection of large evergreens than burlap shields.

Temperature extremes can be only partially guarded against. Where wind chill (lowering of temperature because of the force of wind) is a factor, plants should be located in a protected area. Wrapping the plant in burlap also helps. This has proven to be an effective technique for protecting tender flower buds on otherwise hardy plants.

Certain plants, such as roses, can be cut back in the fall and their crowns mulched heavily to assure insulation against the effects of winter. Likewise, any plant that can be damaged by freezing and thawing of the soil should be heavily mulched after the ground has frozen to insulate against premature thawing.

If the landscaper is trying to prolong the lives of annual flowers in the autumn, and a frost is forecast, the foliage can be sprinkled with water prior to nightfall. This helps to avoid damage caused by a light frost. The water gives off enough warmth to keep the plant tissue from freezing.

Sunscald of young transplants lessens as the plants grow older and form thicker bark. It can be avoided on trees during the first winter of growth by wrapping the trunks of the trees with paper or burlap stripping. For other types of sunscald,

Figure 27-7 A broad-leaved evergreen protected against windburn with a burlap shield

such as that affecting broad-leaved evergreens, the same remedies practiced for windburn and temperature extremes are effective. Wrapping the plants in burlap or the use of antitranspirants gives protection.

Some sunscald can be avoided by the designer. Vulnerable species of plants should not be placed on the south side of a building, nor should they be placed against a reflective white wall that will magnify the sun's effect on the above-ground plant parts.

Heaving of the turf is impossible to prevent completely. The best defense against it is encouraging deep rooting through proper maintenance and good landscaping practices during the growing season.

Bulbs, groundcovers, and other perennials can be protected from heaving through application of a mulch after the ground has frozen. The mulch acts to insulate the soil against surface thawing.

Ice and snow damage to foundation plants can be avoided if the designer is careful not to place plants beneath the overhanging roof line of a building. If the groundskeeper must deal with plants already existing in a danger area, the use of hinged, wooden A-frames over the plants can help to protect them, Figure 27-8. As large pieces of frozen snow and ice tumble off the roof, the frame breaks them apart before they can damage the plants.

To aid plants that have been split or bent because of heavy snow accumulation, the groundskeeper must work quickly and cautiously. A broom can be used to shake the snow off the weighted branches. However, snow removal must be done gently and immediately after the snow stops. If the branches are frozen or the snow has become hard and icy, removal efforts will cause more damage than benefit.

If breakage of plants occurs during the winter, the groundskeeper should prune the damaged parts as soon as possible. This prevents further damage to the plants during the rest of the winter.

Certain plants, such as upright arborvitae and upright yews, become more susceptible to heavy snow and ice injury as they mature. Often the damage cannot be repaired. Large and valuable plants in a landscape can be winterized in the autumn by tying them loosely with strips of burlap or twine. (Do not use wire.) When prepared in this manner, the branches cannot be forced apart by the heavy snows of winter, and splitting is avoided

Animal damage can be prevented by either eliminating the animals or protecting the plants from their feeding. While rats, mice, moles, and voles are generally regarded as offensive, plantings are damaged as much or more by deer, rabbits, chipmunks, and other gentler kinds of animals. Certain rodenticides (substances that poison rodents) may be employed against some of the undesirable animals that threaten the landscape. In situations where the animals are welcome but their winter feeding damage is not, a protective enclosure of fine mesh wire fencing around the plants helps to discourage animals from feeding there, Figure 27-9.

Figure 27-8 Hinged wooden A-frames protect foundation plantings from damage caused by sliding ice and snow.

Figure 27-9 A wire enclosure placed around young plants during the winter helps to protect them from animal damage. Even smaller mesh is needed to discourage rodents.

Salt injury to plants and paving need not be as bad as was illustrated earlier if caution is exercised by the groundskeeper. Salt mixed with coarse sand does a better job than either material used separately. The sand provides traction on icy walks, and a small amount of salt can melt a large amount of ice. Excessive salt has no value; it only kills plants and destroys paving. In very cold temperatures, salt does not melt ice; therefore, in these cases it serves no purpose. Salt can turn compacted snow, which is comparatively safe for traffic, into inches of slush, which is messy and even more slippery. The problem of salt injury can be solved by reducing the amount of salt spread on walks and streets during the winter season. The addition of sand distributes the salt more evenly and provides grit for better traction.

Snowplow damage is to be expected if a designer places plants too close to walks and roadways. Therefore, one obvious solution to the problem begins with the designer. When planning landscapes for areas in which winter is normally accompanied by a great deal of snow, the designer should avoid placing shrubs or other items near intersections or other places where snow is likely to be pushed.

Another type of plow damage is the result of the plow driver's inability to see objects beneath the snow. If possible, all objects such as outdoor furniture and lights should be removed from plow areas prior to the winter season. If it is not possible to move them, low objects should be marked with tall, colored poles that can be seen above the snow.

Whenever possible, snow blowers should be used instead of plows. These machines are much less likely to cause damage.

Rutting of lawns usually results from the practice of permitting individuals to park cars on lawns. The best solution to the problem is to avoid doing so. Otherwise, sawhorses or other barriers offer a temporary solution.

Achievement Review

A. Indicate whether the following types of winter injury are caused by natural conditions (N) or human error (H).

1. crumbled paving resulting from too much salt
2. dry, dead twigs on the windward side of a pine tree
3. dried, blistered bark on the trunk of a recently transplanted tree
4. deep ruts in the lawn in front of a house
5. dead branches in a shrub following an unusually cold winter
6. failure of a shrub to flower in the spring
7. perennials lying on the surface of the soil in the early spring with roots exposed to the drying air
8. dead grass next to a walk heavily salted during the winter
9. an upright evergreen split in the center by snow sliding off a roof
10. a young tree girdled at the base

B. Of the types of winter damage in the following list, which ones could be reduced or prevented by proper winterization of the landscape during the late autumn?

1. crumbled paving
2. sunscald on new transplants
3. breakage of outdoor furniture by snow plows
4. rutting of the lawn by automobiles
5. foundation plants broken by falling snow
6. sunscald on broad-leaved evergreens
7. tree limbs broken off by an ice storm
8. bulbs heaved to the surface of the soil
9. windburn on evergreens
10. flooded basement caused by melting snow

Suggested Activities

1. Look for signs of winter injury in nearby landscapes. Find windy corners where evergreens are planted and check for windburned tips. Visit a shopping center, campus, or park where salt is used on walks and parking lots. Look at the paving and nearby plantings for signs of damage. Note the placement of plants in relation to walk intersections and other places where snow may be piled.

2. Conduct ice melting tests. Freeze four pie plates of water. Apply four different mixtures of salt and/or sand to the surfaces and see which melts first. In the first pan, use all salt; in the second, half salt and half sand; in the third, one-quarter salt and three-quarters sand; and in the fourth, all sand. What conclusions can be drawn from the trials?

   NOTE: Returning the treated pans of ice to the freezer (approximately 20 degrees F) will assure that no natural melting occurs. Check the pans every 15 minutes for observations.

3. Demonstrate the damaging effects of salt upon plant life. Grow some experimental plants in advance. Root each plant in a separate container. Apply only water to some of the plants for a week. To others, apply varying dilutions of a salt-water solution. To a third group, apply water to the soil, but mist the foliage with saltwater several times each day. Record observations of each treatment daily.

4. Demonstrate the reduction of water loss from plant tissue caused by antitranspirants. Purchase a small bottle of antitranspirant from a garden center, or write to a manufacturer and request a complimentary sample. Dilute with water as specified on the label. Using a small pump sprayer, apply the antitranspirant to one of two comparable samples of cut evergreens in vases of water. Be certain to apply it evenly over all needle surfaces. The untreated sample should dry out sooner than the treated one, indicating greater water loss.

5. Repeat the demonstration in #4 using a sun lamp or a fan to simulate the drying effects of sun and wind and the protection offered by antitranspirants.

Pricing Landscape Maintenance

Objectives:

Upon completion of this chapter, you should be able to

❧ describe the values of cost analysis to a maintenance firm

❧ list the features of a landscape maintenance cost analysis

❧ describe unit pricing

❧ prepare a maintenance cost estimate

The Need for Cost Analysis in Landscape Maintenance

An accurate analysis of the costs of specific tasks done by a landscape maintenance firm has several values:

❧ It assures that all costs to the firm are recognized.

❧ It permits a fair price to be charged to the customer.

❧ It allows a comparison of the profitability of different tasks.

❧ It can compare the efficiency of different crews performing the same task.

The first two benefits were discussed in Chapter 15. They are equally important in landscape maintenance. The last two warrant explanation also. A new landscape maintenance firm often believes that all jobs are good jobs. "No job is too big or too small" is the way their advertisements often read. Actually, as a firm grows, not all jobs can be accepted. There may be insufficient time or personnel to respond to every client request. The tasks that return the greatest profit for the labor invested will need to be emphasized over those that return less profit. A carefully prepared cost analysis can illustrate these different profit potentials.

The performance efficiency of several crews or laborers doing the same task can also be compared with a maintenance cost analysis. Assuming the cost of materials and the site conditions to be the same, the labor time required is the only comparable variable. Carefully kept work records documenting the crew size and hours of labor required to complete a task permit the performance comparison.

Features of the Maintenance Cost Analysis

A cost analysis of a landscape maintenance job includes

❦ a listing of all tasks to be performed.

❦ the total square footage area involved for each service

❦ the number of times each service is performed during the year.

❦ the time required to complete each task once.

❦ the time required to complete each task annually.

❦ the cost of all materials required for each task.

❦ the cost of all labor required for each task.

In order for total job costs to be precisely calculated, each individual cost must be described in a form that can be added easily to all other job costs. **Unit pricing** reduces all area dimensions and material quantities to a common measurement, such as a thousand square feet or an acre. It is a necessary first step, since materials are often purchased by the truckload, bale, or other bulk measurement. Unit pricing also permits maintenance cost analysis to be adapted for computer use.

Calculations for Cost Analysis

To apply the technique properly requires practice. Study the following examples and their explanations.

Example 1

Problem:

To calculate the cost of mowing 10,000 square feet of lawn with an 18-inch power mower 30 times each year.

Necessary Information:

It takes 5 minutes to mow 1,000 square feet. The laborer receives $6.50 per hour. There are no material costs.

Solution:

Maintenance Operation: Lawn mowing with 18-inch power mower

Square footage area involved	10,000 sq. ft.
Number of times performed annually	30
Minutes per 1,000 sq. ft.	5
Total annual time in minutes (1)	1,500
Material cost per 1,000 sq. ft.	none
Total material cost	none
Wage rate per hour	$5.00
Total labor cost (2)	$125.00
Total cost of maintenance operation per year	$125.00

Note:

(1) To obtain the total annual time in minutes:

   a. divide the square footage of area involved by 1,000 [10,000 sq. ft. ÷ 1,000 = 10]

   b. multiply by minutes per 1,000 sq. ft. [10 × 5 = 50 minutes]

   c. multiply by number of times performed annually [50 minutes × 30 = 1,500 minutes]

   d. enter answer under total annual time in minutes.

(2) To obtain the total labor cost:

   a. divide the total annual time in minutes by 60 minutes [1,500 minutes ÷ 60 minutes = 25 hours]

   b. multiply by the wage rate per hour [25 hours × $5.00 = $125.00]

   c. enter answer under total labor cost.

Example 2

Problem:

To calculate the cost of mulching 2,000 square feet of planting beds with wood chips, 4 inches deep.

Necessary Information:

The task is done once each year. It requires 30 minutes to mulch 1,000 square feet, 4 inches deep. The laborer receives $6.50 per hour. The wood chips cost $165.00 per 1,000 square feet of coverage.

Solution:

Maintenance Operation: Mulching plantings with wood chips, 4 inches deep

Square footage area involved	2,000 sq. ft.
Number of times performed annually	1
Minutes per 1,000 sq. ft.	30
Total annual time in minutes (1)	60
Material cost per 1,000 sq. ft.	$165.00
Total material cost (2)	$330.00
Wage rate per hour	$5.00
Total labor cost (3)	$5.00
Total cost of maintenance operation per year (4)	$335.00

Note:

(1) To obtain the total annual time in minutes:
   a. divide the square footage of area involved by 1,000 [2,000 sq. ft. ÷ 1,000 = 2]
   b. multiply by minutes per 1,000 sq. ft. [2 × 30 = 60 minutes]
   c. multiply by number of times performed annually [60 minutes × 1 = 60 minutes]
   d. enter answer under total annual time in minutes.

(2) To obtain the total material cost:
   a. divide the square footage of area involved by 1,000 [2,000 sq. ft. ÷ 1,000 = 2]
   b. multiply by material cost per 1,000 sq. ft. [2 × $165.00 = $330.00]
   c. multiply by number of times performed annually [$330.00 × 1 = $330.00]
   d. enter answer under total material cost.

(3) To obtain the total labor cost:
   a. divide the total annual time in minutes by 60 minutes [60 minutes ÷ 60 minutes = 1 hour]
   b. multiply by the wage rate per hour [1 hour × $5.00 = $5.00]
   c. enter answer under total labor cost.

(4) To obtain the total cost of maintenance operation per year:
   a. add total material cost and total labor cost [$330.00 + $5.00 = $335.00]
   b. enter answer under total cost of maintenance operation per year.

The Completed Cost Estimate

A full cost estimate for maintenance is simply an enlargement of the previous examples. For convenience, all of the maintenance operations that deal with the same area of the landscape are grouped together in the estimate. Study the following example and note the calculation of all figures.

Data

I. A landscape requires the following maintenance tasks and equipment:
   a. 24,500 square feet of lawn cut 30 times each year with a power riding mower
   b. 500 square feet of lawn cut 30 times each year with an 18-inch power hand mower
   c. all lawn areas fertilized twice each year
   d. 5,000 square feet of shrub plantings fertilized once each year
   e. shrubs pruned once each year
   f. 400 square feet of flower beds requiring soil conditioning once each spring
   g. flowers planted once each year
   h. flowers hand weeded 10 times each year
   i. flower beds cleaned and prepared for winter once each autumn

II. Calculated time requirements for the maintenance tasks:
   a. a power riding mower cutting 1,000 square feet of lawn in 1 minute
   b. an 18-inch power mower cutting 1,000 square feet of lawn in 5 minutes
   c. lawn fertilization requiring 3 minutes per 1,000 square feet for spreading
   d. shrub fertilization requiring 5 minutes per 1,000 square feet
   e. pruning time for shrubs averaging 60 minutes per 1,000 square feet
   f. soil conditioning for flower beds requiring approximately 200 minutes per 1,000 square feet
   g. flower planting requiring 600 minutes per 1,000 square feet

The Completed Cost Estimate

Maintenance Operation*	Sq. Footage Area Involved*	Number of Times Performed Annually*	Minutes Per 1,000 Sq. Ft.*	Total Annual Time in Minutes**	Material Cost Per 1,000 Sq. Ft.*	Total Material Cost**	Wage Rate Per Hour*	Total Labor Cost**	Total Cost of Maintenance Operation Per Year**
Lawn									
Mowing – rider	24,500 sq. ft.	30	1	735	None		$6.50	$79.63	$ 79.63
Mowing – 18" power	500 sq. ft.	30	5	75	None		$6.50	$ 8.13	$ 8.13
Fertilization	25,000 sq. ft.	2	3	150	$ 5.00	$250.00	$6.50	$16.25	$266.25
Shrubs									
Fertilization	5,000 sq. ft.	1	5	25	$ 6.00	$ 30.00	$6.50	$ 2.73	$ 32.73
Pruning	5,000 sq. ft.	1	60	300	None		$6.50	$32.50	$ 32.50
Flowers									
Soil conditioning	400 sq. ft.	1	200	80	$ 3.00	$ 1.20	$6.50	$ 8.65	$ 9.85
Planting	400 sq. ft.	1	600	240	$125.00	$ 50.00	$6.50	$26.00	$ 76.00
Hand weeding	400 sq. ft.	10	60	240	None		$6.50	$26.00	$ 26.00
Autumn cleanup	400 sq. ft.	1	400	160	None		$6.50	$17.36	$ 17.36
									$548.45

Notes: * All entries in this column came directly from the data given.
       ** All entries in this column were calculated using methods in the earlier examples. Students should practice the calculations to insure their understanding of the methods.

h. weeding of flowers requiring 60 minutes per 1,000 square feet

i. cleanup of flower beds in the autumn requiring 400 minutes per 1,000 square feet

III. All laborers receive wages of $6.50 per hour.

IV. Material costs:

a. lawn fertilizer at $5.00 per 1,000 square feet

b. shrub fertilizer at $6.00 per 1,000 square feet

c. conditioning materials for flower beds at $3.00 per 1,000 square feet

d. flowers for planting averaging $125.00 per 1,000 square feet

V. Overhead and profit. To convert the cost analysis to a cost estimate, additional charges for overhead and profit must be added. As described in Chapter 15, these charges may be figured as a percentage of the total cost of the maintenance operation. In this example, overhead costs of 10 percent ($54.85) and profit allowance of 25 percent ($137.11) could be added to the $548.45 cost to the firm. The price quotation to the client would then be $740.41.

NOTE: Many firms apply different overhead and profit percentages to different parts of a project, rather than using single percentages, as in this example.

Practice Exercise

Complete a cost estimate based upon the following data. Add additional overhead costs of 10 percent and a profit allowance of 25 percent.

I. A landscape requires the following maintenance tasks and equipment:

a. 12,000 square feet of lawn cut twenty-five times a year with a power riding mower

b. 600 square feet of lawn cut twenty-five times each year with a 25-inch power hand mower

c. all lawn areas fertilized twice each year

d. all lawn areas raked once each spring with a 24-inch power rake

e. 700 square feet of shrub plantings cultivated with hoes twice each year

f. shrubs pruned once each year

g. shrubs fertilized once each year

h. 250 square feet of flower beds requiring soil conditioning once each spring

i. flowers in 250 square feet planted once each year

j. flowers in 250 square feet weeded by hand ten times each year

k. flower beds in 250 square feet cleaned and prepared for winter once each autumn

II. Time requirements for the maintenance tasks include:

a. power riding mower cutting 1,000 square feet of lawn in 1 minute

b. the 25-inch power hand mower cutting 1,000 square feet of lawn in 3 minutes

c. lawn fertilization requiring 3 minutes per 1,000 square feet for spreading

d. lawn raking with a 24-inch power rake requiring 10 minutes per 1,000 square feet

e. hand hoe cultivation of the shrubs requiring 60 minutes per 1,000 square feet

f. pruning of shrubs averaging 60 minutes per 1,000 square feet

g. shrub fertilization requiring 5 minutes per 1,000 square feet

h. soil conditioning for flower beds requiring approximately 200 minutes per 1,000 square feet

i. flower planting requiring 600 minutes per 1,000 square feet

j. weeding of flowers requiring 60 minutes per 1,000 square feet

k. cleanup of flower beds in the fall requiring 400 minutes per 1,000 square feet

III. All laborers receive wages of $7.00 per hour.

IV. Material costs include:

    a. lawn fertilizer at $5.00 per 1,000 square feet

    b. shrub fertilizer at $6.00 per 1,000 square feet

    c. conditioning materials for flower beds at $3.00 per 1,000 square feet

    d. flowers for planting at $125.00 per 1,000 square feet

Achievement Review

A. List four ways in which a cost estimate benefits a landscape maintenance firm.

B. What seven items of data are required before beginning a maintenance cost estimate?

C. Figure the total annual time in minutes for a task that is done five times a year, involves 6,000 square feet of area, and requires 7 minutes per 1,000 square feet to accomplish.

D. Figure the total material cost for mulch that is purchased to cover 1,500 square feet of area and costs $125.00 per 1,000 square feet. The mulch is applied once each year.

E. Calculate the total labor cost for a job that requires a total of 420 minutes. The worker assigned to the job is paid $6.00 per hour.

The Final Word

Customer Service

This text began with consideration of the meaning of a word. The word was *landscaper*. Many of the skills and crafts necessary to the practice of landscaping have been discussed in the chapters of the text. It is now appropriate to close the text with another word and an explanation of its meaning. The final word is *service*, and without an acceptance of its meaning and importance as well as knowledge of how and to whom it should be applied, no one is fully prepared to be a landscaper in the 1990s.

Landscaping is a service industry. It is labor intensive and involves the satisfaction of customer needs. A company is only as successful as its clients allow it to be. The key to understanding what service means and requires is acceptance of some new descriptions of old ideas.

Few businesses today have not encountered the concept of total quality management in the popular press as well as in the literature of their industry. The landscape industry is no exception. Total quality management, or TQM as it is frequently abbreviated, is a system of business operation that is driven by the objective of meeting and exceeding client expectations through ongoing improvements in the processes that comprise the business. It replaces the old idea of quality control, as established by the company, with the more factual realization that performance by a company is only qualitative if the client perceives it as such. It matters little if the company praises its own performance. Only praise by the customer who pays the bill truly measures whether the project was done satisfactorily or not. Thus modern landscape companies must do everything possible to ascertain the true desires of their clients at the outset of a project if they are to provide the quality of service that will meet and exceed their clients' expectations. This is no easy task, since each client is different and each expresses or fails to express his or her desires for the project differently. For example, two different clients may each want a patio constructed. Each may have offered nearly similar specifications to the landscaper during the discussion phase of the project; e.g., same size, materials, furnishings, and budget. What was not specified to the company was that one of the clients had in his mind a memory of a patio his family had when he was a child, and he is going to measure the new one against his sentimentalized remembrance. The other client cares more about what her friends and neighbors will think of the patio than anything else about it. She is most concerned that it be visually stunning to impress her guests. Meeting and exceeding the expectations of the two clients will require personal attention to each one by staff members who are able and willing to provide it. If the service meets the quality expectations of the clients, each is likely to return to the company as a future customer. If they are not pleased with the service provided, it is unlikely that they will provide repeat business and highly probable that they will tell their friends and family of their dissatisfaction.

The provision of such personalized service to clients requires all employees of the landscape company to realize that they are in customer service no matter what their job. Even if they seldom or never have direct contact with the bill-paying clients, they must be shown and convinced that the way they perform their jobs affects the ultimate impression of the company's performance by the client. The on-site crew must not only perform their tasks competently, they must look and behave professionally. Dirty, tattered clothing or uniforms that don't match convey a different image than a well-groomed crew dressed in matching uniforms, working with freshly painted equipment, and driving clean trucks. One crew may fit the bad image of landscapers that white-collar professionals often have, while the other crew exceeds the client's expectations, and pleasantly so. Even if the finished project could be done equally well by both crews, the crew perceived as providing the best service would probably be the one that presented the best image. Similar perceptions of professionalism or lack thereof are conveyed by the way crew members behave at the work site. Smoking, loud talk, profanity, wasting time, blaring radios, and spitting can all contribute to client dissatisfaction and make the installation of their new landscape an ordeal rather than a pleasure.

Another instance where clients are sensitive to their treatment by a company occurs when the clients approach a crewperson or supervisor with a question or request and find that employee powerless to respond. While it is difficult for some owners and managers of companies to empower their employees enough to address client concerns when and where they arise, it is an important step toward making employees feel involved with the company, not merely used by it. The level of client satisfaction is raised when clients' questions are regarded as sufficiently important to warrant immediate responses.

In addition to the crewpersons, others in the company affect the client's perception of service and quality. The number of rings before the telephone is answered and the tone of the person's voice all create an impression. Languishing on hold, listening to squeaky recorded music, is certain to frustrate any caller. Voice mail systems and answering machines, unless used carefully and selectively, can also alienate customers. A landscape is a living thing, and clients want to talk about it with living people, not recorded voices.

Treatment by salespersons, the speed and accuracy of billing, and the follow-up concern for client satisfaction are among the many benchmarks by which clients evaluate the services of a landscape company. Since people remember bad experiences, such as rudeness or an unreturned phone call, more than good experiences, one inattentive employee can undermine the superb performances of a dozen other employees. The commitment to total quality management must be company wide. All employees must be taught to understand and value it, then expected to practice it. If they cannot or will not, despite repeated training, then they must be let go. Accepting the philosophy of TQM is as difficult for some executives as it is for others in the company. TQM negates their role as overseers and bosses and instead requires them to be leaders and coaches. TQM is most easily accepted by team players, and most difficult for strong-willed, dominant personalities to accept.

Some would argue that providing a quality of service described as acceptable by the client rather than the company will add greatly to the cost of building landscapes. That has proven to be untrue. It takes about the same amount of time and materials to build a landscape improperly as it does to do it correctly the first time. The defective landscape, which must be fixed after the customer complaint, ends up costing the company more in additional labor expenses and diminished reputation. It also costs a company more to recruit and train new employees to replace departed, disgruntled employees than it does to offer sufficient incentives to retain trained, contented employees.

The original proponents of total quality management also challenged traditional business thinking by redefining the word *client*. The term seems simple: the client is the recipient of goods and services. The new interpretation includes the concept of both external clients and internal clients in a business. The external clients are the ones who order and pay for landscape projects. They are the ones for whom every worker must be convinced to do his or her best every minute of every working day. They make the final judgment about the true quality of the project, and they are the ones whose expectations are to be surpassed. The internal client is the newer expansion of the term. It refers to the company personnel, each of whom is part of one or more systems that deliver some piece of the service and/or product to the external client. If a system or process for the provision of a service or product can be likened to a chain, then each employee can be seen as being linked to those other employees who are responsible for functions or materials that either precede or follow the contribution of that particular employee. In such a situation, each employee within a process or system can be seen as a client of the employee whose step in the process precedes his or hers. Once the client-employee has contributed to the process, it is handed off to the next employee in the chain. Now the client-employee becomes the supplier to the next employee in the sequence, his client. Seen this way, it is easier to accept the proposition that every employee within the landscape company is at times the supplier and at times the client of other suppliers.

The not-so-easy part is gaining an agreement from all employees to accept no defective products or performances from their internal suppliers. If an employee does not pass along work to her internal client in the best possible state, the end product or service will not be as good as it can be. It will wind up either costing the company money to remedy it, or disappointing the external client rather than exceeding expectations. Defective performance must be identified at its source and the internal supplier given the opportunity to correct the defect before returning it to the internal client. To gain acceptance of this important quality factor by employees, the leadership of the company must remove the stigma traditionally attached to errors. Employees must be permitted to make mistakes and correct them without fear of being criticized or reprimanded. This is where the role of the leaders as coaches and cheerleaders for their employees, rather than vigilantes seeking criminals, is critical.

Each process or system within the company must be broken down into its component steps, then studied, made better, the improved process implemented, and the entire process repeated again, and again, and again. TQM is never ending, and it does not happen quickly. Neither can it be accomplished with motivational posters and shiny plaques. It is a management style that calls upon every employee to participate. It uses their expertise to improve its products and services. It rewards their creativity rather than discouraging independent thinking. In the end, it fulfills the expectations of both internal and external clients, and even surpasses those expectations. In a service industry such as the landscape industry, catering to a clientele whose own companies are teaching them about TQM, the expectation level of clients grows higher each year. There are a number of good companies in every major market in the country. The ones that will be most successful will be the ones that keep in mind and excel at the final word . . . *service*.

Appendix A

Examples of Landscape Designs

The illustrations that follow are examples of landscape designs done by professionals currently working in the landscape industry. The designs include residential areas, small commercial sites, and recreational areas. These plans are the work of designers with college preparation ranging from two to five years. The designers are employed as landscape contractors, landscape architects, or recreational planners.

Students will gain several insights by studying the plans. First, each plan can be seen as a graphic explanation of the designer's ideas. Some are mainly concepts, with few precise details, aimed at selling the designer's proposals to a client. Others are detailed instructions of what, where, and how elements of the landscape are to be developed.

Second, notice how the graphics vary. Some of the plans are highly mechanical in appearance, due to the use of lettering guides and waxed press-on letters. Others are less rigid in appearance, due to free-hand lettering and a looser graphic style. All of the designs are of professional quality. The graphic technique used depends upon how much time the project warrants at a particular stage in its development, and how much competition there is among designers to gain a potential client.

(Courtesy of A. J. Tomasi Nurseries, Inc., Pembroke, Mass.)

(Courtesy of A. J. Tomasi Nurseries, Inc., Pembroke, Mass.)

A SMALL COURT

ELEVATION

CHAI POUR

SPACE

DESIGNED BY MICHAEL BOICE
ALBANY, NY

BLOW-UP OF FOUNDATION PLANTING

SCALE - 1' = 10'

- THE SECOND NUMBER AFTER THE PLANT NAME INDICATES THE PRUNING HEIGHT

DESIGNED BY
MICHAEL E. BOICE

MAKE THE BOICE CHOICE!

SCALE - 1' = 10'   BLOW-UP

SCALE - 1' = 20'

- ALL PLANT BEDS CONTAIN PINE BARK MULCH
- AFTER PLANT NAME SHOWS EXISTING PLANTS - EX. WHITE FIR 2 =

SCALE - 1' = 20'

PLANT LIST		
BOTANICAL -SHRUBS-	COMMON	NO. USED
AZALEA MOLLIS	CHINESE AZALEA	2
BERBERIS THUNBERGI ATROPURPUREA	RED JAPANESE BARBERRY	2
CORNUS ELEGANTISSIMA VARIEGATA	VARIEGATED DOGWOOD	1
COTONEASTER APICULATA	CRANBERRY COTONEASTER	3
EUONYMUS ALATUS	BURNING BUSH	1
FORSYTHIA INTERMEDIA 'LYNWOOD GOLD'	LYNWOOD GOLD FORSYTHIA	2
JUNIPERUS CHINENSIS 'HETZI'	HETZ JUNIPER	15
JUNIPERUS CHINENSIS 'HETZI COMPACTA'	DWARF HETZ JUNIPER	3
JUNIPERUS HORIZONTALIS PLUMOSA	ANDORRA JUNIPER	3
JUNIPERUS SQUAMATA MEYERI	MEYER JUNIPER	3
PHILADELPHUS CORONARIUS AUREUS	GOLDEN MOCKORANGE	1
PHILADELPHUS VIRGINALIS	VIRGINAL MOCKORANGE	12
PIERIS JAPONICA	JAPONICA (ANDROMEDA)	4
PINUS MUGHUS	DWARF MUGO PINE	1
RHODODENDRON SPP	RHODODENDRON	8
TAXUS CUSPIDATA	JAPANESE YEW (PYRAMIDAL)	3
WEIGELA FLORIDA VARIEGATA	VARIEGATED WEIGELA	1
-TREES-		
ABIES CONCOLOR	WHITE FIR	8
ACER PALMATUM	JAPANESE MAPLE	1
BETULA PENDULA LACINIATA	CUT LEAF WEEPING BIRCH	1
ELAEAGNUS ANGUSTIFOLIA	RUSSIAN OLIVE	1
MALUS SPP	MacINTOSH APPLES	4
PICEA PUNGENS	COLORADO BLUE SPRUCE	5
PRUNUS SUBHIRTELLA 'PENDULA'	WEEPING CHERRY	1
SORBUS AUCUPARIA	EUROPEAN MOUNTAIN ASH	1
- VINES & GROUND COVER-		
EUONYMUS FORTUNEI ARGENTEO-MARGINATUS	VARIEGATED WINTERCREEPER	3
PACHYSANDRA TERMINALIS	PACHYSANDRA	1400
SEDUM 'DRAGONS BLOOD	'DRAGON'S BLOOD' SEDUM	50
	GROUND MOSS	100

SCALE - 1' = 20'

PLANTING PLAN FOR

MR. & MRS. C. MASICK

COBLESKILL, N.Y.

DESIGNED BY MICHAEL BOICE
ALBANY, NY

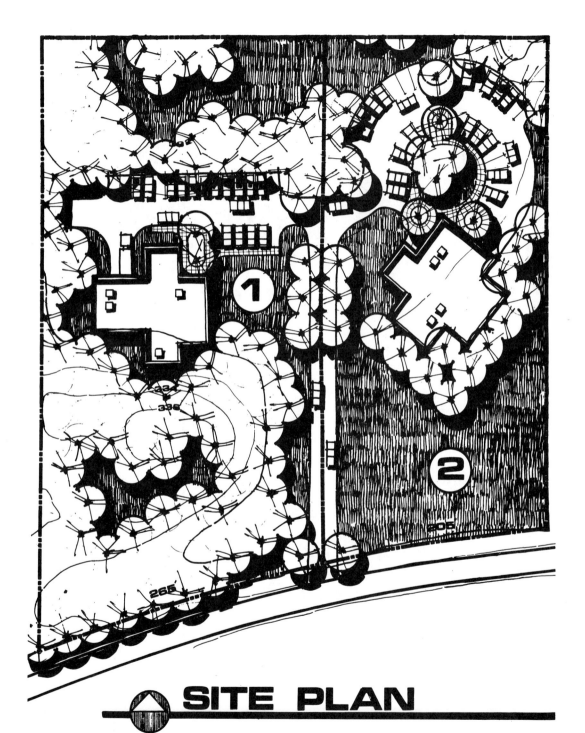

SITE PLAN

DESIGNED BY MARK MAGNONE
SCHENECTADY, NY

PROPOSED SITE PLAN

DESIGNED BY MARK MAGNONE
SCHENECTADY, NY

DESIGNED BY GOLDBERG AND RODLER
HUNTINGTON, NY

DESIGNED BY RUSSELL IRELAND
EAST NORWICH, NY

DESIGNED BY JOHN KRIEG, LANDSCAPE ARCHITECT
ROBT. C. BIGLER ASSOC., ARCHITECTS
PHOENIX, ARIZONA

LANDSCAPE PLAN...

DESIGNED BY JOHN KRIEG, LANDSCAPE ARCHITECT
ROBT. C. BIGLER ASSOC., ARCHITECTS
PHOENIX, ARIZONA.

COAL STREET PARK
WILKES-BARRE, PENNSYLVANIA

DESIGNED BY THE ALLEN ORGANIZATION
GLENS FALLS, NY

MASTER PLAN FOR DEVELOPMENT OF

RECREATION PARK

MIDDLEBURY, VERMONT

ALLEN ORGANIZATION, PARK & RECREATION PLANNERS
DEPARTMENT OF ABSTRACT, ASSOCIATE CONSULTING ENGINEERS
GLENS FALLS, NEW YORK

DESIGNED BY THE ALLEN ORGANIZATION
GLENS FALLS, NY

Appendix B

Tools of the Trade

Specialized Hand Tools and Their Functions

Tool and Name	Function	Tool and Name	Function
Grass Shears	Used to trim grass along walks, roadways, the edge of planting beds, and around trees, posts, etc.	Pruning Saw	Removes any tree or shrub part that cannot be easily cut with the lopping shears. Usually parts are 1 inch or more in diameter.
Pruning Shears	Used to trim tree and shrub twigs up to 1/2 inch in diameter.	Crosscut Saw	Removes large limbs and small trees. The saw has additional general uses.
Lopping Shears	Used to trim tree and shrub twigs from 1 inch to 1 1/2 inches in diameter.	Grass Hook	For reducing the height of overgrown grass areas. It requires the user to bend over.
Hedge Shears	Prunes shrubs grown closely spaced as hedges. These shears are only used on young, tender new growth.		

Specialized Hand Tools and Their Functions (continued)

Tool and Name	Function	Tool and Name	Function
Grass Whip	For reducing grass height without bending over. (Once reduced in height, a lawn mower can be used on the grass.)	Spading Shovel	A combination tool with uses similar to both spades and shovel. It can be used for digging as well as scooping.
Spading Fork	Used for turning over the soil when it is not too hard or compacted. Also used for lifting bulbs in the fall.	Scoop	Good for moving loose materials such as crushed stone, peat moss, soil, etc. Scoops have high sides. They are not used for digging.
Spades	Obvious general uses in digging. Spades have flatter shapes than shovels. They penetrate the soil more easily but have less scooping capability.	Manure Fork	The best tool for moving coarse, lightweight materials such as straw, wood chips, etc.
Shovel	Used for cleaning loose soil from planting holes and other scooping uses. A shovel has sides that a spade does not have.		

Specialized Hand Tools and Their Functions *(continued)*

Tool and Name	Function	Tool and Name	Function
Single-Bit and Double-Bit Axes	Obvious chopping uses. Especially useful in tree removal and for cutting up fallen timber.	Lawn Comb	An excellent rake for collection of leaves and coarse debris from lawn surface.
Weed Cutter	Removes annual weeds by cutting them off at ground level. Not very effective against biennial and perennial weeds.	Shrub Comb	Used for raking debris from small areas between shrubs.
Toothed Rakes	Used for heavy-duty raking that requires a strong tool. Commonly used in preparation of lawn seed beds and cultivation of planted beds.	Bulb Planter	Used to install flowering bulbs.
Broom Rake	Very useful in places where a lightweight, springy rake is needed. Very good for collecting debris and clippings from lawn surface.	Push Hoe	Similar to a scuffle hoe. It is good for rooting out weeds.

Specialized Hand Tools and Their Functions *(continued)*

Tool and Name	Function	Tool and Name	Function
Hand Trowel	Used for transplanting bedding plants into flower beds, borders, and boxes.	Post Hole Digger	Prepares holes for the support posts of fences.
Transplanting Hoe	Uses are similar to those of a hand trowel. It has less adaptability for other types of digging.	Garden Hoe	Widely used for breaking up the soil prior to planting. It is also good for cultivating planted beds and for weed removal.
Scuffle Hoe	Useful in weeding and cultivating in planted beds. It cuts off weeds and loosens surface soil.	Hand Cultivator	Loosens the surface soil in flower beds and around shrubs.

Specialized Hand Tools and Their Functions (*continued*)

Tool and Name	Function	Tool and Name	Function
Pick	Used for breaking up hard rocky soil. It has two pointed ends for gouging into the soil.	Cutter Mattock	Stronger than a grading hoe. Its uses are similar. It has two flat ends.
Sprayer	Needed to apply pesticides, antitranspirants, and other chemicals in liquid form. Sprayers are available in a wide range of sizes.	Spreader	Used for the application of fertilizer, seed, and other dry turf products.
Grading Hoe	Loosens hard or compacted soil during preparation for planting. Has a sharpened flat end.		

Appendix C

Professional and Trade Organizations

Agronomy, Soil

AMERICAN SOCIETY OF AGRONOMY
677 S. Segoe Road
Madison, WI 53711

CROP SCIENCE SOCIETY OF AMERICA
677 S. Segoe Road
Madison, WI 53711

SOIL CONSERVATION SOCIETY OF AMERICA
7515 N.E. Ankeny Road
Ankeny, IA 50021

SOIL SCIENCE SOCIETY OF AMERICA
677 S. Segoe Road
Madison, WI 53711

WEED SCIENCE SOCIETY OF AMERICA
1508 W. University Ave.
Champaign, IL 61821

Botany

AMERICAN ASSN. OF BOTANICAL GARDENS
   AND ARBORETA
P.O. Box 206
Swarthmore, PA 19081

BOTANICAL SOCIETY OF AMERICA
University of Kentucky
Lexington, KY 40506

Bulbs

NETHERLANDS FLOWER-BULB INSTITUTE
Corporate Plaza II
51 Cragwood
S. Plainfield, NJ 07080

Business Management

GARDEN INDUSTRY OF AMERICA, INC.
2501 Wayzata Boulevard
Minneapolis, MN 55440

NATIONAL LAWN AND GARDEN
   DISTRIBUTORS ASSN.
1900 Arch Street
Philadelphia, PA 19103

NATIONAL SMALL BUSINESS ASSN.
1604 K Street NW
Washington, DC 20006

Fertilizers, Chemicals

FERTILIZER INSTITUTE
1015 18th Street NW
Washington, DC 20036

NATIONAL FERTILIZER SOLUTIONS ASSN.
8823 N. Industrial Road
Peoria, IL 61615

POTASH/PHOSPHATE INSTITUTE
2801 Buford Highway NE
Suite 401
Atlanta, GA 30329

Flowers, Plants

ALL-AMERICAN ROSE SELECTIONS, INC.
P.O. Box 218
Shenandoah, IA 51601

AMERICAN CAMELLIA SOCIETY
P.O. Box 1217
Fort Valley, GA 31030

AMERICAN DAHLIA SOCIETY, INC.
345 Merritt Avenue
Bergenfield, NJ 07621

AMERICAN HIBISCUS SOCIETY
206 N.E. 40th Street
Pompano Beach, FL 33064

AMERICAN IRIS SOCIETY
7414 E. 60th Street
Tulsa, OK 74145

AMERICAN PRIMROSE SOCIETY
2568 Jackson Highway
Chehalis, WA 98532

AMERICAN RHODODENDRON SOCIETY
14635 S.W. Bull Mountain Road
Tigard, OR 97223

AMERICAN ROSE SOCIETY
P.O. Box 30,000
Shreveport, LA 71130

BEDDING PLANTS, INC.
P.O. Box 286
Okemos, MI 48864

BROMELIAD SOCIETY
647 S. Saltair Avenue
Los Angeles, CA 90049

HOLLY SOCIETY OF AMERICA, INC.
407 Fountain Green Road
Bel Air, MD 21014

INTERIOR PLANTSCAPE ASSN.
2000 L Street NW
Suite 200
Washington, DC 20036

LIVING PLANT GROWERS ASSN.
1419 21st Street
Sacramento, CA 95814

NORTH AMERICAN LILY SOCIETY, INC.
P.O. Box 476
Waukee, IA 50263

Government Agencies

ANIMAL AND PLANT HEALTH
   INSPECTION SERVICE
6505 Belcrest Road
Hyattsville, MD 20782

SOIL CONSERVATION SERVICE
P.O. Box 2890
Washington, DC 20013

Grass

AMERICAN SOD PRODUCERS ASSN.
4415 W. Harrison
Hillside, IL 60162

THE LAWN INSTITUTE
1501 Johnson Ferry Rd., NE
Suite 206
Marietta, GA 30062

PROFESSIONAL LAWN CARE ASSN.
1000 Johnson Ferry Road, NE
Suite C-135
Marietta, GA 20068

Horticulture

AMERICAN HORTICULTURAL SOCIETY
P.O. Box 0105
Mt. Vernon, VA 22121
and
7931 E. Boulevard Drive
Alexandria, VA 22308

HORTICULTURAL RESEARCH INSTITUTE, INC.
1250 I St., NW
Suite 500
Washington, DC 20005

Landscaping

AMERICAN INSTITUTE OF
LANDSCAPE ARCHITECTS
602 E. San Juan Avenue
Phoenix, AZ 85012

AMERICAN SOCIETY OF LANDSCAPE ARCHITECTS
4401 Connecticut Ave., NW
5th Floor
Washington, DC 20008

ASSOCIATED LANDSCAPE CONTRACTORS
OF AMERICA
12200 Sunrise Valley Drive
Suite 150
Reston, VA 22091

COUNCIL OF TREE AND LANDSCAPE APPRAISERS
1250 I Street NW
Suite 504
Washington, DC 20005

INTERNATIONAL FEDERATION OF
LANDSCAPE ARCHITECTS
Lisbon, Portugal

NATIONAL BARK PRODUCERS ASSN.
301 Maple Avenue W
Tower Suite 504
Vienna, VA 22180

NATIONAL LANDSCAPE ASSN.
1250 I St., NW
Suite 500
Washington, DC 20005

PROFESSIONAL GROUNDS
MANAGEMENT SOCIETY
120 Cockeysville Rd.
Suite 104
Hunt Valley, MD 21031

Nurseries

AMERICAN ASSOCIATION OF NURSERYMEN
1250 I St., NW
Suite 500
Washington, DC 20005

MAILORDER ASSN. OF NURSERYMEN, INC.
210 Cartwright Boulevard
Massepequa Park, NY 11762

WHOLESALE NURSERY GROWERS
OF AMERICA, INC.
2000 L Street NW
Suite 200
Washington, DC 20036

Outdoor Living

CALIFORNIA REDWOOD ASSN.
591 Redwood Highway
Suite 3100
Mill Valley, CA 94941

GARDEN INDUSTRY OF AMERICA, INC.
2501 Wayzata Boulevard
Minneapolis, MN 55440

GOLF COURSE SUPERINTENDENTS ASSN.
OF AMERICA
1617 St. Andrews Drive
Lawrence, KS 66044

KEEP AMERICA BEAUTIFUL, INC.
9 West Broad St.
Stamford, CT 06902

NATIONAL SWIMMING POOL INSTITUTE
2111 Eisenhower
Alexandria, VA 22314

Pest Control

ENTOMOLOGICAL SOCIETY OF AMERICA
4603 Calvert Road
College Park, MD 20740

INTERNATIONAL PESTICIDE APPLICATORS, INC.
a.k.a. Washington Tree Service
20057 Ballinger Road NE
Seattle, WA 98155

NATIONAL PEST CONTROL ASSN.
8100 Oak Street
Dunn Loring, VA 22027

WEED SCIENCE SOCIETY OF AMERICA
1508 W. University Ave.
Champaign, IL 61821

Power Equipment, Parts

OUTDOOR POWER EQUIPMENT INSTITUTE, INC.
1901 L Street NW
Washington, DC 20036

SMALL ENGINE SERVICING DEALERS ASSN., INC.
P.O. Box 6312
St. Petersburg, FL 33736

Seeds

AMERICAN SEED TRADE ASSN.
601 13th St., NW
Suite 570-South
Washington, DC 20005

Soil Conditioners

CANADIAN SPHAGNUM PEAT MOSS
    INFORMATION BUREAU
928 Broadway
New York, NY 10010

PERLITE INSTITUTION, INC.
6268 Jericho Turnpike
Commack, NY 11725

Trees

AMERICAN FOREST INSTITUTION
1619 Massachusetts Avenue NW
Washington, DC 20036

AMERICAN FORESTRY ASSN.
P.O. Box 2000
Washington, DC 20013

AMERICAN SOCIETY OF
    CONSULTING ARBORISTS
5130 W. 101st Circle
Westminster, CO 80030

NATIONAL ARBORISTS ASSN., INC.
P.O. Box 1094
Amherst, NH 03031

NATIONAL CHRISTMAS TREE ASSN.
611 E. Wells Street
Milwaukee, WI 53202

SOCIETY OF MUNICIPAL ARBORISTS
7447 Old Dayton Road
Dayton, OH 45427

Wildflowers

NATIONAL WILDFLOWER RESEARCH CENTER
4801 LaCrosse Ave.
Austin, TX 78739

Glossary

Accent plant A plant that is more distinctive than many plants, but does not attract the eye as much as a specimen plant.

Acclimaticization The preparation of plants for a reduced light setting.

Adobe A heavy soil common to the southwestern United States.

Aeration The addition of air into the soil; it is accomplished during soil conditioning with materials such as sand or peat moss. It can be encouraged in established lawns by the use of machines called aerators.

Aesthetic Attractive to the human senses.

Alkaline Characterized by a high pH.

Ames lettering guide A plastic device used with a T-square to produce guidelines for hand lettering.

Analysis The proportion in which each of three standard ingredients is present: nitrogen, available phosphoric acid, and water-soluble potash.

Angle The relationship between two joined straight lines.

Annual A plant that completes its life cycle in one growing season.

Antitranspirant (*also* antidesiccant) A liquid sprayed on plants to reduce water loss, transplant shock, windburn, and sunscald.

Arid A term used in the description of landscapes where there is little usable water.

Axonometric view A drawing that permits multiple sides of an object to be seen in a single, measurable view.

Balance The even distribution of materials on opposite sides of a central axis. There are three kinds of balance: symmetric, asymmetric, and proximal/distal.

Balled and burlapped A form of plant preparation in which a large part of the root system is retained in a soil ball. The ball is wrapped in burlap to facilitate handling during sale and transplanting.

Bare root A form of plant preparation in which all soil is removed from the root system. The plant is lightweight and easier to handle during sale and transplanting.

Bedding plant An herbaceous plant preseeded and growing in a peat pot or packet container.

Bid A statement of what work, materials, and standards of quality will be provided by a landscape firm in return for the price specified. Once agreed to by a client and the landscape firm, it is legally binding upon both.

Bid bond A security deposit submitted by a contractor at the time of bidding to guarantee that the bid will be honored if selected.

B.L.A. Bachelor of Landscape Architecture. The first professional degree for those who study landscape architecture.

Branching habit A combination of the number of branches, the average size of branches, the flexibility of branches, and the vertical or horizontal direction of their growth.

Broadcast When material such as seed or fertilizer is distributed evenly over a site.

Bulb A flowering perennial that survives the winter as a dormant fleshy storage structure.

Calibration The adjustment of a piece of equipment so that it distributes a given material at the rate desired.

Caliche A highly alkaline soil common to the southwestern United States.

Canopy The collective term for the foliage of a tree.

Capital Money used to finance a business.

Change order An amendment to the original contract to which all parties agree. Once signed, it becomes part of the original contract.

Chemical forms Material that is synthesized and not naturally occurring.

Client The person or organization that owns and provides the financing for a project.

Closed specifications Specifications that do not allow for the substitution of specific products with similar products.

Compaction A condition of soil in which all air has been driven out of the pore spaces. Water is unable to move into and through the soil.

Compass A graphic design tool used for the construction of circles.

Complete fertilizer A fertilizer containing nitrogen, phosphorus, and potassium, the three nutrients used in the largest quantities by plants.

Conditioning Preparation of soil to make it suitable for planting.

Containerized A form of plant preparation for sale and transplanting. When purchased, the plant is growing with its root system intact within a plastic, metal, or tarpaper container.

Contingency Allowance added to insure that unpredictable costs will be covered.

Contour interval The vertical distance between contour lines.

Contour lines Broken lines found on a topographic map. They represent vertical elevation.

Contract An agreement between two parties that is legally binding.

Contractor Party in contract with the client or the client's representatives.

Cool-season grass A type of grass that grows best in temperate regions and during the cooler spring and fall months.

Corporation A form of business operation that makes the business a legal entity, separate from its owners.

Cost Refers to the recovery of expenditures.

Cost estimate An itemized listing of the expenses in an operation. It can be applied to a single task or a total project.

Crotch The point on a tree at which two branches or a branch and the trunk meet.

Crown The point at which aboveground plant parts and the root system meet.

Cultipacker A large lawn seeder pulled by a tractor.

Cut A grading practice that removes earth from a slope.

Deciduous A type of plant that loses its leaves each autumn.

Degree The unit of measurement for angles.

Density The number of leaf shoots that a single grass plant will produce.

Design-build firm A landscape business that provides both design and construction services.

Design specifications All details regarding materials, methods, and performance standards are included in the specifications.

Diameter The distance across a circle as measured through the exact center.

Diazo machine A duplicating machine that makes positive copies from vellum tracings onto heavy paper.

Diazo process Duplication process used to reproduce landscape drawings in large numbers. Makes a direct positive copy using specially treated paper, ultraviolet light, and an ammonia vapor developer.

Dormancy (adj.: dormant) A period of rest that perennial plants experience during the winter season. They continue to live, but have little or no growth.

Downstream With regard to irrigation, describes the direction of water flow away from the source.

Drafting powder An absorbent granular material used to prevent smudging and keep drawings free of skin oil.

Drainage The act of water passing through the root area of soil. Soil is well drained if water disappears in 10 minutes or less from a shrub or tree planting.

Drainage tile A plastic or clay tube buried beneath the soil that collects excess water from the soil and carries it away.

Drop spreader A device for the application of granular materials such as grass seed and fertilizer. The material is dispensed through holes in the bottom of a hamper as the spreader is pushed across the lawn.

Effluent water Treated sewage.

Elevation view A scaled drawing with two dimensions, one horizontal and one vertical.

Enclosure materials Materials that form the walls of the outdoor room.

Enrichment A contribution made to the outdoor room by a landscape item that is not an element of a wall, ceiling, or floor.

Erosion The wearing away of the soil caused by water or wind.

Espalier A form of pruning in which plants are trained against a fence or wall. The effect is vinelike and two dimensional.

Estimate An approximation of the price that a customer will be charged for a landscape project.

Estimator The individual usually assigned to do the landscape take-off.

Evergreen A type of plant that retains its foliage during the winter. There are needled forms (such as pine, spruce, hemlock, and fir) and broad-leaved forms (such as rhododendron, pieris, euonymus, and holly).

Exotic plant A plant that has been introduced to an area by human beings, not nature.

Fauna Animal life.

Fertilization The addition of nutrients to the soil through application of natural or synthesized products called *fertilizers*.

Fertilizer analysis The percentage of various nutrients in a fertilizer product. A minimum of three numbers on the fertilizer package indicates the percentage of total nitrogen (N), available phosphoric acid (P_2O_5), and water-soluble potash (K_2O), in that order.

Fill A grading practice that adds earth to a slope.

Filler material Used to dilute and mix fertilizer, adding weight and bulk.

Flail mower A mower used for turfgrasses that are only cut a few times each year.

Flora Plant life.

Flower bed A free-standing planting made entirely of flowers with no background of shrub foliage.

Flower border A flower planting used in front of a planting of shrubs. The shrubs provide green background for the blossoms.

Foam board Two sheets of thin cardboard bonded to a central sheet of styrofoam. It is commonly used for mounting designs for presentations.

Focal point A point of visual attraction. A focal point can be created by color, movement, shape, size, or other characteristics.

Focalization of interest Principle of design that selects and positions visually strong items in the landscape composition.

Foliage texture The effect created by the combination of leaf size, sunlight, and shadow patterns on a plant.

Foliar burn A reaction to chemicals in fertilizer.

Footcandle The amount of light produced by a candle at a distance of one foot.

Foundation planting The planting next to a building that helps it blend more comfortably into the surrounding landscape.

General contractor The firm directly responsible to the client for the construction of a project.

Girdling The complete removal of a strip of bark around the main stem of a plant. After girdling, the ability of nutrients to pass from roots to leaves is lost, causing the eventual death of the plant.

Grading Changing the form of the land.

Graft A man-made bond between two different plants, one selected for its aboveground qualities (scion) and the other for its belowground qualities (stock).

Grass seed blend A combination of two or more cultivated varieties of a single species.

Grass seed mixture A combination of two or more different species of grass.

Groundcover A low-growing, spreading plant, usually 18 inches or less in height.

Groundskeeper A professional engaged full-time in landscape maintenance.

Hardiness The ability of a plant to survive through the winter season.

Hardscape Design materials that are not living plant materials. The term usually is applied to the constructed materials of a landscape.

Heading back A pruning technique that shortens a shrub branch without totally removing it.

Heaving An action that causes shallowly rooted plants, such as grasses, groundcovers, and bulbs, to be forced to the surface of the soil. The action results from repeated freezing and thawing of the soil surface.

Herbaceous A type of plant that is nonwoody. It has no bark.

Herbicide A chemical used to kill weeds.

Holdfasts Special appendages of certain vines that allow them to climb.

Humus Created by decaying organic matter, it aids the soil in moisture retention.

Hydroseeder A spraying device that applies seed, water, fertilizer, and mulch at the same time.

Hydrozoning Grouping plants on the basis of their water needs.

Incurve The center of a corner planting bed and a natural focal point.

Inert material Filler material that has no purpose other than to carry and dilute active ingredients in a mixture.

Inorganic Consisting of nonliving materials.

Intangible A quality denoting something that cannot be touched.

Irrigation Supplying water to plants through artificial means.

Jump-cut A pruning technique for the removal of large limbs from trees without stripping bark from the trunk. It involves a series of three cuts.

Landscape architect A licensed professional who practices landscape planning, usually on a scale larger than residential properties.

Landscape contractor A professional who carries out the installation of landscapes.

Landscape designer A professional who devotes all or part of a work day to the design of landscapes.

Landscape installation The actual construction of the landscape.

Landscape maintenance The care and upkeep of the landscape after installation.

Landscape management The extended care of existing landscapes, usually under terms of a contract.

Landscape nursery worker A professional who is concerned with the sale and installation of landscape plants and related materials.

Landscaping A profession involving the design, installation, and maintenance of the outdoor human living environment.

Lateral bud Any bud below the terminal bud on a twig.

Latin binomial The unique two-part scientific name that identifies each plant and animal.

Leaching The dissolving of materials (such as nutrients) in the water which is present in soil, causing the material to quickly pass the point at which plant roots can benefit from them.

Lead branch The most important branch of a tree; cannot be removed without destroying the distinctive shape of the tree.

Lettering Machine A device that types an assortment of letter styles and sizes onto a transparent tape. The tape can then be applied to a landscape design.

Lettering template A stencil used to form letters. It is an alternative to free-hand lettering.

Light compensation point The point of exact balance between photosynthesis and respiration by plants.

Light quality The color of light emitted by a particular source.

Lime A powdered material used to correct excess acidity in soil.

Line-width variation A variation of hand-lettering style, where the designer uses an instrument with a chiseled tip, so lines that are thin and wide can be made with the same instrument.

Loam Soil that contains approximately equal amounts of clay, silt, and sand (a desirable condition).

Lux The amount of illumination received on a surface one meter from a standard light source (an international measurement).

Macro-range Viewer sees the landscape from the most distant vantage point.

Marker paper Smooth, hard surfaced paper used for marker renderings.

Market The geographic area from which a business attracts most of its customers.

Masses Solid vertical areas of plantings, buildings, walls, or land forms.

Massing plants Plants that serve to fill large amounts of space both on the ground and in the air.

Masterformat A standardized system of organization for project manuals.

Microclimate Regions that provide atypical growing conditions.

Microcomputer A small computer that adapts well to the business needs of landscape firms.

M.L.A. Master of Landscape Architecture. The second professional degree for those who study landscape architecture.

Morphology Physical structure of plants.

Mulch A material placed on top of soil to aid in water retention, prevent soil temperature fluctuations, or discourage weed growth.

Mylar film Polyester drafting film used for drawing when durability of the original work is important.

Native plant A plant that evolved naturally within a certain locale.

Naturalized plant A plant that was introduced to an area as an exotic plant, but which has adapted so well that it may appear to be native.

Newsprint paper Inexpensive paper used for preliminary sketches and designs.

Non-collusion affidavit A document affirming that bidders have not discussed a project prior to bidding in order to establish a minimum bid among themselves.

Noxious weeds Persistent weeds defined by law in most states. They are perennial and difficult to control. The presence of these weeds in a grass seed mix indicates that the seed is of low quality.

Nutrient ratio A comparison of the proportion of each nutrient in a fertilizer to the other nutrients in the same fertilizer. Example: A fertilizer with a 5-10-5 analysis has a 1-2-1 ratio of ingredients.

Open specifications Specifications that permit substitutions after approval by the designer or owner.

Organic Consisting of modified plant or animal materials.

Outcurve The sides of a corner planting.

Overhead Operational costs of doing business; not individual job costs.

Partnership A form of business operation engaged in by two or more persons.

Pastels Chalky crayons made of powdered pigment and used in color rendering.

Patching Warranted when bare spots are at least 1 foot in diameter; seed, sod, plugs, or sprigs should be selected to match the grasses of the established lawn.

Perennial A plant that lives more than two growing seasons. It usually is dormant during the winter.

Performance bond A security provided by the contractor to provide the owner with sufficient money to have a project completed should the contractor default during the performance of the contract.

Performance specifications Establish the standard of performance or service that must be met by the feature being described.

Perspective view A drawing that permits multiple sides of an object to be seen in a single view. Dimensions are not measurable.

Pesticide A chemical used for the control of insects, plant diseases, or weeds.

pH A measure of the acidity or alkalinity of soil. A pH of 7.0 is considered neutral. Ratings below 7.0 are acidic, above 7.0 alkaline (basic).

Photocopy process Duplication process also used to reproduce landscape drawings in large numbers; operates much like an office copier, using black toner to reproduce the image.

Photosynthesis Production of food by the plant.

Physiology Biological functioning of plants.

Plan view A measurable drawing seen as though the viewer's line of sight is perpendicular to the surface.

Plant list An alphabetical listing of the botanical names of plants used in a landscape plan, their common names, and the total number used.

Plug A small square, rectangle, or circle of sod, cut about two inches thick.

Plugging A method of lawn installation that uses cores of live, growing grass.

Price Refers to an outlay of funds.

Principles of design Standards by which designs can be created, measured, discussed, and evaluated.

Prismacolors (reg. trademark) Colored pencils used in rendering.

Process A sequence of steps, in the form of decisions or activities, that result in the accomplishment of a goal. Also, a sequence of stages during which something is created, modified, or transformed.

Project manual A bound compilation of the documents pertaining to a landscape (or other construction) project.

Propagation The reproduction of plants. It may be sexual or asexual (by vegetative cuttings, layering, etc.).

Proportion Principle of design concerned with the size relationships between all the features of the landscape.

Proprietary specifications Those that require the use of specific products.

Protractor A graphic design tool for measuring angles.

Pruning The removal of a portion of a plant for better shape or more fruitful growth.

Pubescence The presence of fine hairs on the surface of leaves.

Puddling Compaction of soil to such a degree that water will not soak into it.

Purity The percentage, by weight, of the pure grass seed in a mixture.

Quadrille paper Ruled paper divided 4, 5, 8, or 10 squares per inch and used for first-draft drawings.

Quickly available fertilizer A fast-action fertilizer that has its nitrogen in a water-soluble form for immediate release into the soil.

Radius One-half of the diameter of a circle.

Reel mower A mower used for home, recreational, and commercial lawn maintenance. The blades rotate in the same direction as the wheels and cut the grass by pushing it against a nonrotating bedknife at the rear base of the mower.

Respiration The use of food reserves for growth and maintenance by plants.

Rhizome An underground stem. New shoots are sent to the surface some distance out from the parent plant. Each new plant develops its own root system and becomes independent of the parent plant.

Rhythm and line Principle of design. Something is repeated at a standard interval, creating a rhythm. Lines establish the shape and form of a landscape.

Riser The elevating portion of a step.

Root systems Of plants, range from total tap root systems, with a large single root growing straight down into the soil, to full fibrous systems, with thousands of fine, hairlike roots spreading out in all directions.

Rotary mower A mower used for home, recreational, and commercial lawn maintenance. The blades move like a ceiling fan, parallel to the surface of the lawn, cutting the grass off as they revolve.

Rotary spreader A device for the application of granular material such as grass seed and fertilizer. The material drops from a hamper onto a rotating plate and is propelled outward in a semicircular pattern.

Scaffold branch A lateral branch of a tree.

Scale (engineer's) A measuring tool that divides the inch into units ranging from 10 to 60 parts.

Scion Shoot portion of a tree.

Seed analysis A breakdown of the contents of the seed package on which it appears; must appear on every package of seeds sold.

Serif A decorative stroke attached to a letter to create an ornate appearance.

Shrub A multistemmed plant smaller in size than a tree.

Silhouette The outline of an object viewed as dark against a light background.

Simplicity Principle of design that makes the viewer comfortable in the landscape; does not imply simplistic.

Site An area of land having potential for development.

Slope A measurement that compares the horizontal length to the vertical rise or fall of land. The measurement can be determined from a topographic map.

Slow-release fertilizer A slow-action fertilizer in which the nitrogen content is in a form not soluble in water. The nitrogen is released more slowly into the soil for more efficient intake by plants.

Sodding A method of lawn installation that uses strips of live, growing grass. It produces an immediate effect on the landscape, but is more costly than seeding.

Software Supportive material that allows a computer to carry out specific functions.

Soil texture The composition of a soil as determined by the proportion of sand, silt, and clay that it contains.

Species A category of plant classification distinguishing the plant from all others.

Specifications Written requirements for the installation of a landscape. They are usually prepared and made available to contractors before the bidding for a contract begins.

Specimen plant A plant that is highly distinctive because of such qualities as flower or fruit color, branching pattern, or distinctive foliage. Its use creates a strong focal point in a landscape.

Spreader A garden tool used for the even distribution of materials such as grass seed and fertilizer.

Sprig A piece of grass shoot. Sprigs are commonly used to establish warm-season grass plantings.

Sprinkler irrigation Water applied under pressure over the tops of plants.

Stock Root portion of a tree.

Stolon A stem that grows parallel to the ground. New plants develop from it and become independent of the parent plant.

Subcontractor Firm in contract with the prime contract holder to provide selected services for the accomplishment of a project.

Sucker A succulent branch that originates from the root system. The vegetation of suckers is abnormal and undesirable.

Sunscald A temperature-induced form of winter injury. The winter sun thaws the aboveground plant tissue, causing it to lose water. The roots remain frozen, and thereby unable to replace the water. The result is drying of the tissue.

Symbols Drawings that represent overhead views of trees, shrubs, or other features of a landscape plan.

Take-off The calculation of quantities from plans and specifications.

Tangible A quality denoting something that is touchable.

Taxonomist A person who specializes in the classification of plants and animals.

Technical specifications Written descriptions of the technical standards that the work must meet; often used for portions of projects that are not easily verified for correctness after construction has been completed.

Tender A condition of plants that implies their lack of tolerance to cold weather.

Tendrils Special appendages of certain vines that allow them to climb.

Terminal bud The end bud on a twig.

Terrain The rise and fall of the land.

Texture A description of the coarseness or fineness of a plant, compared to other nearby plants.

Thatch Dead, semidecomposed grass clippings on the surface of soil.

Thinning out A pruning technique that removes a shrub branch at or near the crown of the plant.

Topiary A form of pruning in which plants are severely sheared into unnatural shapes such as animals or chess pieces.

Topography A record of an area's terrain.

Trace elements Nutrients essential to the growth of many plants, but needed in far less amounts than the major elements.

Trajectory Path of irrigation water as it is propelled through the air.

Transfer film Paper-thin, transparent plastic sheeting with a clear adhesive backing used for reproduction of selected graphics in desired locations on layout sheets.

Transplant To relocate a plant.

Tread That portion of a step on which the foot is placed.

Triangle A three-sided graphic design tool. It commonly has either a 30°-60°-90° or a 45°-90°-45° combination of angles.

Trickle irrigation Water supplied directly to the root zone of plants.

T-square A long straightedge that takes its name from its shape. It is a graphic design tool.

Twining One method by which certain vines are able to climb.

Unit cost The price of the smallest available form of an item described in a cost estimate.

Unit pricing The reduction of all landscape area dimensions and material quantities to a common measurement, such as thousand square feet or acre.

Unity Principle of design in which all separate parts contribute to the creation of the total design.

Vellum Thin, transparent paper used for tracing designs.

Vertical mowing A technique that requires a power rake or a mower whose blades strike the turf vertically. It is done to break up the soil plugs left by an aerator or to remove excessive thatch.

Voids The horizontal open areas of the landscape.

Warm-season grass A grass that grows best in warmer regions of the country and during the summer months.

Water sprout A succulent branch that grows from the trunk of a tree. The vegetation of water sprouts is abnormal and undesirable.

Weed A plant growing where it is not wanted and with no economic value.

Weep hole A means of preventing water buildup behind a retaining wall.

Windburn Drying out of plant tissue (especially evergreens) by the winter wind.

Winter injury Any damage done to elements of the landscape during the cold weather season of the year.

Working drawing A copy of a landscape design done on heavy paper or plastic film. The working drawing is used repeatedly during actual construction of the landscape and must be very durable.

Wound paint A sealing paint used over plant wounds of 1 inch or more in diameter after pruning.

Xeriscaping® Techniques of landscaping that conserve water.

Zoning The regulations of a community that govern what uses can be made of different areas within the community.

Index

Note: Page numbers in **bold type** reference non-text material